EAST NORFOLK

Flora of
Norfolk

Flora of Norfolk

Dr C P Petch
MA MD FRCP

E L Swann
Past President of the Norfolk
and Norwich Naturalists Society
Honorary Member of the Botanical
Society of the British Isles

NORWICH: JARROLD AND SONS LIMITED

First Published 1968 on the occasion of
the centenary of the Norfolk and Norwich
Naturalists' Society. Founded in 1869

© Text. Dr C. P. Petch and Mr E. L. Swann
Illustrations Jarrold and Sons Limited

Printed and bound in Great Britain by
Jarrold and Sons Limited, Norwich

CONTENTS

ACKNOWLEDGEMENTS

It is without precedent to have a regional *Flora* so lavishly illustrated; the expense alone of modern colour reproduction would put such a work beyond the financial resources of most Natural History societies.

We would like to convey our deep appreciation to the directors of Jarrold and Sons Limited, who without thought of financial gain, have generously allowed us the use of all the resources of their well-known firm to provide an account of the floral richness of our county.

The authors

We are also grateful to the following photographers who have kindly allowed us to use their coloured transparencies for the illustrations.

Paul Banham 53, E. G. Burt 108, Dr J. A. Campbell 70, Dr S. Clark 18, 89, Mrs D. M. Dean 21, 74, 109, Prof. J. H. Fremlin 20, Dr M. George 95, 96, A. H. Hems, 22, 34, 36, 39, 47, 52, 54, 57, 58, 59, 60, 71, 73, 83, 84, 93, 94, 97, 99, 102, 103, 104, 106, J. C. E. Hubbard 24, 43, G. E. Hyde 4, H. John Jarrold 101, Jarrolds 6, 25, 48, 63, A. C. Jermy 41, 111, Reg Jones 5, 9, 32, 40, 49, 50, 51, 77, 85, 88, Miss V. M. Leather 8, 28, 65, 66, 79, Dr Lodge 46, J. E. Lousley 19, 31, 72, 110, 114, Miss D. M. Maxey, 2, 3, 13, 15, 16, 29, 30, 33, 35, 42, 56, 61, 67, 78, 80, 81, 90, 92, 98, 115, D. B. Osbourne 62, W. H. Palmer 44, 113, J. Secker 91, Miss G. Tuck, 1, 7, 11, 12, 23, 26, 27, 37, 38, 45, 55, 64, 68, 69, 75, 82, 100, 105, 107, 112, G. D. Watts 10, 76, 86, 87, K. Wilson 14.

LIST OF ILLUSTRATIONS

7

FOREWORD

C. E. Hubbard, C.B.E., D.Sc., F.L.S.

Norfolk is so very rich in areas of special biological importance that naturalists everywhere will welcome this fine addition to our county *Floras*. I am sure that everyone with an interest in plants will join me in warmly congratulating its enthusiastic authors and their numerous recorders and contributors on this valuable outcome of many years of intensive, yet very enjoyable, field studies.

Within its boundaries may be found a remarkable range of plant-life whether one visits its long and largely unspoilt coastal region of shingle-ridges, sand-dunes, and salt marshes; the sandy heaths and commons, woodlands, and cultivations of the Breckland; the fens and bogs of many parts of the county; the chalk downs; or the Broads. In these varied communities there are fortunately still many rare species and interesting local variants which have been able to persist despite the ever increasing pressure on their habitats by man and his domestic animals. Now with the publication of an up-to-date account of its flora, the wealth of plants and the localities in which many of the rarer kinds grow will be brought to the notice of a new generation. Some will continue as students, at the university, colleges, and schools, whilst many more will make botanical excursions or spend their holidays in the county. It will not be out of place, therefore, to remind them that whether they are using the Norfolk countryside as a source of information or for relaxation, they should endeavour to preserve its plants and their habitats for future generations.

Many noted botanists, and naturalists generally, have lived in the county and others have made it a base for their studies, so it is not surprising to find it was the first county with the wisdom to form a Naturalists' Trust (1926), the purpose being to acquire and preserve areas of special biological importance. Thus at the present time it is responsible for the conservation of numerous reserves, covering several thousand acres. In addition to these reserves, the National Trust owns extensive areas of considerable beauty, all rich in plants, both on the coast and inland, while over all the Nature Conservancy keeps a watchful and helpful eye.

On account of the extensive use of herbicides, and the improved and intensive methods of cultivation, many of the plants which were once the delight of our youth and added beauty to the fields and roadsides, as well as many rare or unusual so-called 'weeds' of arable land, have been very much reduced in numbers, or may even be extinct in some localities. It is, therefore, of increasing importance to all nature lovers to press for the preservation of all those attractive plants that persist wherever they are of no danger to livestock. Furthermore, in a county with countless pits and ponds in which flourish many unusual plants, some at least of those not required for watering stock might be left as habitats for the aquatics which line their margins.

Finally a very special tribute is due to Messrs. Jarrolds, the printers and publishers of this profusely illustrated account of Norfolk plants, and to everyone concerned with its production, for this, the centenary year of the Norfolk and Norwich Naturalists' Society.

PREFACE

Norfolk is a sea-coast county on the shores of the North Sea and the Wash with a coastline of 85 miles. The greatest length from east to west is 67 miles and from north to south about 43 miles. With an area of approximately 1,315,000 acres or about 2,000 square miles it is the fourth largest English county, ranking after Yorkshire, Lincolnshire and Devonshire. It is separated from the neighbouring counties of Suffolk, Cambridgeshire and Lincolnshire by the rivers Waveney, Little Ouse, Great Ouse and Nene.

The first *Flora* of the county was published in 1866 by the Rev. Kirby Trimmer and, although in his introduction he stated, 'it is statistically divided into two nearly equal portions, known as East and West', he made no attempt to show the relation between plant distribution and physiographic features.

In his *Topographical Botany*, 1873, H. C. Watson divided the county arbitrarily into two vice-counties separated by the 1° line of longitude, specified as West Norfolk (v-c 28) and East Norfolk (v-c 27).

For the purposes of the 1914 *Flora*, W. A. Nicholson divided the county into four more or less equal areas in accordance with the plan originated by the Rev. George Munford in his list of Norfolk plants published in White's *Directory*, 3rd edition, 1869. Such a division is unworkable as it makes little or no distinction in regional boundaries and entails far too much reliance on accurate map-reading.

The 1914 *Flora* dealt mainly with the plants of East Norfolk, and in our *West Norfolk Plants Today*, 1962, we attempted to redress the balance of 'the neglected half'.

Geologically, the drift deposits are so complicated that no divisions can be based on surface geology. The size of the county militates against the use of parish boundaries as there are well over 600 of these. Those who favour the use of 10-kilometre grid squares are faced with thirty-nine whole ones and the parts of twenty-seven others with the additional difficulty of making allowance for the inviolate division between East and West bisecting six whole squares and the part of one other. Lastly, no classification can be based on river systems as the larger rivers form the county boundary, although, as Petch states in *West Norfolk Plants Today*, 'the river Nar forms a convenient northern boundary to many Breckland species, and the Great Ouse delimits a tract of intensively cultivated land on its west that is very different from the rest of the vice-county'.

It will be appreciated that the authors have experienced some little difficulty in arriving at a satisfactory system for such a large county but they feel that, aided by both the chapter on the Major Plant Communities and the map giving places of botanical interest, visitors should have no difficulty in locating the floral wealth of Norfolk.

Practically every species is recorded on a 10-kilometre basis, whilst for less common plants with six or fewer records localities are specified together with the initials of the finder; absence of initials implies the joint work of the authors, except in the Bryophytes, where the authority is Burrell, and *Rubus* where it is Linton.

With few exceptions, noted in the text, nomenclature follows Clapham, Tutin & Warburg: *Flora of the British Isles*, 1962. Both Extinctions and Casuals are set apart from the main list.

An unusual feature of a regional *Flora* is the number of coloured illustrations chosen mainly from the photographic work of members of the Norfolk and Norwich Naturalists Society.

THE HISTORY OF BOTANY IN NORFOLK

E. L. Swann

Both by reason of its rich flora and such famous botanists as Sir James Edward Smith, the Hookers, Dawson Turner and John Lindley, Norfolk has for long been in the forefront of botanical science.

One of the difficulties is to find some, perhaps extraneous, event that formed the starting point. Although the written history of British plants began with Turner's *Herbal* in 1551, there was an awakening of interest some two hundred years earlier in Norfolk. It was during the reign of Edward III (1327–77) that Flemish weavers were persuaded to set up their looms in and around Norwich. The abundance of local supplies of wool laid the foundation of the staple industry and Norwich became the second largest city, ranking after London. According to repute, the Flemings and the Walloons introduced many new or little-known plants such as the lilac, carnation, Provence rose, asparagus, carrot, celery and cabbage. Their gardens were usually walled-in enclosures and it is not difficult to imagine them bringing in other plants from the country outside the city walls; nor is it surprising that the taste for gardening spread to the local inhabitants, who formed one of the earliest known horticultural societies.

With the ground already well prepared, the 'Age of the Herbalists' in the sixteenth century found Norwich beginning to build up its reputation. Printing became general and one of the earliest writers was JOHN CAIUS, alias KEY (1510–73), physician to Edward VI, Mary and Elizabeth I, and founder of Caius College, Cambridge. He communicated histories of rare plants to Gesner and was the author of *De Stirpium* (1570).

SIR THOMAS BROWNE (1605–82), the famous physician, writer and antiquarian, contemporary of John Ray, was born in London but spent the last forty-five years of his life in Norwich. To him we owe the first records of *Acorus calamus* and *Eryngium maritimum* in 1643 and 1668 respectively (see plant list).

BENJAMIN STILLINGFLEET (1707–71) was born at Wood Norton. He persuaded William Hudson, a London apothecary, to become the author of *Flora Anglica*, published in 1762, 'the first true attempt to set forth a British *Flora* according to the system and nomenclature of Linnaeus'. He sent Hudson many Norfolk plants and discovered *Deschampsia setacea* which Hudson described under *Aira setacea*. Whilst staying with his friend Robert Marsham, a correspondent of Gilbert White, at Stratton Strawless in 1755, Stillingfleet prepared his *Calendar of Flora* which forms part of his *Miscellaneous Tracts relating to Natural History, Husbandry and Physick*, 1759, and reached four editions. The *Calendar* records nearly 250 plants, many of which are first Norfolk records. By reason of his appearing in London wearing blue worsted stockings he, unwittingly, became the originator of the expression 'blue stocking'.

HUGH ROSE (1717–92), a Norwich apothecary, is chiefly to be remembered for instructing J. E. Smith in botany. He corresponded with Hudson, and four years before becoming afflicted with incurable blindness, with the help of Bryant, published *Elements of Botany* in 1775.

The REV. HENRY BRYANT (1721–99) died at Colby where he was Rector. He discovered *Hammarbya paludosa* and his was the first record of *Crassula tillaea* from Drayton Heath in 1766. He joined Rose in writing *Elements of Botany* and was mainly responsible for the *Appendix* of Norfolk and Suffolk plants.

JOHN PITCHFORD (1737–1803) was a Norwich surgeon and friend of Smith. His favourite genera were *Carex* and *Mentha*. Like most of the Norwich school of botanists he contributed to *English Botany*; and he discovered *Holosteum umbellatum* and *Peucedanum palustre*.

THOMAS JENKINSON WOODWARD (*c.* 1745–1820) was a native of Bungay and a lawyer by profession. He was one of the chief contributors to the second edition of Withering's *Botanical Arrangement of British Plants*, 1787–93; *English Botany*, and, with Goodenough, wrote *British Fuci*, 1793. His was the first record of *Scirpus sylvaticus* in Norfolk.

LILLY WIGG (1749–1828) was born at Smallburgh and died at Great Yarmouth. He was, in turn, shoemaker, schoolmaster and bank clerk. He instructed Dawson Turner in the algae and contributed both to Withering's second edition and *English Botany*. His was the first British record for *Trifolium suffocatum*. A manuscript copy of his *Esculent Plants* is at the Botany Department, British Museum.

The REV. CHARLES SUTTON (1756–1846) was born at Norwich. He was taught botany by Pitchford and became the expert on *Orobanche*. He discovered *O. elatior* and wrote a monograph on the genus which was published by the Linnean Society in 1798.

For more than a hundred years (1723–1825) the living of Fincham in West Norfolk was in the hands of the Forby family. The REV. JOSEPH FORBY was interested in willows and discovered *Salix Forbyana*, a fine basket osier, which was named by Smith to commemorate both Joseph and his nephew Robert. Whilst living at Barton Bendish, the REV. ROBERT FORBY took in pupils and among them was Dawson Turner who contributed records from that parish to the *Botanist's Guide*.

JAMES CROWE (1750–1807) of Lakenham near Norwich was closely associated with Smith for many years. Together they paid particular attention to the willows which were grown in Crowe's large garden. In addition to the Lakenham property, Crowe also had estates at Leziate and Saham Toney, both in West Norfolk. He studied the mosses, lichens, fungi and algae and to him are due the first records of *Vaccinium oxycoccus* at Dersingham, where it still grows, and *Aceras anthropophorum*. *Hieracium maculatum*, still frequent in the Norwich area, owes its introduction to him from seeds brought back from Westmorland in 1781. His name is commemorated in *Salix Croweana*, a variety of *S. phylicifolia*.

The REV. ROBERT BRANSBY FRANCIS (1768–1850) held various

livings in Norfolk but it was whilst Rector of Holt that he did most of his botanical work, chiefly with the cryptogams. His wife found *Dryopteris cristata* at Holt Lowes, its first British record, but his is the only Norfolk record for *Campanula patula*. Like many of his contemporaries, he contributed to *English Botany*.

DAWSON TURNER (1775–1858), born at Great Yarmouth, was a banker. He was mainly interested in algae, producing *British Fuci* in 1802, *Fuci* in 1808–19, and in 1805, with Dillwyn, published the *Botanist's Guide*. He contributed largely to *English Botany* and his collections are at Kew.

The impetus given to binomial nomenclature in 1753 by the first edition of the *Systema Naturae* of Linnaeus helped to resolve the chaos which had previously prevailed and to stimulate the pioneer work being carried out in the county. Although Ray had prepared botanists for this change, the most strenuous supporter was a Norfolk man, JAMES EDWARD SMITH (1759–1828). From his earliest years he was interested in botany and his main reason for taking a medical degree at Edinburgh was because it was the only university at that time which included botany in its medical course. Whilst continuing his medical studies in London he was persuaded by Sir Joseph Banks to buy 'the whole collection of Linnaean books, manuscripts and Natural History specimens'. His father advanced him the sum of £1,000 for this purpose. In 1787 he established the Linnean Society and continued as its President until his death. He spent the last thirty years of his life at Norwich and was knighted in 1814. He stoutly maintained the Linnaean arrangement, the possession of Linnaeus' Herbarium lending force to his utterances. His many contributions included the text of Sowerby's *English Botany* (1790–1814), *Flora Britannica* (1800–4), and *The English Flora* (1824–8). After his death the Linnaean collections were acquired by the Linnean Society for 3,000 guineas.

WILLIAM SKRIMSHIRE (1766–1830) lived at Wisbech in Cambridgeshire, but he was in active correspondence with the Norfolk botanists and collected mainly in the adjoining vice-county of West Norfolk. There is a large collection of his plants, together with his *Catalogue of Plants,* preserved at the Wisbech Museum which include the first Norfolk records of *Cladium mariscus, Polygonum bistorta* and *Utricularia vulgaris*.
Four hundred years after the introduction of the Flemish weavers we still find their influence, for it was in 1779 that JOSEPH FOX, a journeyman weaver, succeeded for the first time in raising *Lycopodium* from spores.

SIR WILLIAM JACKSON HOOKER (1785–1865) was born at Norwich. Ornithology and entomology first attracted his attention but 'being happily the discoverer of a rare moss (*Buxbaumia aphylla*) which he took to Smith, he received from that botanist the bias which determined his career. Henceforth, botany was his sole aim'. His tastes were catholic as he became interested in flowering plants, mosses, hepatics, lichens and fresh-water algae. His love of gardening aided him in becoming Director of Kew Gardens (1841–65). Smith dedicated to him the genus *Hookeria*, a moss which still grows at Holt where he discovered it. He published upwards of fifty volumes of descriptive botany which include *British Jungermanniae* (1816), *Muscologia Britannica* (1817), *Species Filicum* (1846–64), *The British Ferns* (1861), and he edited the *Botanical Magazine* from 1826.

SIR JOSEPH DALTON HOOKER (1817–1911) was the second son of W. J. Hooker. Born at Halesworth in Suffolk, his connection with Norfolk botany is mainly indirect. He succeeded his father as Director of Kew in 1865, a position he held for twenty years. His published work covers more than sixteen pages of the Kew Bulletin, 1912. In 1868 he was President of the British Association in Norwich.

The REV. GEORGE MUNFORD (1794–1871) was born at Great Yarmouth and died at East Winch in West Norfolk where he was vicar in 1849. He wrote the 'Flowering Plants of West Norfolk' in *Ann. & Mag. Nat. Hist.,* 1841 and *Botany* in White's *History of Norfolk*, 1863.

JOHN LINDLEY (1799–1865) was born at Catton and, from his earliest days, became interested in horticulture. In 1841 he established the *Gardeners' Chronicle*; in 1858 he became Secretary to the Horticultural Society (later, the R.H.S.). His *Outlines of Botany* first appeared in 1830 but was enlarged and published as *Elements of Botany* in 1841.

JOHN DREW SALMON (?1802–59), botanist and ornithologist, lived at Thetford from 1835 until 1848. He amassed a large herbarium which is preserved at the Castle Museum, Norwich.

The REV. KIRBY TRIMMER (1804–87) was the author of the first *Flora of Norfolk* which was published in 1866 whilst holding the livings of Burnham Norton and Stanhoe in West Norfolk. The genus *Mentha* was his chief interest as is shown by the *Supplement* to his first work, appearing in 1885. His collection of mints is in the British Museum herbarium. He discovered *Potamogeton trichoides*.

CHARLES JAMES PAGET (1811–44) and his brother JAMES (1814–99) were born at Great Yarmouth, the latter being created a baronet for his eminent services to surgery. They were associated with Dawson Turner, and fully maintained, in Great Yarmouth, the renown of the Norfolk school of botanists, producing the *Sketch of Natural History of Yarmouth* in 1834. Sir James Paget's herbarium is preserved at the Castle Museum, Norwich.

SAMUEL PICKWORTH WOODWARD (1821–65) was born at Norwich. He became Professor of Natural History at the Royal College of Agriculture, Cirencester, where his herbarium and drawings are kept. His 'Flora of Central Norfolk' appeared in the *Mag. Zool. Bot.,* in 1841. The hybrid thistle, *Carduus Woodwardii* (*C. acaulon × dissectum*), was named after him by H. C. Watson.

HAMPDEN GLEDSTANES GLASSPOOLE (1825–87) was born at Ormesby St Michael. He collected a large herbarium bequeathed to H. D. Geldart. He botanised chiefly in the Yarmouth area, adding one species to the British list in *Carex trinervis*, since considered to be a hybrid; *Ammophila baltica*; and *Sparganium neglectum* Beeby. Of a shy and retiring nature, his friends, and these included men like Arthur Bennett, James Britten, Daydon Jackson and the Rev. W. W. Newbould, would say, 'If only we could get Glasspoole to tell us all he knows, we should learn something.'

HERBERT DECIMUS GELDART (1831–1902), a wine merchant, belonged to an old Norwich family. The frequent references to his finds in the *Flora* of 1914 testify to his activities in the field, whilst his many papers in the Transactions of the Norfolk and Norwich Naturalists Society confirm both his wide knowledge and critical accuracy. He compiled the article on *Botany* in Mason's *History of Norfolk*, 1883; a paper on *Botany in Norfolk* in W. A. Dutt's *Norfolk*, published in 1900; and the botanical portion for the county in the Victoria History, 1901. His herbarium is at the Castle Museum, Norwich.

DR FREDERICK LONG (1840–1927) carried on his practice at Wells-next-the-Sea until 1899 when he retired and lived at Norwich. His was a lifetime devoted to botany, in the course of which he added several plants to the county list, including *Sonchus arvensis* var. *angustifolius*, then (1885) new to the British flora, *Poa subcaerulea*, *Bromus erectus* var. *villosus* and *Trifolium pratense* var. *parviflora*. His collection of 2,000 plants is at the Castle Museum, Norwich.

ARTHUR BENNETT (1843–1929), although neither native nor resident of the county, joined the Norfolk and Norwich Naturalists Society in 1883, and the Norfolk Broads and Breckland provided him with rich hunting grounds. His favourite genera were *Potamogeton, Najas* and *Carex*. His advice on taxonomy and nomenclature was sought by professional and amateur alike; he revised the genus *Potamogeton* in J. D. Hooker's *Student's Flora of the British Isles* and added notes to the *Carices*. Whilst visiting the Broads in 1880 he discovered *Chara stelligera* and, three years later, added *Najas marina*, new to both Norfolk and Britain. His papers, some twenty-seven in all, the *Transactions* of the Society represent a valuable addition to our knowledge of Norfolk plants.

The REV. EDWARD FRANCIS LINTON (1848–1928) accepted the living of Sprowston near Norwich in 1878. During his ten years in Norfolk he devoted considerable time to the flora, specialising in *Rubus, Rosa* and *Salix*. In conjunction with the Revs. W. Moyle Rogers, R. P. Murray, and his brother, W. R. Linton, he edited a *Set of British Rubi* which were eagerly taken up. With his brother, he later issued a *Set of British Willows* on which genus he wrote a monograph (*J. Bot.*, 1913, *Supp.*) and, again with his brother, did outstanding work on the *Hieracia*. His wide knowledge was drawn on by both Exchange Clubs and his notes on plants were invaluable for the identification of critical species. His name is commemorated by *Rubus Lintoni*, an endemic species he found at Sprowston.

WILLIAM A. NICHOLSON (1858–1935), born at Birkenhead, joined the staff of Gurneys Bank (now Barclays) in Norwich and became interested in botany. In 1913 the Norfolk and Norwich Naturalists Society accepted his offer of the MS. of the *Flora of Norfolk* on which he had been working for thirty years. It was published in 1914.

WILLIAM GEORGE CLARKE (1877–1925), Yorkshire-born, lived at Norwich, where he took up journalism. First devoting considerable time to geology and archaeology, he later took up botany and his many papers on Breckland gained him a more than local reputation. The continual mention of 'WGC' in the *Flora* shows the fruits of his tireless energy and it was his pen that coined 'Breckland', a name which has passed into the language. With his friend, W. H. Burrell, the well-known bryologist, he cycled all over the county, adding both species and stations to the *Flora*. A few months before his fatal illness he completed his well-known book *In Breckland Wilds.*

DR JOHN LOWE, born at Sleaford in Lincolnshire, began his medical practice in King's Lynn in 1857. Whilst resident in Norfolk he devoted much of his scanty leisure to cryptogamic botany, but also compiled a paper on *The Flora of Lynn and Neighbourhood* (Bot. Soc. Edinb., 1866). He died in 1902.

FRED ROBINSON lived at Watton in West Norfolk where he practised as a lawyer. He was a friend of G. C. Druce and an active supporter of the Botanical Exchange Club, contributing no fewer than 3,700 sheets from 1913 to 1924. In 1911 he discovered *Calamagrostis stricta* on the site of the former Hockham Mere; in 1914 he found *Cucubalus baccifer* at Merton, claiming it as a native plant; his was the first record for *Cerastium arvense* var. *latifolia* in 1909.

WILLIAM CHARLES FRANK NEWTON (1895–1928) studied the flora around Watton where he spent several holidays prior to 1913. He joined the staff of the John Innes Horticultural Institute in 1922 but his early death removed one of the most promising research students. His earliest paper, 'Notes on South-West Norfolk Plants', appeared in the *Journal of Botany* in 1913.

H. DIXON HEWITT, an anlytical chemist, lived at Thetford for many years and botanised in the local Breckland. His annotated copy of the 1914 *Flora* at the Castle Museum, Norwich, contains many notes and records.

Of the many well-known botanists who visited Norfolk but who have now passed away, mention must be made of G. C. Druce; J. E. Little, who stayed with relatives at Wallington Hall near Downham Market; C. E. Salmon and J. W. White, authors respectively of the *Flora of Surrey* and *The Bristol Flora.*

In recent times, Dr A. S. Watt of Cambridge has devoted considerable time to the many problems of Breckland, and the results of his researches appear in a long series of papers in the *Journal of Ecology.*

We have been fortunate in having Dr C. E. Hubbard's expert help with the grasses, both in the field and herbarium. He is a West Norfolk man and spent his early life in the vice-county before joining the staff at Kew, where he has recently retired from the post of Keeper of the Herbarium.

J. E. Lousley has often visited Norfolk. In 1949 he found *Vicia villosa*; he was probably the last botanist to see *Viola stagnina* and, until recently refound by Mr E. T. Daniels, recorded by photograph the last few flowers of *Crocus purpureus* at its former station at Harleston in 1937.

Dr Francis Rose visits the county from time to time and is mainly interested in the many calcareous valley fens. Thanks to his field-notes we have been enabled to fill several gaps.

Whilst spending many holidays in Norfolk, E. S. Edees, the *Rubi* referee, worked on a *Flora of Swaffham* which was not published. He passed the manuscript to us and has generously allowed us to make full use of it.

A. C. Jermy of the British Museum (Natural History), a Norfolk man, has worked chiefly from his home near the Broads, adding *Luronium natans* and *Nitella mucronata* var. *gracillima* to the county list.

From time to time, works have been published dealing with the Norfolk flora as a whole or on certain aspects of it. Sir Edward J. Salisbury was President of Norfolk and Norwich Naturalists Society, 1931–2, and his Presidential Address on *The East Anglian Flora* proved to be so important that it was later issued in book form and enjoyed a wider circulation.

In 1934, the first edition of *Scolt Head Island* was published for the Norfolk and Norwich Naturalists Society. Edited by Professor J. A. Steers, it contains a full account of the plant and animal life of the dunes and marshes of this classic locality, with chapters on pollen analysis by Professor H. Godwin, and ecology and floral list by V. J. Chapman.

In 1960, Dr Joyce M. Lambert and her team of workers published *The Making of the Broads* (R.G.S. Research Series 3), in which it was proved beyond all doubt that 'these sheets of water are not the relict of a former great estuary as once supposed' but represent the sites of medieval peat pits.

Many of the professional botanists at Kew have visited the county and both their records and assistance have been acknowledged in the *Flora*. They include J. P. M. Brenan, R. D. Meikle, R. Melville, N. Y. Sandwith and V. S. Summerhayes.

Our own work goes back some forty years, and the results of our field-work in West Norfolk were published by the Botanical Society of the British Isles in 1962 as *West Norfolk Plants Today*.

Bibliography

BRITTEN, Jas. & BOULGER, G. S. 1888 *et seq.* Biographical Index of British and Irish Botanists. *Journal of Botany.*

GELDART, A. M. 1913–14. 'Sir James Edward Smith and some of his friends', Presidential Address, Norfolk and Norwich Naturalists Society. *Transactions*, IX, **5**.

—— 1930–1. 'The Hookers in Norfolk and Suffolk', Presidential Address, Norfolk and Norwich Naturalists Society. *Transactions*, XIII, **2**.

SOUTHWELL, Thos. 1906–7. Some Old-time Norfolk botanists, Norfolk and Norwich Naturalists Society. *Transactions*, VIII, **3**.

TREVELYAN, G. M. 1949 *Illustrated English Social History.*

CLIMATE

Norfolk is situated in one of the driest areas of the British Isles and the average annual rainfall nowhere exceeds 26 inches (659 mm.); as the county is generally flat the rainfall is more or less evenly spread although there is a tendency for a slight increase in the west. The wettest period of the year is from October to December. There are greater extremes of temperature than in most of lowland Britain, ranging from the mean of 16 °C. (62 °F.) in July to 3 °C. (38 °F.) in January. In the south-west Norfolk 'Breckland' temperatures are more extreme than elsewhere; it has the lowest rainfall, a high summer temperature, and, combined with its coarse sandy soil, approximates to the continental type of hot summers and cold winters. Late frosts are frequent, and occasionally a temperature of some 15 degrees lower than other places within the county has been recorded. During February, March and April north-easterly winds blow with some force and their parching effects result in Spring emergence being later than elsewhere ('Blackthorn winter'). Along the coastal fringes, sea-mists known as 'frets' sometimes blanket the land and penetrate inland to a depth of about five miles. The north and north-east coasts appear to have a more Oceanic type of climate and this is reflected by the distribution of several plant species.

THE GEOLOGY AND FOSSIL FLORAS OF NORFOLK

G. P. Larwood, B.Sc., Ph.D., M.A., F.G.S.

The following account of the geology of Norfolk emphasises the nature and succession of fossil floras and concentrates on the important sequence of Pleistocene deposits which are particularly well developed in the county and in East Anglia generally. The earlier geological history of the county and the post-glacial sequences are treated briefly, but the list of references includes several major works with extensive bibliographies on these and other horizons.

Geological formations outcropping at the surface in Norfolk range in age from the Kimmeridge Clay of the Upper Jurassic to most recent deposits such as Fenland silts, river alluvium and coastal dunes. Older formations are concealed at depth, but their nature is known from borehole evidence as at North Creake, Southery and Lowestoft. Details of these concealed rocks are given by Pringle (1923), Kent (1947) and Thurrell (1961).

The stratigraphy of Norfolk is summarised in the following table, the most recent deposits first.

SERIES		STAGES	DEPOSITS & FORMATION NAMES
HOLOCENE		Recent	Most recent deposits, still accumulating
		Flandrian	Alluvium, peat, estuarine clays and beach deposits
PLEISTOCENE	Upper	Weichselian	Solifluxion deposits
			Hunstanton Till
		Ipswichian	Wretton Interglacial deposits
		Gippingian	Gipping Till
	Middle	Hoxnian	Nar Valley and St Cross South Elmham Interglacial deposits
		Lowestoftian	Lowestoft Till
			Marine sands
			North Sea Drift = Cromer Till and Norwich Brickearth
		Cromerian	Arctic Freshwater Bed
			Cromer Forest Bed Series – Interglacial
	Lower	Baventian	Icenian Crag — Weybourne Crag
		Antian	Icenian Crag — Norwich Crag
		Thurnian	
		(Ludhamian)	(Scrobicularia Crag and Red Crag)[1]
Neogene Erosion			
PALAEOGENE	EOCENE AND PALAEO-CENE	(Ypresian)	(London Clay)
		(Sparnacian)	(Reading Beds)
		(Thanetian)	(?Thanet Sands)
post-Cretaceous Erosion			
CRETACEOUS	Upper	Maastrichtian	Upper Chalk
		Campanian	
		Santonian	
		Coniacian	
		Turonian	Middle Chalk
		Cenomanian	Lower Chalk
	Lower	Albian	Red Rock and Gault Clay
		Aptian	Carstone
		Neocomian	Snettisham Clay
			Sandringham Sands
JURASSIC		Kimmeridgian	Kimmeridge Clay

[1] Names in brackets refer to deposits concealed at small depth.

It is evident that Pleistocene and Cretaceous deposits make up the bulk of the succession. The Pleistocene sequence is particularly well developed in Norfolk and Suffolk; glacial deposits alternating with temperate· interglacial deposits and corresponding to successive advances and recessions of the ice during very marked climatic fluctuations. The pollen and other plant remains from these Pleistocene deposits, reflecting the climatic variations, are of special interest and warrant the Pleistocene sequence being considered in some detail.

The pre-Pleistocene deposits of Norfolk

All the pre-Pleistocene deposits outcropping in Norfolk are marine sediments. The oldest, the Kimmeridge Clay, outcrops in western Norfolk but is largely concealed by fenland silts and peats. The blue-grey shaly clays of the Kimmeridgian accumulated in a muddy sea which probably inundated the whole region towards the end of Jurassic time. Fossil molluscs, fish and marine reptiles occur in the Kimmeridge Clay together with occasional plant debris. Although ferns, conifers and cycads characterised Jurassic floras generally, only occasional fragments of carbonised and mineralised coniferous wood, which must have drifted far from land, have been recovered from the Kimmeridge Clay in Norfolk.

After slight earth movements and erosion towards the end of the Jurassic the area was again submerged by a shallow sea spreading from the north-'east. The Lower Cretaceous sands and clays which now outcrop in the west of the county accumulated in this sea. Details of the Lower Cretaceous sequence in Norfolk are discussed by Larwood (1961). Molluscs and fossils of other marine groups together with marine vertebrate remains are locally abundant in the Lower Cretaceous sequence. Drifted fragments of coniferous wood and unidentifiable carbonised plant debris from various Lower Cretaceous horizons suggest that shorelines were not too distant at that time.

The Upper Cretaceous is characterised particularly by the Chalk which reaches a thickness of 1,350 feet in East Norfolk and outcrops widely in the county and underlies much of the later Pleistocene deposits. The formation dips gently eastward and its lower beds outcrop at the surface southward from Hunstanton to Lakenheath, and increasingly higher zones are exposed to the east. Along the east coast of Norfolk the eroded surface of the, Chalk is concealed beneath younger deposits. Details of the fauna and lithology of Upper Cretaceous strata in Norfolk are given by Peake & Hancock (1961).

At the beginning of the Cenomanian, in the early Upper Cretaceous, there was a very extensive marine submergence and chalk accumulated in this widespread sea until the end of the Cretaceous period. The Norfolk area was very distant from any shoreline and, after Cenomanian times, very little land-derived sediment contaminated the deposition of pure white limestones in the Chalk sea.

The more pure forms of chalk, some may contain as much as 98 per cent calcium carbonate, are spectacular examples of the way in which the lower plants may contribute significantly to the formation of sediment. The coarser fraction of chalk is composed of the minutely broken remains of the shells of invertebrates, and the finer fraction is formed from the calcareous skeletons of coccoliths which are microscopic marine plank-tonic algae. Black (1953) has demonstrated, using electron-microscopy, the importance of coccoliths in the constitution of the Chalk and has shown subsequently that a succession of distinctive coccolith assemblages may be used to subdivide the Chalk succession. Conditions in the clear waters of the Chalk seas were particularly favourable to coccolith development and the late Cretaceous is a period of marked abundance of these algae.

In general during the Cretaceous period important changes occurred in the composition of land floras. Throughout the Lower Cretaceous ferns and cycadophytes were abundant and conifers common. This association was effectively a continuation of Jurassic floras. During Upper Cretaceous times essentially modern types of vegetation developed. The dominant cycadophytes of the earlier Mesozoic were replaced by flowering plants, and it is significant that there was marked evolutionary adaptation of pollinating insects at the time of the development of the Angiosperms in the late Cretaceous.

At the end of the Cretaceous further earth movements uplifted, tilted and warped the deposits forming a land area which was extensively eroded. Flint nodules which occur in the Upper Chalk provided much material which was derived to form pebbles in later deposits.

The first-formed post-Cretaceous deposits in Norfolk are concealed at depth in the eastern part of the county and are of Palaeocene and Eocene age. These lower Tertiary, Palaeogene, sediments are marine and fluvia-tile sands and clays. They have been penetrated by a number of boreholes as at Cantley and Great Yarmouth. The basal Tertiary glauconitic sands may be equivalent to the Thanet Sands of south-eastern England, which they resemble lithologically, and the overlying clays with lignite may be fluviatile correlatives of the Reading Beds. Ypresian London Clay is also known from borehole evidence in East Norfolk. Downing (1959) described the concealed western limits of this deposit which is over 300 feet thick beneath Great Yarmouth and consists of brown-grey sandy marine clays. No fossils are recorded from the clay penetrated by the boreholes in Norfolk, but elsewhere in south-eastern England the London Clay has a rich invertebrate and vertebrate fauna and abundant drifted plant remains.

In the floras of the Palaeogene Angiosperms were dominant and ferns and conifers common, but the cycads had almost disappeared. Warm, moist, temperate conditions are indicated by the earliest Tertiary land plants. Rivers flowing from the west into the Eocene sea covering south-eastern England carried down much driftwood and many fruits and seeds which have been preserved in detail. Fruits of the palm *Nipa burtini* are most common and many other species of palms and other plants are represented. Many of these are now exclusively or mainly tropical in distribution. Somewhat later in the Eocene, deposits in southern England demonstrate the increasing abundance of flowering plants.

Throughout most of the Tertiary the flora of western Europe con-tained a high proportion of plants comparable with those of present south-east Asian tropical communities. During the Pliocene, towards the end of the Tertiary, temperatures in the East Anglian region were lower, since molluscs with Mediterranean affinities are a prominent element of the

23

Upper Pliocene Coralline Crag (marine shelly sands) of the Aldeburgh-Orford area in Suffolk.

No Upper Tertiary (Neogene) deposits of Miocene and Pliocene age are preserved at depth in Norfolk. The region was subjected to considerable erosion during Neogene times and shelly sands, part of the Lower Pleistocene Red Crag, rest directly on the eroded surface of the London Clay.

The Pleistocene deposits of Norfolk

In the marine shelly sands of the Red Crag boreal cold water mollusca appear among the fauna for the first time together with the remains of boreal mammals. These faunal elements indicate colder conditions than those of the later Pliocene and are taken as the first signs of the climatic deterioration resulting ultimately in the first advance of ice sheets into the region in post-Cromerian times.

The Lower Pleistocene sequence

Lower Pleistocene formations in Norfolk consist of a series of marine littoral shelly sands (the Red, Norwich and Weybourne Crags) and associated gravels, sands and clays. The mollusca of these deposits were studied extensively by Harmer (1896, 1899, 1900, 1902, 1914–24). He noted a marked increase in the frequency of boreal colder mollusca in the later parts of the Crag sequence. Recently, Funnell (1961) has examined the foraminifers from Crag sequences in Norfolk, and West (1961) has studied the associated pollen. Their detailed conclusions on successive climatic stages in the Crag sequence were based particularly on the analysis of samples from a borehole at Ludham. At this locality the borehole penetrated deposits of Red Crag from −90 to −165 feet O.D. Foraminifers recovered from these coarse shelly sands indicate temperate climatic conditions (Funnell, 1961). About 20 feet of grey silty clays succeed the Red Crag in the Ludham borehole. Funnell compares the temperate foraminiferal assemblage from these beds with that from the Lower Division of the Norwich Crag in Suffolk. From −70 to −40 feet O.D. the foraminifers recovered from grey shelly sands and clays indicate a climatic deterioration leading to very cold glacial conditions which prevailed during the deposition of the succeeding 10 feet of clays. The foraminifers from these clays are comparable with those from the Weybourne Crag. Pollen recovered from the same sequence supports this evidence of climatic fluctuations during the Lower Pleistocene (West, 1961).

The vegetational history of the Lower Pleistocene, based on analysis of pollen from the Ludham borehole sequence, indicates two stages (the Thurnian and the Baventian) with high nonarboreal pollen frequencies characterised by abundant *Ericales*, representing oceanic heath developed in cold to glacial conditions, and two temperate stages of interglacial type (the Ludhamian and the Antian) with higher frequencies of thermophilous tree pollen representing mixed coniferous and deciduous forest which included the Tertiary relic genera *Tsuga* and *Pterocarya*. A third temperate stage at the top of the sequence may correlate with part of the succeeding Cromer Forest Bed Series at the base of the Middle Pleistocene.

The Middle Pleistocene sequence

The early temperate phase of the Middle Pleistocene in Norfolk is represented by the Cromer Forest Bed Series which is famous for its abundant mammalian remains. A recent concise account of fossil vertebrates from the Cromer Forest Bed Series is given by McWilliams (1967). The Series consists of Lower and Upper Freshwater Beds (carbonaceous clays and peats) separated by an Estuarine Bed (sands and gravels with associated peaty deposits and clay lenticles). The Estuarine Bed encloses drifted tree stumps and much macroscopic plant debris. Reid (1882, 1890) described the Cromer Forest Bed Series in detail. Pollen analysis of samples from the Upper Freshwater Bed (Duigan, 1963) has indicated that there were at the time of its deposition climatic fluctuations from open vegetation to boreal coniferous forest (with birch and pine) to temperate mixed oak forest conditions (with alder, oak, elm, lime and some hazel) returning to boreal forest (pine and birch), in the later phases. The water fern *Azolla filiculoides* occurs commonly in the Upper Freshwater Bed which is characterised by *Picea* and by the scarcity of *Abies*, *Corylus* and *Carpinus* pollen. The evidence of true temperate conditions establishes the Cromer Forest Bed Series as in part truly interglacial between the cold Baventian and the ensuing Lowestoftian glacial stage.

The Arctic Freshwater Bed, which locally succeeds the Cromer Forest Bed Series, comprises thin sands and laminated clays with arctic plants such as *Betula nana* and *Salix polaris* reflecting the very cold conditions preceding the deposition of the Cromer Till from the first advance of ice sheets into the Norfolk region in the early Lowestoftian Stage.

The first of the two ice advances of the Lowestoftian, the Cromer Advance, passed south-eastward over Norfolk depositing the North Sea Drift (the Cromer Till and Norwich Brickearth). Scandinavian rocks are characteristic erratics of these deposits. The second, Lowestoft Advance, passed eastward across Norfolk depositing clayey till with many Chalk and Jurassic erratics. Sands with marine mollusca occur between the tills of the two advances but there is no evidence of a return to fully interglacial climatic conditions during the deposition of the sands.

The late Middle Pleistocene in Norfolk concluded with an amelioration of climate accompanied by retreat of the Lowestoft Ice and plant colonisation of the till surface. Lake deposits accumulated in depressions on the surface of the Lowestoft Till at various localities in Suffolk and Norfolk. The vegetation of this Hoxnian Interglacial is best known from the type locality but also from equivalent deposits in the Nar Valley (Stevens, 1960) and at St Cross South Elmham (West, 1961). In the Hoxnian Interglacial deposits a sequence of four climatic plant zones is recognisable. An early glacial zone with high nonarboreal pollen frequencies and *Hippophae* precedes an early temperate zone in which four subzonal assemblages are present. In the early temperate zone an early *Betula-Pinus* subzone is succeeded by *Betula-Quercus-Pinus*, *Alnus-Quercus-Betula-Tilia* and *Alnus-Quercus-Taxus-Ulmus-Tilia* subzones. This sequence is followed by a later temperate zone with an *Alnus-Carpinus-Picea-Abies* assemblage. A glacial zone with arctic plants preludes the succeeding Gippingian Glaciation of the early Upper Pleistocene.

25

The Upper Pleistocene sequence

During the Gipping Glaciation, at the beginning of the Upper Pleistocene, ice sheets advanced south-eastward across Norfolk, and a chalky till with northern igneous erratics was deposited over much of the county.

During the succeeding post-Gipping temperate stage – the Ipswichian Interglacial – peat deposits and silts accumulated in valley sites as near Ipswich in Suffolk and near Wretton in Norfolk. Pollen analysis of samples from such sites indicates an early glacial zone with high nonarboreal pollen frequencies, reflecting open vegetation, followed by a sequence of zones characterised by different assemblages of arboreal pollen: a *Betula-Pinus* zone, a *Pinus-Betula-Ulmus* zone, a *Pinus-Betula-Quercus-Ulmus* zone, a *Quercus-Pinus-Corylus-Acer-Alnus* zone, a *Carpinus* zone and a *Pinus* zone. Early abundant *Corylus* and *Acer* pollen and late abundant *Carpinus* pollen are thus characteristic of the Ipswichian Interglacial. Macroscopic plant remains include several species not now native, and certain water plants indicate that summers were warmer than those of the post-Pleistocene Flandrian stage.

The last advance of ice into Norfolk impinged only on the north Norfolk coast depositing the brown Hunstanton Till during the early Weichselian Stage and causing extensive solifluxion (freeze-thaw disturbances) of the ground beyond. Stone stripes and polygons of the Breckland region, selectively colonised by *Calluna* and other plants, are well-formed examples of such periglacial phenomena. Towards the end of the Weichselian, following the final retreat of the ice, park tundra conditions were developed widely and colonisation by *Betula* was common.

It is evident that there were extreme fluctuations of climate during Pleistocene times in the Norfolk area. Three main phases of ice advance into the region and three intercalated interglacial phases of temperate climate occurred. Such marked fluctuations profoundly altered the distribution of both faunas and floras. A lowering of sea level accompanied each expansion of the ice sheets, which at their maximum advance spread south to the Thames–Severn line. In the periglacial tundra area beyond the southern limits of the ice land-connections were formed by the fall in sea level and such connections were severed only by the rise of sea level concomitant with ice recession. Geographical alterations of this kind greatly affected the speed and mode of plant and animal recolonisation of pioneer soils developing in recently ice-covered areas.

The post-glacial sequence

The main features of Flandrian (post-glacial) vegetation in England are summarised in the table opposite.

West (1967) in an excellent summary of the history of British Quaternary stratigraphy, faunas and floras indicates that early abundant *Corylus* pollen and the occurrence of *Tilia*, *Carpinus* and *Fagus* and the absence of *Picea* and *Abies* distinguish Flandrian vegetational sequences from those of interglacial times. At the beginning of the Flandrian there was a marked rise in arboreal pollen, and the sequence of pollen zones IV–VIII demonstrates the establishment of temperate deciduous forests and the effects of deforestation by man from the marked *Ulmus* decline at the beginning of Sub-Boreal times.

STAGE	VEGETATIONAL PERIODS	CHARACTERISTIC ASSEMBLAGES
	Modern Afforestation	
Flandrian	Sub-Atlantic (Zone VIII) (back to 500 BC)	*Alnus-Quercus-Betula-Fagus-Carpinus assemblage*
	Sub-Boreal (Zone VIIb) (3000–500 BC)	*Alnus-Quercus-Tilia assemblage*
	Atlantic (Zone VIIa) (5300–3000 BC)	*Alnus-Quercus-Ulmus-Tilia* assemblage
	Boreal (Zone VI) (7000–5300 BC)	*Pinus-Corylus* assemblage
	Boreal (Zone V) (7600–7000 BC)	*Corylus-Betula-Pinus* assemblage
	Pre-Boreal (Zone IV) (8300–7600 BC)	*Betula-Pinus* assemblage

Important investigations into the late and post-glacial vegetational history of Norfolk include those of Godwin (1940, 1968), Godwin and Tallantire (1951) and of Tallantire (1953, 1954). These investigations relate chiefly to Fenland deposits and to those of some Norfolk meres at Hockham, Old Buckenham and Lopham. Godwin (1956) discusses vegetational sequences of the Flandrian in detail in his *History of the British Flora*.

During post-glacial times peats and marine silts and clays accumulated in the Fenland and Broadland areas. The two main horizons of marine clay deposition in the Fenland sequence correlate approximately with Lower and Upper Clays of Neolithic and Romano-British marine incursions in the Broadland sequence.

The Broads, formed by the flooding of ancient excavations, are the best known post-glacial lakes in Norfolk. Their detailed vegetational history was investigated by Drs Lambert and Jennings, whose researches revealed a sequence of peats and clays which correlate with the Flandrian vegetational zones as follows.

e. Upper Peats (Zone VIII)
d. Upper Clay (Zone VIII – Romano-British marine incursion)
c. Middle Peats (high Zone VII and transition Zone VII–VIII)
b. Lower Clay (Zone VII – Neolithic marine incursion)
a. Basal Peats passing laterally inland into detrital muds (Zone V to early Zone VII).

A detailed account of these deposits and of the associated historical investigations is given in the R.G.S. Research Series Publication No. 3, *The Making of the Broads* (Lambert et al., 1960).

Bibliography

BLACK, M. 1953. The Constitution of the Chalk. *Proc. Geol. Soc. Lond.*, No. 1499, lxxxi, lxxxv.

DOWNING, R. A. 1959. A Note on the Crag in Norfolk. *Geol. Mag.*, **96**, 81.

DUIGAN, S. L. 1963. Pollen Analysis of the Cromer Forest Bed Series in East Anglia. *Phil. Trans. Roy. Soc., Lond.*, **246**, 149.

FUNNELL, B. M. 1961. The Paleogene and Early Pleistocene of Norfolk. *in* The Geology of Norfolk. *Trans. Nfk. and Norw. Nat. Soc.*, **19** (6), 340.

GODWIN, H. 1940. Studies of the Post-glacial history of British Vegetation. III. Fenland pollen diagrams. IV. Post-glacial land- and sea-level in the English fenland. *Phil. Trans. Roy. Soc., Lond.*, **230**, 239.

—— 1956. *The History of the British Flora.* Cambridge.

—— 1968. Studies of the Post-glacial history of British Vegetation. XV. Organic Deposits of Old Buckenham Mere, Norfolk. *New Phytol.*, **67**, 95.

—— & TALLANTIRE, P. A. 1951. Studies of the Post-glacial history of British Vegetation. XII. Hockham Mere, Norfolk. *New Phytol.*, **39**, 285.

HARMER, F. W. 1896. On the Pliocene Deposits of Holland and their Relation to the English and Belgian Crags. . . . *Quart. J. Geol. Soc. Lond.*, **52**, 748.

—— 1899. On a proposed new Classification of the Pliocene Deposits of the East of England. *Rep. Brit. Assoc. Dover*, 751.

—— 1900. The Pliocene Deposits of the East of England – Part II. The Crag of Essex (Waltonian) and its relation to that of Suffolk and Norfolk. *Quart. J. Geol. Soc. Lond.*, **56**, 705.

—— 1902. A Sketch of the Later Tertiary History of East Anglia. *Proc. Geol. Assoc. Lond.*, **17**, 416.

—— 1914–24. The Pliocene Mollusca of Great Britain. *Mon. Pal. Soc.*, 2 vols.

KENT, P. E. 1947. A Deep Boring at North Creake, Norfolk. *Geol. Mag.*, **84**, 2.

LAMBERT, J. M., JENNINGS, J. N., SMITH, C. T., GREEN, C. & HUTCHIN-SON, J. N. 1960. *The Making of the Broads.* R.G.S. Res. Ser. 3.

LARWOOD, G. P. 1961. The Lower Cretaceous Deposits of Norfolk. *in* The Geology of Norfolk. *Trans. Nfk. and Norw. Nat. Soc.*, **19** (6), 280.

McWILLIAMS, B. 1967. *Fossil Vertebrates of the Cromer Forest Bed in Norwich Castle Museum.* City of Norwich Mus. Publ.

PEAKE, N. B. & HANCOCK, J. M. 1961. The Upper Cretaceous of Norfolk. *in* The Geology of Norfolk. *Trans. Nfk. and Norw. Nat. Soc.*, **19** (6), 293.

PRINGLE, J. 1923. On the Concealed Mesozoic Rocks in South West Norfolk. *Summ. Prog. Geol. Surv. for 1922*, 126.

REID, C. 1882. The Geology of the Country around Cromer. *Mem. Geol. Surv. Eng. Wales.*

—— 1890. The Pliocene Deposits of Britain. *Mem. Geol. Surv. U.K.*

STEVENS, L. A. 1960. The Interglacial of the Nar Valley, Norfolk. *Quart. J. Geol. Soc. Lond.*, **115**, 291.

TALLANTIRE, P. A. 1953. Studies of the Post-glacial history of British

Vegetation. XIII. Lopham Little Mere, a late glacial site in central East Anglia. *J. Ecol.*, **41**, 361.

—— 1954. *Ibid.* XIV. Old Buckenham Mere. Data for the Study of Post-glacial History. *New Phytol.*, **53**, 131.

THURRELL, R. G. 1961. The Sub-Cretaceous Rocks of Norfolk. *in* The Geology of Norfolk. *Trans. Nfk. and Norw. Nat. Soc.*, **19** (6), 271.

MAJOR
PLANT COMMUNITIES

C. P. Petch, M.D., F.R.C.P.

The natural vegetational climax of Norfolk, as of most temperate regions, would be forest, and it must be presumed that at some time after the last ice age most of the county was covered by woodland. From Neolithic times, however, human activity has tended to destroy trees, and to prevent their regeneration, starting on the lighter soils which were readily converted to heathland, extending gradually, as methods of cultivation became more effective, to the richer ones. This process has continued up to the present day, so that most of the county is now intensively farmed. Those areas which particularly attract the botanist, the remaining woods, heaths and fens, are best described as semi-natural. A claim can be made on the other hand for the maritime plant communities as natural ones, and as they are of great interest they will be considered first. All three formations – salt marsh, shingle and sand – are well represented.

Salt Marsh

This forms the shore of the Wash from Wisbech to Wolferton, and lines the estuaries of all the lesser rivers. On the north coast sandy salt marshes shelter behind Scolt Head Island, Blakeney Point and the shingle spit at Holme. The dominant species is the grass *Puccinellia maritima*, which forms a pure sward in the Wash, but the north-coast marshes show a more interesting mixture of species, including *Salicornia perennis*, *Frankenia laevis*, and four species of Sea Lavender (*Limonium*). During the past thirty years the annual *Salicornia* species have been displaced from the role of pioneer colonist of mud by the invasive hybrid *Spartina townsendii*, first introduced into the Lynn Cut in 1909. Along the Wash coast the salt-marsh level is being gradually raised by the deposition of silt, and, as soon as it is high enough, is enclosed for cultivation. Large areas have been reclaimed in this way since 1850 and the pace has accelerated since 1949. The northern salt marshes have been the subject of much ecological research, including the important work of Chapman at Scolt Head, and that of Oliver at Blakeney. Holme salt marsh has been studied by several Cambridge workers, most recently by Conway in 1933.

Sand dune

Dunes extend from Old Hunstanton to east of Wells, and again from Winterton to Great Yarmouth. They form the island of Scolt Head, and the headland of Blakeney Point, each on a shingle foundation. The stages of dune formation can be studied on the foreshore, beginning with the strand flora – *Cakile*, *Salsola* – passing through the early dunes of *Agropyron junceiforme*, to the grey dunes dominated by *Ammophila*, *Elymus* and *Festuca arenaria*. In the spring these dunes are gay with ephemerals – *Cerastium atrovirens*, *Erophila verna*, *Valerianella locusta* – and in a wet summer bee orchids may be frequent. Pyramidal orchids are also sometimes seen. Rare dune grasses found along the coast are *Vulpia mem-*

branacea, Poa bulbosa, Corynephorus canescens, and the inter-generic hybrid *Ammocalamagrostis baltica.*

Shingle

The six-mile bank joining Blakeney Point to the cliffs at Weybourne is one of the best examples of a shingle habitat in the British Isles. Here F. W. Oliver worked on the autecology of *Suaeda fruticosa,* and demonstrated its close adaptation to these conditions. A similar shingle bank extends southwards along the shore of the Wash from Snettisham to Wolferton, and shingle forms the curving 'laterals' of Blakeney Point and Scolt Head Island. Characteristic plants are *Glaucium flavum, Suaeda fruticosa* and *Silene maritima.*

Heath

Kirby Trimmer tells us that in the reign of Charles II a heath extended 'with little interruption' from south of Norwich to Lynn. By Trimmer's time much of this had been enclosed and cultivated, but much remained. During the lifetime of the present authors many heaths have been ploughed – Massingham Heath, Coxford Heath, and heathland adjacent to Roydon Common. Others, particularly in the Breckland, have been planted with conifers. It seems unlikely that *Calluna* heath was ever a natural climax. Observation today suggests that in the absence of burning, birch trees invade these soils. Grazing animals may have played a part in preventing tree growth in the past, but now are given better pasture.

The best known Norfolk heaths are those of the Breckland, brought to the public eye by the writings of the late W. G. Clarke. This area, some 400 square miles in extent, lies half in West Norfolk and half in West Suffolk, centred upon Thetford and Brandon. The word 'breck' from which the name derives is in common use throughout West Norfolk for a field on marginal land, ploughed in times of agricultural prosperity, allowed to revert to grass heath in times of depression. Such grass heath may persist for 100 years. An early investigation of the causes of plant distribution in Breckland was undertaken by E. Pickworth Farrow (1925), who stressed the importance of rabbit grazing to the virtual exclusion of other factors. Later work by A. S. Watt (1936, 1940) has pointed out the significance of soil differences, particularly in respect of chalk content. Although since the time of Clarke and Farrow much of the Breckland has been planted with conifers by the Forestry Commission, the original plant community can still be seen on Weeting Heath, a reserve belonging to the Norfolk Naturalists Trust, as well as at East Wretham, Gooderstone, Cockley Cley and elsewhere. Much of the Stanford battle area is in a similar state. Rainfall here is the lowest in the British Isles and a unique group of so-called steppe species characterises the Breck flora – *Artemisia campestris, Silene otites, Medicago minima, Scleranthus perennis, Phleum phleoides, Veronica spicata, V. triphyllos, V. verna.* Professor Salisbury's map ('East Anglian Flora', *Trans. Nfk. and Norw. Nat. Soc.,* 1932, p. 235) show the close relation between the distribution of three of these and the rainfall. In the spring the ephemeral flora may be more striking than that of the sand dunes.

The Greensand ridge of West Norfolk gives rise to even poorer and more acid soils than the Breck, and on them have developed the extensive pure *Calluna* heaths of Dersingham, West Newton, North Wootton, Roydon and Grimston. In the last few years Grimston Warren has been afforested. *Crassula tillaea* is abundant on tracks on the Greensand. Heaths on glacial gravels are scattered throughout the county and are particularly striking on the hills behind Salthouse and Kelling on the north coast – the outwash plains of a glacier. Here *Calluna* is mixed with *Erica cinerea* and dwarf gorse. Of other isolated heaths on gravel, Mousehold, near Norwich, is perhaps the best known.

Bog

This denotes the vegetation of acid peat developing in areas of restricted drainage on poor soil, usually close to heath and not to be confused with fen. Characteristic are *Sphagnum* species, *Rynchospora alba*, *Eriophorum angustifolium*, *Vaccinium oxycoccos*, and three species of sundew, but plants of wet heath are also found, e.g. *Erica tetralix*, *Trichophorum caespitosum*, *Carex panicea*, *Genista anglica*. This association is the most vulnerable, through drainage, of all those described, and has completely disappeared from the adjacent county of Cambridgeshire. The extensive bog at Roydon Common, home of *Hammarbya paludosa*, is now a nature reserve of the Norfolk Naturalists Trust. Bogs are also found on the Greensand at Dersingham and North Wootton, and small fragments are scattered throughout the county. That on East Winch Common provides our only site for *Deschampsia setacea*, as well as an abundance of *Gentiana pneumonanthe*.

Fen

Fen vegetation, developed on peat where drainage of water from basic soils prevents an acid reaction, formerly occupied the Great Level in North Cambridgeshire, South Lincolnshire and West Norfolk. Today drainage has abolished it, and apart from specially preserved fragments such as Wicken Fen, no trace remains. Cultivation is causing steady wastage of the drained peat exposing the underlying silt. The river basins of East Norfolk have been managed on different lines, peat having been extracted for burning rather than cultivated *in situ*. Here fen still marks a stage in the succession from the aquatic vegetation of the Broads to fen carr (see below). Throughout the rest of Norfolk, valley fens border the smaller rivers and streams – South Lopham Fen by the Waveney; Blo' Norton Fen by the Little Ouse; Swangey and Rockland Fens and Cranberry Rough in the Thet Valley; Caldecote and Gooderstone Fens and Foulden Common in that of the Wissey; Scarning, Kettlestone Fen and Sculthorpe Moor in that of the Wensum; Marham Fen by the Nar; and lastly, at least until recently, Leziate, Derby and Sugar fens by the Gaywood river.

The dominant species of true fen is *Cladium mariscus*, but where peat is shallower *Juncus subnodulosus* may replace it. *Schoenus nigricans* and *Molinia caerulea* suggest slightly more acid conditions. *Glyceria fluitans* often borders the rivers, *Phragmites communis* and *Schoenoplectus lacustris* the Broads. Transition to carr is marked by *Salix cinerea, Alnus, Frangula,*

Rhamnus, Prunus padus and *Betula pubescens*. Of rarer species, *Liparis loeselii* has been seen recently on South Lopham Fen and East Walton Common. *Carex appropinquata, Lathyrus palustris* and *Peucedanum palustre* are fairly widely distributed. *Calamagrostis stricta*, another rare species in Britain, is confined to Cranberry Rough, a fen carr developed on the site of the former Hockham Mere, and to the adjacent Thompson Common.

Chalk Grassland

Although chalk underlies almost the whole of Norfolk it appears at the surface mainly in the west, and its slope is rarely steep enough to favour typical chalk grassland. Ringstead Downs, a dry valley in the middle chalk about a mile long to the east of Hunstanton, is an exception, as are also on a smaller scale artificial chalk earth works such as Warham Camp and the Devil's Dykes. One of the latter used to run from Narborough to Beecham-well, crossing the dry land between the Nar and Wissey valleys, but is now almost destroyed. Another still runs from Cranwich to Weeting between the Wissey valley and Little Ouse and bears an interesting chalk flora including *Thesium humifusum, Phleum phleoides, Silene otites, Astragalus danicus* and *Carex ericetorum*. A fragment of a third Devil's Dyke at Garboldisham bears *Veronica spicata*.

Chalk grassland, maintained by grazing, also borders Caldecote and Marham fens, and forms large parts of Foulden Common, Lamb's Common, East Walton Common and the now dry Beechamwell Fen. Patches of chalky boulder clay occur in many places, often in sandy areas such as Roydon Common, and this explains the occurrence of chalk-loving species such as *Ophrys apifera* in such sites. Elsewhere leaching has allowed the development of heath over chalk as at Massingham Heath and in parts of Breckland. Mention has already been made of the peculiar flora of the Breckland heaths, and it should be noted that some of its components are chalk lovers – *Silene otites, Phleum phleoides, Dianthus deltoides, Artemisia campestris*. Watt has drawn attention to the high chalk content of some types of Breckland soil.

Woodland

Most Norfolk woods have been adapted to the preservation of game, especially round such great estates as Sandringham, Houghton, Raynham and Holkham. Many of them were planted during the last 100 years, some on the site of previous woods that had been felled. The custom of clear felling and replanting woodland is one that comes naturally to a people familiar with arable farming, but does not accord with ecologically based ideas of forestry, and has a destructive effect on the native shade-loving species. This, often combined with a poor soil, makes the ground flora of such woods limited in the extreme. *Pteridium* and *Rubus* species are the commonest, with occasional *Mercurialis perennis, Silene dioica* or *Circaea lutetiana*. Improvement in the soil is shown by *Anemone nemorosa, Galeobdolon luteum, Allium ursinum* and *Melica uniflora*. The most exacting woodland plants – *Paris quadrifolia, Melampyrum pratense, Galium odoratum, Platanthera chlorantha* – are only found in old-established woods on the clay of central Norfolk. Some of these – Horning-toft Wood, Rawhall Wood, parts of Swanton Novers Great Wood – are on good land which has probably borne forest since prehistoric times, and are of infinitely greater interest than the more usual plantations on sandy soil.

Aquatics

The long discussion over the origin of the Norfolk Broads has been finally settled by the brilliant and painstaking work of Miss Lambert and colleagues (*The Making of the Broads*, R.G.S. Research Series 3, 1960). She has shown by boring that the Broads are man-made medieval turf pits, dug at a time when the land stood higher in relation to the sea than at present. If a fault of British plant ecologists has been their failure to appreciate the influence of earlier human activity on our vegetation, Miss Lambert's work serves as a corrective. It is now easy to understand why the Broads are steadily shrinking in size (though Tansley in 1939 was able to describe them as the largest fresh-water lakes in the South of England), but they are still of great botanical interest. They show a hydrosere leading from open water to fen and eventually fen carr, with local variants depending upon water movement, salinity and similar factors. One aquatic confined to this part of the British Isles is *Najas marina*, found in a few of the north-eastern broads, but now apparently spreading (Ellis, 1966). Other rarities, which may be abundant in Broadland, are *Stratiotes aloides* and *Sonchus palustris*.

Open water is scarce in the rest of Norfolk. The meres of Breckland are a striking feature of an otherwise arid region, but are prone to dry out in a succession of dry years. When wet they show *Potamogeton lucens*, *P. gramineus* and their hybrid *P. zizii*, and one – Home Mere – is our only remaining locality for *Littorella lacustris*. Artificial lakes in parks and the artificial waterways of the fens often show *Acorus calamus*. Small streams draining the chalk are characterised by *Ranunculus pseudofluitans* and *Oenanthe fluviatilis*. Brackish water shows *Ranunculus baudotii*, *Ruppia maritima*, *R. spiralis* and *Potamogeton pectinatus*.

Bibliography

CHAPMAN, V. J. 1934. *in* Steers, A. J. (editor), *Scolt Head Island*, Cambridge.

CONWAY, V. J. 1933. *J. Ecol.*, **21**, 263.

ELLIS, E. A. 1965. *The Broads*, London.

FARROW, E. Pickworth. 1925. *Plant Life on East Anglian Heaths*, Cambridge.

LAMBERT, J. M., JENNINGS, J. N., SMITH, C. T., GREEN, C., and HUTCHINSON, J. N. 1960. *The Making of the Broads*. R.G.S. Research Series 3, London.

OLIVER, F. W. and SALISBURY, E. J. 1913. *J. Ecol.*, **1**, 249.

SALISBURY, E. J. 1932. East Anglian Flora. *Trans. Nfk. and Norw. Nat. Soc.*, **13**, 191.

WATT, A. S. 1936. *J. Ecol.*, **24**, 117.

WATT, A. S. 1940. *J. Ecol.*, **28**, 42.

THE COMPOSITION OF THE FLORA

In assessing the status of many species, no two botanists would ever agree; judgment is so much a matter of personal opinion and many of the terms used are themselves arbitrary. In our attempt to classify the plants of our county we have used the system proposed by J. E. Lousley in his paper *The Recent Influx of Aliens into the British Flora* (Bot. Soc. Brit. Is., 1952 Conference).

NATIVES Species believed to have been in Britain before man, or to have immigrated without his aid by using their natural means of dispersal, or to have arisen *de novo* here.

DOUBTFUL NATIVES Species with a long history in Britain but which are suspected of having been introduced by human agency.

DENIZENS Species growing in natural or semi-natural communities and not dependent for their persistence on human disturbance of the habitat.
COLONISTS Species which grow only in habitats created and maintained by human activities. These are mainly weeds of cultivation restricted to arable fields and disturbed ground.

ALIENS Species believed to have been introduced by the intentional or unintentional agency of man.

NATURALISED ALIENS Introduced species which are naturalised in natural or semi-natural habitats.
ESTABLISHED ALIENS Introduced species which are established only in man-made habitats.
CASUALS Introduced species which are uncertain in place or persistence, i.e., not naturalised or established.

Applying this scheme to Norfolk plants we arrive at the following census:

Natives	982 or	62·8 per cent
Denizens	19	1·2 per cent
Colonists	77	4·9 per cent
Naturalised aliens	79	5·0 per cent
Established aliens	84	5·3 per cent
Casuals	218	14·0 per cent
Doubtful status	20	1·2 per cent
Hybrids	84	5·3 per cent
	1,563	

List of Recorders

In addition to those botanists whose names are quoted in full, the following list gives the names of those whose initials only appear in the text.

Addington, Mrs S.	SA	Hodgson, J.	JH
Allen, G. O.	GOA	Hoff, W. G.	WGH
Armstrong, Miss M. B.	MBA	Howitt, R. C. L.	RCLH
Bagnall-Oakeley, R. P.	RPB-O	Hubbard, C. E.	CEH
Barnes, Miss R. M.	RMB	Jermy, A. C.	ACJ
Bennett, A.	AB	Leadbitter, E.	EL
Bishop, Mrs E.	EB	Leather, Miss V. M.	VML
Bitton, E. Q.	EQB	Libbey, R. P.	RPL
Bourne, P. J.	PJB	Linton, E. F.	EFL
Brenan, J. P. M.	JPMB	Lousley, J. E.	JEL
Brown, Miss M. I.	MIB	Marks, Miss K.	KM
Bull, A. L.	ALB	Maxey, Miss D. M.	DMM
Bull, K. H.	KHB	Melville, R.	RM
Bullock-Webster, G. R.	GRB-W	Moore, P. D.	PDM
Burrell, W. H.	WHB	Newton, W. C. F.	WCFN
Chandler, J. H.	JHC	Noble, Miss E. R.	ERN
Clarke, W. G.	WGC	Palmer, R. C.	RCP
Copping, A.	AC	Pankhurst, J. S.	JSP
Coker, P.	PC	Richardson, F. D. S.	FDSR
Cross, R. S.	RSC	Riddelsdell, H. J.	HJR
Daniels, E. T.	ETD	Robinson, F.	FR
Danvers, Miss M. V.	MVD	Rocke, G. H.	GHR
Day, F. M.	FMD	Rose, F.	FRo
Ducker, A. C.	ACD	Rumbelow, P. E.	PER
Ducker, B. F. T.	BFTD	Ryves, T. B.	TBR
Durrant, K.	KD	Sanderson, H. A.	HAS
Edees, E. S.	ESE	Sandwith, N. Y.	NYS
Ellis, A. E.	AEE	Sell, P. D.	PDS
Ellis, E. A.	EAE	Silverwood, J. H.	JHS
Evans, J. B.	JBE	Simon, P. H.	PHS
Fitter, R. S. R.	RSRF	Smith, T. C.	TCS
Forrest, Miss C.	CF	Stirling, A. M.	AMS
Garrod, G.	GG	Taylor, B.	BT
Gurney, Miss C.	CG	Townsend, C. C.	CCT
Halligey, P.	PH	Tuck, Miss G.	GT
Harley, R. M.	RMH	Tutin, T. G.	TGT
Hewitt, H. D.	HDH	Willé, Mrs P. A.	PAW

Among the many local botanists who have given us considerable help, we would make special mention of Mr E. T. Daniels of Norwich who has scrutinised the typescript, generously allowed us to make full use of his field-work on the *Centaurea nigra-nemoralis* complex and added innumerable records for East Norfolk; Mr A. L. Bull of Foxley sent us yearly lists of finds in mid- and south Norfolk; Mrs S. Addington from her home area of Tasburgh, south of Norwich; Miss D. M. Maxey for her help over many years and her records from central Norfolk; Mr J. H. Silverwood who added many records from north, south and East Norfolk; Miss G. Tuck from the Stanhoe-Docking area of north-west Norfolk; and Miss C. Forrest of Pulham St Mary.

Bibliography

ARNELL, S. 1956. *Illustrated Moss Flora of Fennoscandia* (Hepaticae).

BOREAU, A. 1857. *Flore du Centre de la France.*

BRAITHWAITE, R. 1887–1905. *British Moss Flora.*

CHAPMAN, V. J. 1934. Botany in Steers, J. A. (editor), *Scolt Head Island.*

CLAPHAM, A. R., TUTIN, T. G. & WARBURG, E. F. 1952, 1962. *Flora of the British Isles,* 1st and 2nd editions.

CLARKE, W. G. 1937. *In Breckland Wilds,* 2nd edition, rewritten by R. R. Clarke.

CRUNDWELL, A. C. & NYHOLM, E. 1964 The European Species of the *Bryum erythrocarpum* Complex, (*Trans. Brit. bryol. Soc.,* IV, 597).

DIXON, H. N. 1954. *The Student's Handbook of British Mosses,* 3rd reprint.

DUCKER, B. F. T. & WARBURG, E. F. 1961. *Physcomitrium eurystomum* in Britain (*Trans. Brit. bryol. Soc.,* IV, 95).

DUNCAN, U. K. 1961–2. Illustrated Key to Sphagnum Mosses (*Trans. Bot. Soc. Edin.,* XXXIX).

FERNALD, M. L. 1950. *Gray's Manual of Botany.*

FRYER, A. & Bennett, A. 1915. *The Potamogetons of the British Isles.*

GILPIN, F. W. 1888. *Flowering Plants of Harleston.*

GREENE, S. W. 1957. The British Species of the *Plagiothecium denticulatum-P. silvaticum* Group (*Trans. Brit. bryol. Soc.,* III, 181).

HUBBARD, C. E. 1954. *Grasses,* Pelican Book No. A295.

MacVICAR, S. M. 1912. *The Student's Handbook of British Hepatics.*

MOSS, C. E. 1914. *Cambridge British Flora.*

NICHOLSON, W. A. 1914. *Flora of Norfolk.*

NYHOLM, E. 1956 *et seq. Illustrated Moss Flora of Fennoscandia* (Musci).

PATON, J. A. 1965. *Census Catalogue of British Hepatics,* 4th edition.

PERRING, F. H. & WALTERS, S. M. (editors). 1962. *Atlas of the British Flora.*

PETCH, C. P. & SWANN, E. L. 1962. West Norfolk Plants Today (*Proc., Bot. Soc. Brit. Isles, Supp.*).

PUGSLEY, H. W. 1930. Revision of the British Euphrasiae (*Journ. Linn. Soc.,* XLVIII).

—— 1919. Revision of Fumaria (*Journ. Linn. Soc.,* XLIV).

—— 1948. Prodromus of the British Hieracia (*Journ. Linn. Soc.,* LIV).

ROUY, G. & FOUCAUD, J. 1893. *Flore de France.*

SUMMERHAYES, V. S. 1951. *Wild Orchids of Britain.*

SWANN, E. L. 1955. An Annotated List of Norfolk Vascular Plants (*Trans. Nfk. and Norw. Nat. Soc.*).

SYME, J. T. B. (editor). 1863. *Sowerby's English Botany,* 3rd edition.

TANSLEY, A. G. 1939. *The British Isles & their Vegetation.*

TRIMMER, K. 1866. *Flora of Norfolk;* Supplement, 1885.

WARBURG, E. F. 1963. *Census Catalogue of British Mosses,* 3rd edition.

WATSON, W. C. R. 1958. *Handbook of Rubi of Great Britain & Ireland.*

Journal of Botany.

Journal of Ecology.

Reports of the Watson's Bot. Exc., Club.

Reports of the Bot. Soc. and Exc. Club.

Reports of the Bot. Soc. of the Brit. Isles.

Watsonia.

Transactions of the Nfk. and Norw. Nat. Soc.

Transactions of the Brit. bryol. Soc.

CHARACEAE

Stoneworts are found in still or slowly flowing water. They flourish in ponds, gravel, clay and peat pits where the water is clean and usually alkaline. The larger areas of water such as the Norfolk Broads have in the past provided ideal conditions and attracted such well-known botanists as Canon G. R. Bullock-Webster, H. & J. Groves, Arthur Bennett, C. E. Salmon and Rev. E. F. Linton.

This difficult family has been somewhat neglected of late but we are indebted to Mr G. H. Rocke for several recent records, whilst Mrs S. T. Phillips led an excursion of the Botanical Society of the British Isles to Hickling Broad in 1960.

We are grateful to Miss Sims and Mr R. Ross of the British Museum (Natural History) for their advice on nomenclature, and, on their suggestion, we have used the revised names given by the late Mr G. O. Allen in his *British Stoneworts* (Haslemere Nat. Hist. Soc., 1950). A recent treatment on a world scale has been published by R. D. Wood & K. Imahori (*Revision of the Characeae*, 2 vols., 1964–5, J. Cramer, Weinheim). The authors take a much wider view of the species than most earlier workers, a consequence of their having seen a much wider range of material. As Mr Ross points out, the treatment by Wood & Imahori is more likely to stand the test of time, and, in our synonymy, we indicate how these authors treat the various entities.

Our account incorporates the many records given in the 1914 *Flora*.

NITELLEAE

Nitella opaca (Bruz.) Agardh

East: 21, St Faiths, 1885, EFL; 31, Acle, AB; 32, Barton Broad, 1902, AB; Ingham, 1966, GHR; 50, Gt. Yarmouth, J. J. Owles; 51, Caister-on-Sea, LW in *Bot. Guide*.

N. flexilis (L.) Agardh
(*N. flexilis* (L.) Agardh subsp. *flexilis*)

West: 69, Southery, 1899, GRB-W; 88, Fowlmere, 1898, GRB-W.

N. translucens (Pers.) Agardh

East: 'E. Norfolk', no localities specified, 1880, H. & J.G.

N. mucronata (A. Braun) Miquel var. *mucronata*
(*N. furcata* subsp. *mucronata* (A. Braun) R. D. Wood)

West: 'St Johns, Lit. Ouse', 1897, GRB-W.

Var. *gracillima* Groves & Bullock-Webster
(*N. furcata* subsp. *mucronata* (A. Braun) R. D. Wood)

East: 30, Rockland Broad, *c.* 1953, ACJ.

N. tenuissima (Desv.) Kütz.

West: 70, Foulden Common, 1950–61, GHR; 07, Lopham Fen, 1897, GRB-W, its first Norfolk record.

Tolypella intricata (Trent. ex Roth) Leonh.
(*T. intricata* (Trent. ex Roth) Leonh., forma *intricata*)
East: 49, Gillingham, GHR & C. West, 1950.

T. prolifera (Ziz ex A. Braun) Leonh.
(*T. intricata* forma *prolifera* (Ziz ex. A. Braun) R. D. Wood)
West: 59, Welney, 1899, GRB-W; 1956, ELS confd. GOA; 69, Southery, 1899, GRB-W.

T. glomerata (Desv.) Leonh.
(*T. nidifica* var. *glomerata* (Desv.) R. D. Wood)
Usually in shallower margins of brackish dykes by tidal rivers.
West: 88, Fowlmere, 1899, GRB-W.
East: 04, Cley, DT; 31, Acle, GHR; 41, Stokesby; Runham, GHR.

CHAREAE

Nitellopsis obtusa (Desv.) J. Groves
Locally common. Under *Lychnothamnus stelliger* Braun in 1914 *Flora*.
East: 30, Rockland Broad, AB; 41, Filby, AB; 42, Hickling Broad, 1953, GHR; Horsey Mere, 1965, GHR, 'abundant and increasing'; Heigham Sounds; Blackfleet Broad; Martham; Old Meadow Dike; 32, Stalham to Barton Broad, 1899, GRB-W.

Chara canescens Desv. & Lois.
East; 42, Hickling Broad, 1900, GRB-W; 1954, GHR, 'has decreased considerably and have not seen it for twelve years'.

C. vulgaris L., var. *vulgaris*
(*C. vulgaris* L., forma *vulgaris*)
The most abundant and widespread species; localities too numerous to specify.

Var. *papillata* Wallr.
(*C. vulgaris* forma ?)
East: 30, Hassingham, GHR, 'not rare; always with the typical plant'; 31, Acle, GHR; 42, Hickling Broad, 1950, C. West.

Var. *longibracteata* (Kütz.) Groves & Bullock-Webster
(forma *longibracteata* (Kütz.) H. & J. Groves)
East: 41, Ormesby St Michael, 1883, Glasspoole, in *Herb. Mus. Brit.*

Var. *refracta* (Kütz.) Groves & Bullock-Webster
(*C. vulgaris* L., forma *vulgaris*)
East: 41, Martham, 1960, GHR.

C. rudis A. Braun
(*C. hispida* forma *rudis* (A. Braun) R. D. Wood)
East: 30, Hassingham, 1954, GHR, its first Norfolk record; 42, Potter Heigham, GHR, 1960.

C. tomentosa L.

East: 42, Hundred Stream near Potter Heigham, 1881, AB in *Herb. Mus. Brit.*

C. hispida L.
(*C. hispida* L., forma *hispida*)

Common.

West: 69, Wretton Fen, 1917, J. E. Little; 70, Foulden Common, ELS.
East: 09, Old Buckenham Fen, GHR; 31, Acle; Upton, GHR; 'too common in E. Norfolk to require localities'; 42, Hickling Broad, 1960, S. T. Phillips.

C. contraria A. Braun ex Kütz.
(*C. vulgaris* forma *contraria* (A. Braun ex Kütz.) R. D. Wood)

West: 88, Fowlmere, 1899, GRB-W.
East: 32, Barton Broad, 1910, Miss Pallis; 41, Rollesby, CES & AB; 42, Potter Heigham, 1885, H. & J. G.; Hickling Broad, 1965, GHR, 'abundant'.

C. contraria × *hispida*

Not mentioned by Wood and Imahori.

East: 41, Somerton, 1890, C. Cotton; 42, Potter Heigham, H. Groves; Hickling Broad, 1965, GHR, 'plentiful most years'; Horsey Mere, GHR.

C. baltica var. *rigida* Groves & Bullock-Webster

Not mentioned by Wood and Imahori who treat *C. baltica* as *C. hispida* forma *baltica* (Bruz.) R. D. Wood.

East: 42, Hickling Broad, 1898, its first record; 1956–65, GHR.

C. aculeolata Kütz.
(*C. hispida* L., forma *hispida*)

West: 70, Foulden Common, 1958, ELS det. GOA; 71, Marham Fen, 1919, J. E. Little; 09, Swangey Fen, GHR.
East: 41, Martham Broad, Salmon and White, 1915 (as *C. polyacantha* Braun).

C. aspera Deth. ex Willd.
(*C. globularis* forma *aspera* (Deth. ex Willd.) R. D. Wood)

Prefers brackish water.

East: 41, Somerton, 1965, GHR; Runham, GHR; 42, Hickling Broad, 1960, S. T. Phillips.

C. desmacantha (H. & J. Groves) Groves & Bullock-Webster
(*C. globularis* forma *curta* (Nolte ex Kütz.) R. D. Wood)

West: 70, Foulden Common, 1961, GHR det. GOA, 'grows well in several of the pits'; 98, Langmere, 1897, GRB-W; 07, Lopham Fen, GRB-W.

C. connivens Salzm. ex A. Braun
(*C. globularis* forma *connivens* (Salzm. ex A. Braun) R. D. Wood)

In brackish water. Rare.

East: 41, Martham Broad, Salmon & White, 1915; 42, Heigham Sounds, 1889, J. Bidgood; Hickling Broad, 1900, GRB-W; 1954, GHR.

C. globularis var. *capillacea* (Thuill.) Zanev.
(*C. globularis* Thuill., forma *globularis*)

As this plant has been confused with *C. fragilis* Desv., we have been unable to apportion the older records. It is probable that some are referable to the next species.

West: 88, Croxton, Burrell & Clarke.

East: 08, Roydon near Diss, EFL; 14, Weybourne, J. Saunders; 19, Flordon, Burrell & Clarke; 28, Harleston, *Fl. Pl. H.*; 32, Barton Broad; 42, Potter Heigham; Hickling; Blackfleet Broad, all by AB; Hickling, 1965, GHR.

C. delicatula sensu Agardh non Desv.
(*C. globularis* forma *virgata* (Kütz.) R. D. Wood)

West: 70, Foulden Common, 1950, GHR; 1958, ELS confd. GOA; 09, Swangey Fen, GHR.

East: 31, Acle; Upton; 32, Ingham, GHR confd. GOA; 42, Potter Heigham; Hickling, GHR.

BRYOPHYTA

E. L. Swann

At the time of the publication of the last *Flora of Norfolk* in 1914, W. H. Burrell, who was responsible for the section dealing with bryophytes, wrote, 'The frequency with which newly discovered plants were based on material from East Anglia is a measure of the enthusiasm of local bryologists at the close of the eighteenth and beginning of the nineteenth centuries.' Burrell (1865–1945) ascribed his interest in bryology to the time when E. M. Holmes requested his help to find a *Sphaerocarpos* 'known to occur in certain turnip fields in Norfolk and which he successfully located' (obituary notice, *The Naturalist*, 1945). He added the hepatic *Lophozia Schultzii* var. *laxa* to the British flora.

Thanks to his extensive field-work, he was able to increase the total of mosses to 190 whilst the hepatics amounted to sixty species. Norfolk workers cannot but be grateful for the very sure foundation he laid.

In such a large county with so few resident bryologists no account can be looked upon as exhaustive; negative evidence must be accepted on its merits. As recent work shows only too clearly, additonal new records continue to be made. *Physcomitrium eurystomum*, a moss new to the British Isles, was found by B. F. T. Ducker on the margin of one of the Breckland meres; *Campylopus introflexus, Orthodontium lineare* and *Isopterygium seligeri* are probably new arrivals; whilst the British Bryological Society's excursion in 1967 resulted in the finding, among others, of *Cryptothallus mirabilis, Riccia rhenana, Hypnum imponens* and *Schistostega pennata*.

Doubtful and missing records

In common with other counties there have been many losses due to drainage, felling of woodland, building, and more intensive cultivation, including the loss of natural chalk grassland, but it is doubtful if *Swartzia montana* ever occurred in Norfolk, whilst *Rhacomitrium lanuginosum, Amblyodon dealbatus* and *Antitrichia curtipendula*, accepted by Burrell very hesitatingly, are certainly extinct and the same may apply to *Splachnum ampullaceum* although this has been found recently in a neighbouring Suffolk fen.

The surprising absence of some records, such as *Fissidens incurvus* and *F. exilis*, may be due to their critical status, whilst the several records for species of *Ulota* and *Orthotrichum* made by earlier workers and now no longer seen, suggest the effect of increasing atmospheric pollution. *Bartramia pomiformis, Zygodon viridissimus* and *Leucodon* are far less frequent.

On the other hand, species such as *Tetraphis pellucida, Pohlia delicatula, Mnium rugicum, Leskea polycarpa, Campylium chrysophyllum* (curiously omitted from the 1914 *Flora*), *Camptothecium nitens, Cinclidium stygium* and *Brachythecium rivulare* appear to be more widespread than before, perhaps due to a greater concentration on hunting in fens. The reported rarity in Burrell's day of such plants as *Homalia trichomanoides* and *Isothecium myurum* reflects the concentration of botanists in the eastern part of the county, a disparity which is also apparent in the phanerogams.

Modern taxonomic work has shown that old records of *Calypogeia trichomanis*, *Mylia taylori* and *Plagiochila asplenioides* s.s., cannot be upheld but their places are taken by *C. muellerana*, *M. anomala* and *P. asplenioides* var. *major*, the last surely worthy of specific rank.

Census for Norfolk and adjoining counties

The figures for mosses are taken from the 1963 **Census Catalogue** and those for hepatics from the 1965 **Census Catalogue**, both published by the British Bryological Society.

The richness of the Norfolk flora is undoubtedly due to the persistence of many of its fens and bogs; those species peculiar to Cambridgeshire reflect the greater interest taken in critical taxa, whilst in the case of Lincolnshire there is evidence for the intrusion of the more northerly element.

	Sphagna	Mosses	Hepatics	Total
Norfolk	16	267	78	361
Suffolk	11	226	67	304
Lincolnshire	12	222	57	291
Cambridgeshire	nil	199	33	232

Norfolk only

Sphagnum magellanicum
S. contortum
Ditrichum heteromallum
Dichodontium pellucidum
Tortula virescens
Pottia wilsoni
Leptodontium flexifolium
Funaria obtusa
Physcomitrium eurystomum
Schistostega pennata
Bryum turbinatum
Cinclidium stygium
Philonotis caespitosa
Ulota bruchii
U. ludwigii
Neckera pumila

Hookeria lucens
Drepanocladus vernicosus
Camptothecium nitens
Eurhynchium schleicheri
Hypnum imponens
Petalophyllum ralfsii
Lophozia porphyroleuca
Leiocolea rutheana
Nardia geoscyphus
Cephaloziella elachista
Cephalozia pleniceps
C. macrostachya
Cladopodiella francisci
C. fluitans
Scapania curta

Suffolk only

Sphagnum robustum
Dicranella crispa
Tortula cunefolia
Splachnum ampullaceum
Pottia crinita
Bartramia ithyphylla
Pterogonium gracile

Eurhynchium praecox
Riccia beyrichiana
Pallavicinia lyellii
Trichocolea tomentella
Leiocolea turbinata
Cephaloziella integerrima
Odontoschisma denudatum

43

Cambridgeshire only

Tortella inflexa
Weissia rostellata
W. multicapsularis
Rhacomitrium heterostichum

Bryum creberrimum
Scleropodium caespitosum
Lejeunea cavifolia

Lincolnshire only

Tetraplodon mnioides
Bryum marratii
B. uliginosum
B. warneum
Mnium stellare
Breutelia chrysocoma
Neckera crispa
Hygroamblystegium fluviatile

Brachythecium plumosum
Pylaisia polyantha
Hypnum lindbergii
Anthoceros punctatus
Blepharostoma trichophyllum
Plagiochila asplenioides s.s.
Scapania aspera

List of Authorities and Contributors quoted

Of the 1914 Flora

Bodger, J. W.	JWB	Newton, W. C. F.	WCFN
Burrell, W. H.	WHB	Paget, James	JP
Clarke, W. G.	WGC	Sherrin, W. R.	WRS
Crowe, James	JC	Stirling, J.	JS
Francis, R. B.	RBF	Thompson, W. E.	WET
Holmes, E. M.	EMH	Turner, Dawson	DT
Hooker, W. J.	WJH		

After 1914

Appleyard, Mrs J.	JA	Paton, Mrs J. A.	JAP
Brit. bryol. Soc. 1967		Petch, C. P.	CPP
excursion	BBS	Pigott, C. D.	CDP
Bull, A. L.	ALB	Proctor, M. C. F.	MCFP
Catcheside, D. G.	DGC	Richards, P. W.	PWR
Crundwell, A. C.	ACC	Rocke, G. H.	GHR
Dickson, Jas.	JD	Rose, F.	FR
Ducker, B. F. T.	BFTD	Swinscow, T. V. D.	TVDS
Duckett, J. G.	JGD	Townsend, C. C.	CCT
Duncan, Miss U. K.	UKD	Wallace, E. C.	ECW
Jones, E. W.	EWJ	Warburg, E. F.	EFW
Little, E. R. B.	ERBL	Watson, E. V.	EVW
Manning, S. A.	SAM	Whitehouse, H. L. K.	HLKW
Marquand, C. V. B.	CVBM		

In the absence of initials the records are those made by Burrell; where the station is followed by an exclamation mark it denotes the author's work.

Acknowledgments

The writer wishes to record his thanks to Mr G. H. Rocke who has devoted much time to bryophytes, particularly in the east of the county and, either alone or in the company of the writer, has added many additional stations and new records; to Dr Francis Rose who has made his field-notes freely available; to Mr C. C. Townsend of Kew; and to many of the BBS Referees for their invaluable help with critical species, particularly Miss U. K. Duncan, Mrs J. W. Fitzgerald, Mr R. D. Fitzgerald, Mrs J. A. Paton, Dr M. C. F. Proctor, Mr F. A. Sowter and Mr E. C. Wallace.

Arrangement

Both the names and arrangement follow the two *Census Catalogues* (*British Hepatics,* Mrs J. A. Paton, 1965, and *British Mosses,* E. F. Warburg, 1963). The names of the species are followed by their frequency and habitat and then by the stations in West and East Norfolk.

The localities are named on the One-inch Ordnance Survey maps but to facilitate location these are preceded by the number of the relative 10-km. grid square. Finally, the distribution of the species in the neighbouring counties is given (S = Suffolk, C = Cambridgeshire and L = Lincolnshire).

Hepatics

ANTHOCEROTACEAE

Anthoceros punctatus L.

No recent records. Reported by Burrell as rare on bare earth.

East: 03, Gunthorpe; 11, Drayton. L.

SPHAEROCARPACEAE

Sphaerocarpos michelii Bellardi

Formerly locally common in turnip and clover fields on clay and loam. Traced to the west of Swaffham by W. G. Clarke up to 1910.

West: 83, stubble field (mustard) on loam, S. Creake, 1958, !

East: 01, near Elsing, UKD, 1967. Burrell recorded some eighteen stations. S.

S. texanus Aust.

There is a specimen in Manchester Museum collected by W. J. Hooker in 1845 from 'Vere', a locality unknown. Burrell has two further records from East Norfolk. S.

REBOULIACEAE

Reboulia hemisphaerica (L.) Raddi

Rare. In hedgerows.

East: 02, Guestwick; 03, Edgefield, RBF; 12, Booton; 14, Beeston Regis; 21, Frettenham; St Faiths and Spixworth; 22, on a sandy bank near Felmingham, FR, 1951; 23, Aldborough. S.

45

MARCHANTIACEAE

Conocephalum conicum (L.) Underw.

Widespread and abundant on exposed soil of dykesides and brickwork of sluices in shade.

West: 62, Babingley, BBS; 70, Cockley Cley ! 72, Hillington ! 99. Thompson Water, CCT & ELS.   S. C. L.

Lunularia cruciata (L.) Dum.

Frequent. Sides of dykes, damp peat banks and moist compacted soil of old gardens in shade.

West: 62, Whin Hill Covert, Wolferton ! Sandringham Warren ! forecourt of old garden, King's Lynn ! 73, Sedgeford, BBS; 82, Tatterford, JA; 89, bank of R. Wissey, Bodney !
East: 12, Booton Common, GHR & ELS; 19, Flordon.   S. C. L.

Preissia quadrata (Scop.) Nees

West: 91, base-rich fen, Scarning, FR, 1956; 93, Kettlestone Fen, FR.
East: 22, Bryants Heath near Felmingham, FR.   S. L.

Marchantia polymorpha L., var. *polymorpha*

Formerly regarded as rather rare. Burrell gives several stations in East Norfolk. Now far more frequent on damp soil of streamsides; abundant on damp garden paths and occasionally on moist sand in gravel pits; sometimes abundant on sugar-beet sludge heaps.

West: 60, Crimplesham ! 62, King's Lynn ! N. Wootton, CPP; 73, Sedgeford, BBS; 88, Two Mile Bottom near Thetford ! 99, Hockham Rough, BBS.   S. C. L.

RICCIACEAE

Riccia warnstorfii Limpr.

West: 03, Swanton Novers Great Wood, 1967, BBS, JAP.   C.

R. glauca L.

Burrell regarded it as locally common in turnip and clover fields on clay and loam in East Norfolk. It appears to be far from common today when fields are rarely left unploughed for long periods. Has probably been confused with *R. sorocarpa*.

West: 03, Swanton Novers Great Wood, 1967, BBS, JAP.
East: 01, Hockering, BBS, ERBL; near Elsing, 1967, UKD.   S. C. L.

R. sorocarpa Bisch.

Occasional. Less frequent than formerly (see previous species). Grows with *R. glauca* in similar situations.

West: 61, E. Winch ! 64, Ringstead Downs, BBS; 98, Ringmere ! 99, Cranberry Farm, Hockham, CCT & ELS; 03, Swanton Novers Great Wood, BBS.
East: in *Census Catalogue*, but no localised records traced.   S. C. L.

R. fluitans L.

Frequency variable. Free-floating or creeping on mud of pond edges. Sometimes abundant in some of the Breckland meres.

West: 60, Shouldham Thorpe; 62, N. Wootton, CPP; 88, Fowlmere, abundant in 1962, ! Home Mere, CPP & ELS; 98, pool near Ringmere, BBS; 99, Stow Bedon ! Hockham ! Breckles Heath, ALB.

East: 32, E. Ruston.   S. C. L.

R. rhenana Lorberr ex K. Müll.

Its discovery in 1967, in the moat at Hockering Wood, by Mrs J. W. Fitzgerald and Dr H. L. K. Whitehouse was the fourth British record.

West: 82, very sparse, Tatterford Common, 1967, BBS, JGD & JA.

East: 01, Hockering, very abundant, BBS, JWF & HLKW.   C.

R. cavernosa Hoffm.

West: 98, Ringmere, 1958, ECW; Langmere, abundant, JGD.

East: 01, Sparham; 20, Holverston; under *R. crystallina*.

Ricciocarpus natans (L.) Corda

Occasional. Floating in ponds, ditches and meres.

West: 60, Shouldham Thorpe, WGC; 62, N. Wootton, CPP; 88, Fowlmere, abundant and growing with *Riccia fluitans*, 1962, !

East: 10, Little Melton, 1909; 11, Weston; 12, Booton and Heydon; 19, Flordon; 32, Sutton; 41, Burgh Common ! 51, Caister.   S. C. L.

RICCARDIACEAE

Riccardia multifida (L.) Gray

Occasional in bogs and fens.

West: 62, Roydon Common, FR; 71, E. Walton Common ! 09, Swangey Fen, BBS, 1953.

East: 03, Holt Lowes, FR; 12, Buxton Heath, JAP, 1967; 19, Flordon; 40, Acle, FR.   S. C. L.

R. sinuata (Dicks.) Trev.

In base-rich fens.

West: 09, Swangey Fen, H. J. B. Birks and others, 1966.

East: 12, Buxton Heath, JAP, 1967; 21, gravel pit, west of Coltishall, JAP, 1967; 22, Bryants Heath near Felmingham, FR, 1959.   S. C. L.

R. latifrons (Lindb.) Lindb

In acid bogs.

West: 62, Sandringham Warren, FR, 1959; Roydon Common, in *Sphagnum*, FR; BBS.

East: 12, Buxton Heath, 1967, JAP.

R. pinguis (L.) Gray

In base-rich fens in which it is constant but never abundant.

West: 62, Grimston Warren ! Roydon Common, FR; Sugar Fen, BBS; 84, in a hollow near the well in Norton Hills, Scolt Head, CVBM; 91, Scarning Fen, FR; 93, Kettlestone Fen, FR; 09, Swangey Fen, FR.

East: 03, Holt Lowes, FR; 09, Old Buckenham Fen, FR; 12, Buxton Heath, BBS, 1953; 22, Bryants Heath, FR; 40, Acle, FR.   S. C. L.

Cryptothallus mirabilis Malmb.

Found by Mr Jas. Dickson during the Brit, bryol. Soc. Exc., in April 1967. It was growing in peat litter under *Sphagnum recurvum* and associated with *Molinia, Betula* and *Polytrichum commune.* In view of similar habitats elsewhere in the county its occurrence may be expected in other stations.

West: 62, Sandringham Warren, BBS, 1967, JD.   S. C. L.

PELLIACEAE

Pellia epiphylla (L.) Corda

Widespread and abundant in both vice-counties. On the exposed soil of streamsides and always on acid soil.   S. C. L.

P. endiviifolia Dicks.
(*P. fabbroniana* Raddi)

Locally common. Base-rich fens, ditch-sides on calcarous soil and chalk detritus. More conspicuous in autumn and winter by reason of its seasonal furcate growth. Usually associated with *Dicranella varia.*

West: 62, Sedgeford, BBS; 70, old marl pit, Gooderstone Warren ! Cockley Cley Warren ! 80, Saham Toney; 82, Tatterford, JA; 91, Scarning Fen, FR; 93, Kettlestone Fen, FR; 99, Rockland All Saints Fen, CPP & ELS; 08, source of the R. Waveney, S. Lopham !

East: 11, Horsford; 12, Buxton Heath and Booton, FR; 19, Flordon; 21, Newton St Faiths; 22, Bryants Heath, FR; 23, Southrepps; 40, Acle, FR.   S. C. L.

METZGERIACEAE

Metzgeria furcata (L.) Dum., var. *furcata*

Frequent on trees, irrespective of species.

West: 61, Cranberry Wood, E. Winch ! 64, Ringstead Downs, BBS; 70, on *Pinus sylvestris*, Gooderstone Fen ! 88, on *Salix alba*, Two Mile Bottom near Thetford ! 91, Scarning Fen, BBS; 92, on *Acer campestre*, Big Wood, N. Elmham ! 99, Wayland Wood near Watton; 03, Swanton Novers, FR & ELS.   S. C. L.

M. fruticulosa (Dicks.) Evans

Only one station given by Burrell, Narford, W71, *c.* 1913.

West: 61, on *Salix fragilis*, Cranberry Wood, E. Winch ! 82, Tatterford Common, JAP, 1967; 91, Scarning Fen, BBS.
East: 30, on *Salix* bushes, Wheatfen Broad, BBS, 1953.   S. C.

PALLAVICINIACEAE

Moerckia flotoviana (Nees) Schiffn.

Not uncommon in fens.

1 ROYAL FERN *Osmunda regalis* *Miss Tuck*

2 MOONWORT *Botrychium lunaria* *Miss D. M. Maxey*

3 ADDER'S TONGUE   *Ophioglossum vulgatum*   *Miss D. M. Maxey*

4 WINTER ACONITE   *Eranthis hyemalis*   *G. E. Hyde*

5 GREAT SPEARWORT   *Ranunculus lingua*   *R. Jones*

West: 62, Sugar Fen, JAP, 1967; 93, Kettlestone Fen, FR; 07, Blo'
Norton Fen, FR; 09, Swangey Fen, FR.
East: 02, Whitwell; 12, Buxton Heath, BBS, 1953; 19, Flordon. S.

BLASIACEAE

Blasia pusilla L.

West: 62, on a sandy track over Roydon Common kept moist by seepage
from a spring and associated with *Centunculus*, *Radiola* and *Isolepis*
setacea, June 1959, !

FOSSOMBRONIACEAE

Petalophyllum ralfsii (Wils.) Nees & Gottsche

West: 74, with *Bryum bicolor* and *Sagina nodosa* in a damp hollow west of
the pine clump on the dunes, Holme-next-the-Sea, C. J. Cadbury, 1957;
could not be found by the BBS in 1967.

Fossombronia foveolata Lindb.

East: 11, Horsford, 1936, SAM, as *F. dumortieri*, confirmed by A. Mayfield
and A. Wilson. L.

F. pusilla (L.) Dum., var. *pusilla*

Rare. On heath tracks.

West: 62, Roydon Common, 1962, ! 03, Swanton Novers Great Wood,
BBS.
East: 14, Sheringham; 21, Newton St Faiths. S. C. L.

F. wondraczekii (Corda) Dum.

West: 03, Swanton Novers Great Wood, BBS, 1967, JAP.

PTILIDIACEAE

Ptilidium ciliare (L.) Hampe

Rare on heaths but may be locally abundant in the microhabitat on the
north side of tall *Calluna* where the addition of *Carex arenaria* and
Agrostis canina ssp. *montana* augment the humidity.

West: 60, Shouldham Warren ! 62, N. Wootton ! Sandringham Warren !
78, Weeting Heath, BBS; 82, Massingham Heath ! 83, Syderstone; 88,
Santon ! Two Mile Bottom, Thetford ! 89, E. Wretham Heath, ALB; 98,
Ringmere !

East: 11, Horsford; 21, Newton St Faiths; 22, Stratton Strawless; 39,
Broome; 42, Winterton. S. L.

P. pulcherrimum (Weber) Hampe

In each of the stations it was only seen on one tree.

West: 62, N. Wootton, on *Quercus robur*, BBS; 03, Swanton Novers Great
Wood, on an elm, March 1966, TVDS, FR & ELS.
East: 01, Hockering Wood, on a birch tree, BBS, ECW. C. L.

LEPIDOZIACEAE

Lepidozia reptans (L.) Dum.

Rare. On humus in woodland and on rotting stumps.

West: 79, Mundford Covert ! 91, Scarning Fen, BBS; 03, Swanton Novers, abundant on rotting stumps, living oaks and tussocks of *Leucobryum*, FR & ELS.

East: 01, Hockering, BBS; 12, Blickling; 03, Edgefield, RBF; 11, Felthorpe; 14, Beeston Regis; 22, Stratton Strawless.   S. L.

L. setacea (Weber) Mitt.

Although described by Burrell as rare, it is a frequent species growing in *Sphagnum* on wet acid heaths and in bogs, sometimes very abundant.

West: 61, Leziate; E. Winch Common, BBS; 62, Sandringham Warren ! Roydon Common ! Grimston Warren ! 63, Dersingham Fen !

East: 03, Holt Lowes, abundant in *Sphagnum rubellum* ! 11, Swannington; 12, Buxton Heath, JAP; 22, Bryants Heath, FR.   S. L.

CALYPOGEIACEAE

Calypogeia muellerana (Schiffn.) K. Müll.

Given by Burrell, under *C. trichomanis*, as common on woodland banks and in swamps. The true *C. trichomanis* is unlikely to be a Norfolk species; all our material has colourless, compound oil-bodies and is abundant on acid peat, leaf litter in carrs, bogs and on sandstone exposures in shade.

West: 62, Sandringham Warren ! Roydon Common and Grimston Warren, 1960, ! N. Wootton, CPP & ELS; 63, Dersingham Fen ! 71, E. Walton Common !

East: 03, Holt Lowes, 1957, !   S. L.

C. fissa (L.) Raddi

Although not recorded until about 1914 and its distribution not fully worked out, there seems little doubt that it is far more common in acid bogs growing amongst *Sphagnum*.

West: 61, E. Winch Common, BBS; 62, Ling Common, N. Wootton, BBS; Sandringham Warren, BBS; Roydon Common, FR; 71, E. Walton Common, CPP & ELS; 82, Tatterford Common, BBS; 03, Swanton Novers, FR & ELS.

East: 01, Hockering Wood, BBS; 02, Bawdeswell, FR; 03, Holt Lowes ! 12, Buxton Heath; 32, Barton Turf !   S. C. L.

C. sphagnicola (Arn. & Pers.) Warnst. & Loeske

Distribution imperfectly known as it is not mentioned by Burrell. It appears to be a rare species restricted to *Sphagnum* in bogs.

West: 62, Sandringham Warren, lf–la on *Sphagnum recurvum*, FR & ELS, 1959.

C. arguta Nees & Mont.

Rare.

West: 62, Greensand cliff near Wolferton station, BBS, JGD, 1967.

East: 32, by the side of a ditch, Catfield, WHB.

LOPHOZIACEAE

Lophozia ventricosa (Dicks.) Dum., var. *ventricosa*

Tolerates a wide range of substrata but is more frequent on acid soils of damp heaths, alder carrs and damp sand pits; rarely in the alkaline soils of Breckland.

West: 62, Sandringham Warren, BBS; old sand pits, Roydon Common ! N. Wootton, CPP; 63, Dersingham Fen, amongst *Aulacomnium*, MCFP; 78, depressions on heath near Snake Wood, Brandon ! 99, Hockham, CCT & ELS; 08, Garboldisham Heath.
East: 12, Buxton Heath, JAP; 20, Thorpe St Andrew; 21, Newton St Faiths.   S. L.

L. porphyroleuca (Nees) Schiffn.
Very rare. No recent records.
East: 03, Holt Lowes, RBF.

L. excisa (Dicks.) Dum.
Rare. Wet heathy places; in depressions in chalk grassland.
West: 78, Weeting Heath, BBS; 08, Garboldisham Heath, !
East: 03, Holt and Edgefield, RBF; 20, Norwich, WJH; 50, Gt. Yarmouth, WJH.   S. L.

L. incisa (Schrad.) Dum.
Very rare. Wet heaths. First recorded from Holt by Rev. R. B. Francis.
East: 03, Holt, FR; 22, Bryants Heath, FR.   S.

L. bicrenata (Schmid.) Dum.
Rare. Heaths.
West: 62, Sandringham Warren, BBS 1967, JAP.
East: 03, Holt and Edgefield, RBF; 20, Norwich, WJH; 50 Gt. Yarmouth, St Faiths.   S. L.

Leiocolea rutheana (Limpr.) K. Müll.
First discovered in Britain by Burrell at Flordon, E19, 1909. Was at first taken to be *Lophozia Muelleri* but material was submitted to Schiffner who detected the paroicous inflorescence and placed it under L. *Schultzii* as a new variety '*laxa*'. Locally abundant in base-rich fens. Confined to Norfolk.
West: 91, Scarning Fen, 1955, ! 09, Swangey Fen, 1952, GHR & ECW.
East: 03, Holt Lowes, FR; 12, Buxton Heath, FR, 'in great profusion and in large swelling tufts mixed with *Philonotis calcarea, Drepanocladus vernicosus, Aulacomnium palustre, Bryum pseudotriquetrum, Ctenidium molluscum, Acrocladium giganteum*, etc., on the sides of hummocks in highly calcareous fen'; 1967, JAP, 'now very sparse'.

Barbilophozia attenuata (Mart.) Loeske
Rare on sandy banks in shade.
West: 62, Sandringham Warren, 1952, MCFP; 63, Dersingham Fen !
East: 03, Holt, RBF.   L.

B. hatcheri (Evans) Loeske
Apparently rare; so far detected only in two adjacent stations.
West: 62, Sandringham Warren, 1960, ! base of Greensand cliff, Wolferton, BBS.   S.

B. barbata (Schmid.) Loeske
West: 78, at base of an ant-hill, Weeting Heath, BBS 1967, Miss M. Dalby.
                                                                                    S. L.

Tritomaria exsectiformis (Breidl.) Schiffn.
Rare. Moist heaths. No records since the first in 1911 under *Sphenolobus*, Moss Exc. Club.
East: 03, Holt and Edgefield, RBF; 11, Felthorpe, RBF; Horsford; 20, Norwich, WJH.   S. L.

Gymnocolea inflata (Huds.) Dum., var. *inflata*
Widespread and abundant throughout the county. Colonises bare, wet hollows in heath and bog.   S. L.

JUNGERMANNIACEAE

Solenostoma crenulatum (Sm.) Mitt.
Considered by the early bryologists as locally common with *Nardia scalaris* on heaths.
West: 63, Dersingham Fen, WJH (under *Aplozia*).
East: 01, Hockering Wood, BBS; 03, Holt Lowes, in valley bog, 1959, ! 11, Horsford; 20, Thorpe St Andrew; 33, Witton, GHR.   S. L.

Plectocolea hyalina (Lyell ex Hook.) Mitt.
East: 01, Hockering Wood, BBS, 1967, ERBL.

Nardia scalaris (Schrad.) Gray
Occasional. On damp heathland and woodland banks.
West: 62, N. Wootton, CPP & ELS; 03, Swanton Novers, FR & ELS.
East: 03, Briston; Holt Lowes, RBF; 14, Sheringham; 20, Sprowston and Thorpe St Andrew; 22, Scottow and Westwick; Bryants Heath, FR.
                                                                                    S. L.

N. geoscyphus (De Not.) Lindb.
West: 62, near King's Lynn, 1930, H. H. Knight.

PLAGIOCHILACEAE

Mylia anomala (Hook.) Gray
Although the two species of *Mylia* are given in the 1914 *Flora*, recent work has shown that this species is the only one found in Norfolk. It is frequent on wet heaths.
West: 62, Sandringham Warren, 1959, FR & ELS; Roydon Common, 1948, ! Grimston Warren !
East: 03, Holt Lowes, FR; 12, Buxton Heath, 1967, JAP.   S.

Plagiochila asplenioides (L.) Dum., var. *major* Nees

Burrell recorded the true species as rare in marshes. Our recent records are from dykesides in woods on boulder clay and all come under the variety which appears to be worthy of specific rank.

West: 82, Tatterford, JA; 91, Horse Wood, Mileham, CPP & ELS; 03, Swanton Novers, GHR.

East: 01, Hockering Wood, BBS; 12, Buxton Heath, BBS, 1953, UKD.

HARPANTHACEAE

Lophocolea bidentata (L.) Dum.

Widespread and abundant throughout the county. On the ground in woodland, alder carrs, fens, Breckland grass heaths and shaded lawns.

S. C. L.

L. cuspidata (Nees) Limpr.

Equally widespread and abundant throughout the county on tree boles, both living and rotten, in woodland; occasionally on floors of pits and in shade on the older grey sand-dunes of the coast. S. C. L.

L. heterophylla (Schrad.) Dum.

Widespread and abundant throughout the county in similar habitats to the last but not seen in coastal areas.

Chiloscyphus pallescens (Ehrh.) Dum.

West: 62, Roydon Common, FR; BBS; 91, Scarning Fen ! 93, Kettlestone Fen, FR; 07, Blo' Norton Fen, FR; 09, Swangey Fen !

East: 03, Holt Lowes ! 12, Buxton Heath, JAP; 14, Beeston Bog ! 40, Acle, FR. S. C. L.

CEPHALOZIELLACEAE

Cephaloziella elachista (Jack) Schiffn.

East: 12, in boggy ground amongst *Sphagnum*, Buxton Heath, BBS 1953; 1967, JAP.

C. rubella (Nees) Warnst.

West: 62, raw humus on damp heath, Sandringham Warren, 1952, MCFP. S. C.

C. hampeana (Nees) Schiffn.

West: 61, E. Winch Common, BBS, 1967, JAP; 62, near Wolferton, 1967, JAP.

East: 12, on *Sphagnum*, Buxton Heath, 1967, JAP. S. C.

C. starkei (Funck) Schiffn., var. *starkei*

Apparently Burrell realised the difficulties in naming species of this genus as he queried the distribution when quoting records for *C. byssacea*, now sunk in *C. starkei*.

West: 70, Gooderstone Warren, on humus in the 'breck', 1956, ! 62, Grimston Warren, under *Calluna*, ! Roydon Common, BBS; Sugar Fen,

JAP; Sandringham Warren, BBS; 78, Weeting Heath ! 84, Scolt Head, CVBM; 98, Langmere, BBS.

East: 03, Holt and Edgefield, RBF; 12, Buxton Heath, 1967, JAP; 41, Hemsby.   S. C. L.

CEPHALOZIACEAE

Cephalozia bicuspidata (L.) Dum., var. *bicuspidata*

Abundant locally on peat banks in fens and woodland.

West: 60, Mow Fen, Shouldham ! 61, E. Winch Common, BBS; 62, N. Wootton, CPP & ELS; Roydon Common, FR; Sandringham Warren, FR; 63, Dersingham Fen ! 71, E. Walton Common.

East: 01, Hockering Wood, BBS; 03, Holt Lowes ! 12, Buxton Heath, FR; JAP; 22, Bryants Heath, FR.   S. C. L.

Var. *lammersiana* (Hub.) Breidl.

West: 62, Roydon Common, BBS, 1967, JAP.

East: 03, Holt Lowes, FR; 11, Swannington; 22, Swanton Abbot; 32, E. Ruston.   S. C. L.

C. pleniceps (Aust.) Lindb.

Not recorded by Burrell.

East: 03, damp peat with *Erica tetralix*, Holt Lowes ! Edgefield Heath, H. H. Knight; 12, Buxton Heath, 1967, JAP.

C. connivens (Dicks.) Lindb.

Locally abundant in *Sphagnum* bogs, on damp heaths and in fen scrub.

West: 61, E. Winch Common, BBS; 62, in tufts of *Sphagnum compactum*, Sandringham Warren ! Roydon Common, FR; base of *Molinia* tussocks, Grimston Warren !

East: 11, Swannington Common, GHR & ELS; 12, Buxton Heath, BBS, 1953; 22, Bryants Heath, FR.   S. C. L.

C. macrostachya Kaal., var. *macrostachya*

Not distinguished in the 1914 *Flora*.

West: 62, Sandringham Warren, FR & ELS; Roydon Common, f–la, FR; 63, Dersingham Fen, DGC, *c*. 1959.

East: 12, valley sides, Buxton Heath, BBS, 1953.

Cladopodiella francisci (Hook.) Buch

The specific epithet commemorates the Rev. R. B. Francis who first discovered the plant in wet heaths at Holt and Edgefield. It was described by Hooker under *Jungermannia* in his *British Jungermanniae*, 1816. Its persistence at Holt cannot be confirmed as recent searching has been unsuccessful.

West: 61, Bawsey Wood, 1930, H. H. Knight.   S.

C. fluitans (Nees) Buch

Few stations but locally abundant in bog pools and wet heaths.

West: 62, Sandringham Warren, FR & ELS; Roydon Common, !
East: 03, Holt Lowes, EMH; 12, Buxton Heath !

Odontoschisma sphagni (Dicks.) Dum.

Occasional in *Sphagnum* in bogs and fens. Tall and erect forms with distant leaves occur on Sandringham Warren.

West: 61, E. Winch Common, BBS; 62, Sandringham Warren in *Sphagnum rubellum* ! Roydon Common, FR.

East: 01, Bawdeswell; 03, Holt Lowes, FR; 12, Buxton Heath, 1967, JAP; 22, Stratton Strawless. S. L.

SCAPANIACEAE

Diplophyllum albicans (L.) Dum.

Occasional on woodland banks but not as common as formerly.

West: 62, N. Wootton Woods, CPP & ELS; Roydon Common, BBS; 82, Tatterford, JA; 91, Scarning Fen, BBS; 99, Wayland Wood near Watton, CPP & ELS; 03, Swanton Novers, FR & ELS.

East: 01, Hockering Wood, BBS; 03, Holt Lowes ! 22, Bryants Heath, FR; 29, Fritton Decoy ! S. L.

Scapania irrigua (Nees) Dum.

Rare. Wet heaths and woodland rides.

West: 81, Litcham.

East: 03, Holt Lowes, GHR & ELS; 11, Swannington; 12, Buxton Heath, FR; Hevingham; 21, Newton St Faiths. S. L.

S. nemorea (L.) Grolle
(*S. nemorosa* (L.) Dum.)

Rare. Wet heaths.

East: 03, Holt Lowes; Briston, GHR, 1958; 12, Buxton Heath, BBS, 1953; 22, Bryants Heath, FR. S. L.

S. undulata (L.) Dum.

East: 03, Edgefield Heath, H. J. B. Birks and others, 1966.

S. compacta (Roth) Dum.

West: 62, sandy bank of railway cutting, Wolferton, BBS, 1967, ACC.

East: 03, Holt Lowes and Edgefield, RBF. S. L.

RADULACEAE

Radula complanata (L.) Dum.

Occasional on woodland trees and always in small patches.

West: 70, Gooderstone Warren ! 78, Weeting Fen, on *Salix cinerea* ! 92, Horningtoft ! 99, Wayland Wood near Watton ! Thompson Water, CCT & ELS; 09, Swangey Fen, FR. S. C. L.

PORELLACEAE

Porella platyphylla (L.) Lindb.

Recorded by Burrell from trees and walls, chiefly in East Norfolk, and regarded as rather rare (under *Madotheca*).

West: 62, walls of castle, Castle Rising ! and on an ash, same station ! 81, walls of Castleacre Priory and ruins of castle ! 92, base of an ash, Horningtoft !

East: 09, New Buckenham; 10, E. Carleton and Ketteringham; 19, Hapton and Flordon; 29, Newton Flotman; 31, Upton.   S. C. L.

LEJEUNEACEAE

Lejeunea ulicina (Tayl.) Tayl.

Intensive searching has so far failed to confirm its occurrence here which has been reported by M. C. F. Proctor. The record given in *Trans.*, BBS, 1934, for 'Burnham Beeches' as vice-county 28 is incorrect; this station is in v-c 24.

FRULLANIACEAE

Frullania tamarasci (L.) Dum.

West: 78, Weeting Heath, BBS, 1967, JA.   S.

F. dilatata (L.) Dum.

Frequent on trees.

West: 61, Cranberry Fen, E. Winch ! 62, Sandringham Warren, BBS; 69, W. Dereham Fen ! 70, Barton Bendish ! 78, Gravel Pit Wood, Weeting ! 99, Cranberry Rough, Hockham ! 03, Swanton Novers, FR & ELS; 09, Swangey Fen !

East: No localised records traced.   S. C. L.

Musci

SPHAGNACEAE

Sphagnum palustre L.

One of our most abundant species. Common and widespread in bogs, fens, carrs and wet pockets on damp heaths.

West: 62, Sandringham Warren ! N. Wootton, CPP & ELS; Roydon Common and Derby Fen, FR; 81, Litcham Common ! 82, Tatterford, BBS; 99, Hockham, WGC; Stow Bedon, WCFN.

East: 02, Bawdeswell; 03, Holt Lowes, GHR & ELS; Briston, FR; 11, Swannington, GHR & ELS; 12, Buxton Heath, FR; 14, Beeston Regis ! 22, Bryants Heath, Felmingham, FR; 31, Horning; 32, Smallburgh, FR; 40, Acle, FR.   S. L.

S. magellanicum Brid.

Rare. This robust species forms very dense dark-red cushions in un-disturbed bogs of high-water table. Associated with *Narthecium* and *Rhynchospora alba* in West Norfolk.

West: 62, Sandringham Warren ! Roydon Common, FR.

East: 03, Holt Lowes, FR; 12, Buxton Heath, GHR.

Not recorded for neighbouring counties.

S. papillosum Lindb.

Although regarded as rare by Burrell, this species is a frequent component of bogs and wet acid heaths where its yellowish-brown hummocks are conspicuous features. Locally dominant in bog at Roydon and Sandringham Warren. Succeeds *S. cuspidatum* on the margins of water-filled depressions.

West: 62, Sandringham Warren, MCFP; N. Wootton, CPP & ELS; Derby Fen, l–d in bog, FR; Roydon Common.

East: 02, Bawdeswell; 03, Holt Lowes, GHR & ELS; 11, Swannington ! Felthorpe; 12, Buxton Heath, FR; 22, Bryants Heath, FR; Swanton Abbot; 32, Smallburgh. S. L.

S. compactum DC.

Locally abundant. Associated with *Erica tetralix* and *Trichophorum caespitosum* on wet heaths.

West: 61, E. Winch Common ! 62, Sandringham Warren, MCFP; Roydon Common, la–ld on wet heath, FR; 90, Scoulton Heath, CPP & ELS.

East: 02, Bawdeswell; 03, Holt Lowes, GHR & ELS; Briston, FR; 11, Swannington, GHR & ELS; 12, Buxton Heath, FR; 21, Newton St Faiths; 22, Bryants Heath, FR. L.

S. teres (Schp.) Angstr.

Apparently rare; only one recent record.

West: 31, plentiful in Broadland carr, Woodbastwick, 1962, GHR.

East: 03, Holt Lowes, as var. *imbricatum*. S.

S. squarrosum Pers. ex Crome

Frequent in fen carr, damp woods and fens. Tolerates a wide range of soils.

West: 61, Cranberry Fen, E. Winch ! 62, Roydon Common, FR; Derby Fen, MCFP; 91, Scarning Fen, GHR & ELS.

East: 03, Holt Lowes, GHR; 12, Buxton Heath, FR; 30, Blofield, GHR & ELS; 32, E. Ruston and Smallburgh Fen, FR; 42, reed-bed at Horsey Mere, GHR & ELS. S.

S. recurvum P. Beauv.

Widespread and abundant. Forms extensive ochreous-coloured carpets in bogs and fens; fringes bog pools. Associated with *Drosera rotundifolia* and *Narthecium* at Roydon.

West: 60, Mow Fen, Shouldham ! 61, Cranberry Fen, E. Winch, FR; 62, N. Wootton ! Sugar Fen ! Sandringham Warren, MCFP: Roydon Common ! Derby Fen, FR; 71, E. Walton, CPP & ELS; 82, Tatterford, BBS.

East: 03, Holt Lowes! 11, Swannington, GHR; 12, Buxton Heath, FR.
S. L.

S. tenellum Pers.

Locally frequent on wet heath but nearer the bog than *S. compactum*. This species and *S. plumulosum* are the most frequent in fruit.

West: 61, Leziate; 62, Sandringham Warren ! Roydon Common and Grimston Warren, FR.

East: 02, Bawdeswell; 03, Holt Lowes and Briston, GHR; 11, Swanning-ton, GHR & ELS; Drayton and Horsford; 12, Buxton Heath, FR; 21, Horsham St Faiths; 22, Bryants Heath, FR.   L.

S. cuspidatum Ehrh.

Widespread and common in bog pools, frequently free-floating or submerged.

West: 61, E. Winch Common ! 62, Sandringham Warren ! Roydon Common, FR.

East: 03, Holt Lowes ! 12, Buxton Heath, FR; 22, Bryants Heath, FR; 32, Callow Green, Barton Turf !   S. L.

S. contortum Schultz

Burrell regarded this as common and one of the free-floating or sub-merged species. It appears to be rare or overlooked.

West: 62, Roydon Common, H. J. B. Birks & D. A. Ratcliffe, 1966.

East: 03, Briston. 1959, FR; Holt Lowes, FR; 12, Buxton Heath, FR. Restricted to Norfolk.

S. subsecundum Nees var. subsecundum

The restricted species appears to be rare in Norfolk.

West: 61, E. Winch Common, BBS; 62, Roydon Common and Derby Fen, FR.

East: 03, Holt Lowes and Briston, FR; 12, Buxton Heath, 1953, BBS; 22, Bryants Heath, FR; 32, E. Ruston, FR.   L.

Var. inundatum (Russ.) C. Jens.

Occasional. Submerged in bog pools.

West: 61, E. Winch Common ! 62, Sandringham Warren, FR; Roydon Common, FR.

East: 03, Holt Lowes ! Briston, GHR; 11, Swannington, GHR & ELS; 12, Buxton Heath, FR; 22, Swanton Abbot and Bryants Heath, FR. Not recorded for neighbouring counties.

Var. auriculatum (Schimp.) Lindb.

This appears to be the most frequent taxon of the aggregate species.·

West: 61, E. Winch Common, BBS; 62, Sandringham Warren, MCFP; Roydon Common ! 82, Tatterford, BBS; 03, Swanton Novers, FR & ELS.

East: 03, Holt Lowes ! Briston, GHR; 12, Buxton Heath !   S.

S. fimbriatum Wils.

Frequent in fens and carrs, relatively tolerant of a wide range of soil types.

West: 62, Sugar Fen, MCFP; N. Wootton, CPP & ELS; Castle Rising; Roydon Common and Grimston Warren, FR; 63, Dersingham Fen ! 71, E. Walton Common, CPP & ELS; 99, Hockham ! Thompson Common !

East: 11, Swannington, GHR & ELS; 12, Buxton Heath, FR; 29, Hard-wick near Tasburgh ! 32, Ingham, Stalham and Brumstead; 40, Acle, FR; 42, Winterton, GHR.   S. L.

S. rubellum Wils.

Two stations only given by Burrell but it appears that there may be some confusion with a variety of *S. capillaceum*. It is both widespread and abundant, forming hummocks in bogs and fens.

West: 61, E. Winch Common, BBS; 62, Sandringham Warren, MCFP; Roydon Common, FR; 63, Dersingham Fen !

East: 03, Holt Lowes, GHR & ELS; 11, Swannington, GHR & ELS; 12, Buxton Heath, FR; 22, Bryants Heath, FR.   S.

S. capillaceum (Weiss) Schrank
(*S. nemoreum* Scop.)

It appears to be less common than stated by Burrell but see note on previous species.

West: 61, E. Winch Common, BBS; 62, Roydon Common, FR; Derby Fen, FR; 63, Dersingham Fen ! 71, E. Walton Common ! 82, Tatterford, BBS. East: 02, Bawdeswell, FR; 03, Holt Lowes, GHR & ELS; 12, Buxton Heath, FR; 19, Flordon Common !   S.

S. plumulosum Röll.

Common in the less acid habitats such as the numerous more or less calcareous valley fens. This species and *S. tenellum* are the most frequent in fruiting.

West: 62, Roydon Common, FR; Sandringham Warren, BBS; 71, E. Walton Common ! 91, Scarning Fen, GHR; Potter's Fen ! 08, S. Lopham Fen ! 09, Swangey Fen !

East: 03, Holt Lowes ! Briston, GHR; 11, Swannington ! 12, Buxton Heath, FR; 22, Bryants Heath, FR; 32, Smallburgh Fen, FR; 40, Acle, FR; 41, Thurne, FR; 42, Hickling, FR.   S. L.

S. molle Sull.

Apparently rare. Wet heaths. No authentic records until 1966.

West: 61, E. Winch Common, BBS, 1967; 62 Wolferton Bog, J. Birks & D. Ratcliffe, 1966.

POLYTRICHACEAE

Atrichum undulatum (Hedw.) P. Beauv., var. *undulatum*

Common and widespread in woodland, particularly in the boulder-clay region; occassionally in sandy tracks in shade. Forms simulating var. *minus* (Hedw.) occur occasionally in hard compacted soil in some of the forest rides in Breckland.   S. C. L.

Polytrichum nanum Hedw.

Apparently less common than formerly when thirteen stations in East Norfolk were given by Burrell. Bare soil on heaths.

East: 13, Bodham, 1964, GHR; 33, Witton, 1966, GHR.   S. C.

P. aloides Hedw., var. *aloides*

Although Burrell states it to be locally common this hardly accords with recent observations. Thin turf on heaths.

West: 62, Sandringham Warren, BBS; Roydon Common, FR.
East: 03, Holt, H. T. Petch det. ECW; 12, Oulton Street near Aylsham, GHR.   S.

P. urnigerum Hedw.

No recent records but at one time found to be abundant on banks at Gillingham, E49, DT. S. L.

P. piliferum Hedw.

Widespread and common but appears to be less so than the next species. Pioneer coloniser of bare, sandy places, both inland and maritime.

S. C. L.

P. juniperinum Hedw.

Abundant and in more or less similar habitats to *P. piliferum* but appears to favour more stable soil including the grey dunes of both Scolt Head Island and Blakeney Point. S. C. L.

P. alpestre Hoppe

Rare in *Sphagnum* swamps.

West: 61, Leziate; 62, Roydon Common, FR.
East: 03, Holt Lowes; 22, Bryants Heath, FR. S.

P. aurantiacum SW.
(*P. gracile* Sm.)

Rare. Heaths.

West: 61, Leziate; 71, Westacre; 81, Southacre; 88, Croxton.
East: 12, Booton and Marsham; 14, W. Runton, WRS. S. C. L.

P. formosum Hedw.

Frequent in woods and nearby sandy banks.

West: 61, E. Winch Common, BBS; 62, N. Wootton, CPP; Sandringham Warren ! 70, Gooderstone ! 82, Tatterford, JA; 83, S. Creake ! 99, Hockham, BBS.
East: 01, Hockering Wood, BBS; 11, Ringland Hills, WGC; 12, Booton Common ! 22, Bryants Heath, FR. S. C. L.

P. commune Hedw., var. *commune*

Widespread and abundant in bogs, wet heaths, carrs and open woods on acid soil. Fruits freely. S. C. L.

BUXBAUMIACEAE

Buxbaumia aphylla Hedw.

Some ten plants were found by W. J. Hooker in a plantation at Sprowston, E21, in 1805, since when the plant has not been found. Not recorded for neighbouring counties.

FISSIDENTACEAE

Fissidens viridulus Wahlenb.

Rather rare on the heavier soils.

West: 99, Hockham, BBS.
East: 01, Hockering Wood, BBS; 11, Weston; 19, Bunwell; 39, Ditchingham. S. C. L.

F. bryoides Hedw.

Common in bare places in woods and on banks of boulder clay.

West: 62, Bawsey brick-yard ! 82, Eastfield Wood, Tittleshall ! 93, Bulfer Grove near Fakenham ! 99, Wayland Wood near Watton ! 03, Swanton Novers, FR & ELS.

East: 01, Hockering Wood, BBS; 12, Booton Common !   S. C. L.

F. crassipes Wils.

Apparently overlooked by the early bryologists. Locally abundant on lower brickwork of bridges along the upper reaches of the River Wissey where it is frequently submerged; also grows permanently submerged on stones in the river bed.

West: 80, Gt. Cressingham, CPP & ELS; 82, Tatterford Common, JA; 89, bridge over the R. Wissey, Bodney, 1960 ! 99, bridge at Mt Pleasant near Rockland All Saints, CPP & ELS.

East: 01, Lyng, BBS; bridge over R. Wensum, 1967, BBS, its first East Norfolk record.   S. C.

F. taxifolius Hedw.

Widespread and abundant on chalky boulder clay in woods, dykesides, marl and chalk pits.   S. C. L.

F. cristatus Wils.

An uncommon and somewhat critical species sometimes confused with *F. adianthoides* with which it occasionally grows in chalk grassland.

West: 73, 74, Ringstead Downs, BBS; 78, Weeting Heath, FR; 08, Garboldisham Heath, GHR & ELS.

East: 14, Beeston Regis (under *F. decipiens*) with fruit.   S. C. L.

F. adianthoides Hedw.

Widespread and abundant. Constant in base-rich fens of south-west Norfolk, fruiting occasionally in early winter; also in chalk grassland.
          S. C. L.

DICRANACEAE

Pleuridium acuminatum Lindb.

Far less frequent than formerly when Burrell regarded it as common on heaths (under *P. subulatum* Rabenth.)

West: 61, Bawsey! 92, heath near N. Elmham, CPP & ELS; 93, damp, sandy bank near Kettlestone!

East: 01, Hockering Wood, BBS; 11, sides of old cart track, Swannington, GHR & ELS.   S. C. L.

P. subulatum (Hedw.) Lindb.

New to Norfolk.

West: 03, Swanton Novers Great Wood, 1967, BBS.

East: 21, gravel pit, east of Coltishall, 1967, JAP, its first record.

Ditrichum cylindricum (Hedw.) Grout

No mention in 1914 *Flora*; easily overlooked. Occasional in stubble fields with the hepatic *Riccia sorocarpa* on acid soils.

West: 61, E. Winch, 1956, the first record! 81, fallow field near Litcham Common, EWJ; 99, Hockham, CCT & ELS. S. C. L.

D. heteromallum (Hedw.) Britt.

Very rare. Only one old record.

East: 14, Sheringham, 1909, as *D. homomallum*.

Not recorded for neighbouring counties.

D. flexicaule (Schleich.) Hampe

Rare. Confined to chalk grassland and pits in Breckland.

West: 70, limekiln near Swaffham! 70, rail embankment near Swaffham, GHR & ELS; 74, Ringstead Downs, BBS; 78, Weeting Heath, FR; 89, chalk pit near Hilborough, GHR & ELS; near Grimes Graves, Weeting, FR.

East: 14, chalk pit, Sheringham. S. C. L.

Ceratodon purpureus (Hedw.) Brid., var *purpureus*

Widespread and abundant on heaths, in woodland clearings and on sand-dunes. Var. *conicus* (Hampe) Husn., is given for East Norfolk in *Census Catalogue*.

Seligeria calcarea (Hedw.) B., S. & G.

The only moss confined to the chalk in West Norfolk where it grows on exposures of the Lower Chalk in pits.

West: 63, Heacham! 70, Barton Bendish! 72, Hillington! 73, Shernbourne! 74, Ringstead Downs! 80, Ashill! 88, Santon! S. C. L.

Pseudephemerum nitidum (Hedw.) Reim.

Rare. Sandy soil. Under *Pleuridium axillare* Lindb., in 1914 *Flora*.

West: 03, Swanton Novers Great Wood, 1967, BBS.

East: 14, Sheringham; 50, Gt. Yarmouth. S. L.

Dicranella schreberana (Hedw.) Dix.

Rare. Exposed places on clay soil.

West: 69, bank of New Cut, Wretton, March 1967, F. G. Bell & J. M. Lock; 71, exposed clay subsoil of dyke, Marham Fen, May 1967!

East: 01, stubble field, Elsing, April 1967, EWJ; 14, Beeston and Runton (under *D. schreberi*). S. C. L.

D. varia (Hedw.) Schimp., var. *varia*

Frequent throughout West Norfolk. On detritus of old chalk quarries, bare places on the boulder clay, damp clay slopes and on bare surfaces of the Lower Chalk.

West: 63, Heacham chalk pit! 70, Cockley Cley and Gooderstone Warren! 78, Hockwold to Weeting! 80, Hilborough, a small plant with male inflorescences, ! 91, Rawhall ! 99, Hockham, BBS; 08, Garboldisham.

East: 01, damp sand pit, Lyng, ACC. S. C. L.

D. rufescens (With.) Schimp.

Rare. On detritus of disused chalk quarry. Not given in *Census Catalogue*, but West Norfolk gathering determined by E. V. Watson.

West: 63, Heacham, 1956!

East: 14, wet clay bank at Beeston Regis and Runton; 24, Overstrand. S.

D. cerviculata (Hedw.) Schimp.

Regarded as locally common by Burrell on wet peaty soil, particularly on dykesides.

West: 61, Bawsey! 62, Reffley! Grimston Warren!

East: 12, Buxton Heath, FR.   S. L.

D. heteromalla (Hedw.) Schimp.

Very common. Woodland banks, sides of old rabbit burrows and epiphytic on trees.   S. C. L.

Dichodontium pellucidum (Hedw.) Schimp., var. *pellucidum*

Only once recorded.

East: 18, Diss, WRS.

Not recorded for neighbouring counties.

Dicranoweisia cirrata (Hedw.) Lindb.

Common on trees. The occurrence of gemmae on the leaves of the comal tuft observed by Crundwell (BBS, *Trans.*, 1, **4**, 1950) is confirmed for Norfolk material. They are calvate and jointed, up to 160 μ in length, with up to 5 transverse septa and 2–3 longtitudinal, brownish green in colour but are not abundant.   S. C. L.

Dicranum majus Sm.

Rare or overlooked. Not recorded by Burrell. Bogs and damp woods.

West: 01, Swanton Morley, GHR, abundant; 03, Swanton Novers Great Wood, FR & ELS.

East: 01, Hockering Wood! 03, Holt Lowes, 1957, first record,! 30, Hassingham, GHR.   L.

D. bonjeani De Not.

Frequent in both base-rich fens and bogs.

West: 61, E. Winch Common! 62, Sandringham Warren, BBS; Roydon Common, FR; 70, Gooderstone Fen! 71, E. Walton Common! 82, Tatterford Common, JA; 98, Ringmere, BBS; 99, Cranberry Rough, Hockham, CCT & ELS.

East: 02, Bawdeswell, FR; 03, Holt Lowes, FR; 11, Swannington, GHR & ELS; 12, Booton Common, GHR & ELS; Buxton Heath, FR.   S. L.

D. scoparium Hedw.

One of the common mosses of heathland, woodland rides, tree bases, acid grassland and older grey sand-dunes. The var. *orthophyllum* is recorded from maritime habitats but appears to be a habitat modification. S. C. L.

D. polysetum Michx.

(*D. rugosum* Brid.)

Rare. Not recorded by Burrell.

West: 61, E. Winch Common, 1963, GHR; 88, Santon, ALB; 88 or 89, 'among grass in a clearing in pine plantation beside the A134 between Thetford and Mundford', 1960, S. J. P. Waters in BBS, *Trans*, IV, **2**, 1961, its first record.   S. L.

D. spurium Hedw.

Rare. In *Calluna-Erica tetralix* association.

West: 61, E. Winch Common, 1958, GHR; 62, Sandringham Warren, 1952, MCFP; Roydon Common! (Burrell's only record for 'Grimston' most likely refers to this station.)

East: 03, Briston Common, 1961, GHR. L.

Campylopus fragilis (Brid.) B., S. & G.

Rare. Heaths. No recent records.

West: 62, Roydon; 92, Brisley.

East: 02, Bawdeswell; 11, Ringland Hills, 1917, WGC; Drayton; Weston, WHB; 12, Booton. S. L.

C. pyriformis (Schultz) Brid.

Widespread and abundant, particularly on the Greensand heaths, colonising bare areas of peat; occasionally on decaying wood. Fruiting on wet peat at Holt Lowes, 1959. S. C. L.

C. flexuosus (Hedw.) Brid., var. *flexuosus*

Rare. Heaths and woods.

West: 61, E. Winch Common, BBS; 62, Roydon Common, BBS; 71, Pentney; 82, Tatterford Common, JA; 98, Langmere, BBS; 03, Swanton Novers, FR.

East: 03, Holt Lowes and Briston, FR; 11, Felthorpe and Horsford; 12, Buxton Heath, CCT & ELS; Hercules Wood, Blickling, GHR; 21, Newton St Faiths; 22, Bryants Heath, FR; 23, Roughton; 30 Blofield. S. C. L.

C. introflexus (Hedw.) Brid.

Apparently a new arrival to the county. Fruiting occasionally.

West: 61, forming mats under birch trees, Bawsey, 1966,! 62, Roydon Common, BBS; Sandringham Warren, BBS; Ling Common, N. Wootton, BBS; 63, Dersingham Heath, 1964, Birks, Duckett & Lees in BBS, *Trans.*, IV, **5**, 962; 81, Litcham Common, ALB; 84, Holkham, ALB; GHR.

East: 42 (not '50' as recorded), in damp sand, Winterton Ness, J. K. Marshall, its first Norfolk record. S.

C. brevipilus B., S. & G.

Rare. *Erica tetralix* heaths.

West: 61, E. Winch Common! 62, Wolferton Fen, 1960,! 62, Roydon Common, FR; 78, Weeting Heath, FR.

East: 03, Holt Lowes, GHR & ELS; 11, Horsford; 12, Buxton Heath, BBS, 1953; 21, Newton St Faiths; 22, Bryants Heath, FR; 23, Roughton. S.

Leucobryum glaucum (Hedw.) Schimp.

Common. Heaths, woods and fens. In woods, detached balls with little distinction between upper and lower surfaces are to be found as a result of human interference and scratching by birds; in fens, large spreading tufts as much as four feet in width occur. Does not fruit in Norfolk. S. C. L.

ENCALYPTACEAE

Encalypta vulgaris Hedw.

Frequent in chalk grassland and chalk pits.

West: 78, Weeting Heath! 88, Santon Warren, ECW; 99, Stow Bedon!
S. C. L.

E. streptocarpa Hedw.

Apparently rare. Not mentioned by Burrell. In similar habitats.

West: 61, E. Winch Common, BBS; 78, Weeting Heath, 1957, FR; 89, heath near Grimes Graves, Weeting, FR. C. L.

POTTIACEAE

Tortula ruralis (Hedw.) Crome

Widespread and abundant. Trees, thatch, sands (both inland and coastal).
S. C. L.

T. ruraliformis (Besch.) Rich. & Wall.

Abundant on coastal sands and shingle but also occurring inland in Breckland.

West: 62, 63, 74, 84, along the coast from Wolferton to Holkham! 73, Boiler Common, Bircham! 78, Weeting Heath! 80, Hilborough!. 98, Ringmere, BBS.

East: 04, Blakeney. C. S.

T. intermedia (Brid.) Berk.

Regarded as common by Burrell. In the mortar joints of walls.

West: 78, Hockwold church wall! 81, Castleacre; 93, Binham. S. C. L.

T. laevipila (Brid.) Schwaegr., var. *laevipila*

Frequent on trees.

West: 61, E. Winch! 70, Barton Bendish! 71, Lambs Common, E. Walton! 73, Shernbourne! 74, Ringstead Downs! 78, Weeting, BFTD; Weeting Fen! 84, Burnham Overy, BBS; 90, Carbrooke Fen, GHR & ELS.
S. C. L.

Var. *laevipiliformis* (De Not.) Limpr.

West: 63, Heacham; 70, Beechamwell.

East: 19, Flordon. S.

T. virescens (De Not.) De Not.

West: 62, trunk of elm near Sandringham House, BBS, 1967, its first Norfolk record. Confd. ACC.

Not recorded for neighbouring counties.

T. papillosa Wils. ex Spruce

Appears to be less frequent than formerly when it was regarded as common on trees.

West: 61, Cranberry Fen, E. Winch! 62, Ling Common, N. Wootton, BBS; 84, Burnham Overy, BBS; 99, Hockham, BBS.

East: 01, Lyng, EWJ. S. C. L.

65

T. latifolia Hartm.

Rare. On brickwork of bridges over rivers and particularly on piles in some of the Broadland rivers.

West: 89, on more or less submerged brickwork of bridge over R. Wissey, Bodney, 1960, its first West Norfolk record,!

East: 01, revetment of bridge, Lyng, EWJ, 1967; 20, Bramerton; Postwick; Trowse and Whitlingham (as *T. mutica*); 29, Shotesham. S. C. L.

T. subulata Hedw., var. *subulata*

Not as common as formerly. Sandy hedgebanks, boulder-clay banks and maritime shingle.

West: 74, Ringstead Downs, BBS; 78, Weeting Heath, BBS; 81, Washpit Lane, Southacre! 82, Eastfield Wood, Tittleshall, 1960,! Tatterford, BBS.

East: 04, high shingle, Blakeney, 1927, PWR. S. C. L.

T. muralis Hedw., var. *muralis*

Very common on walls; also recorded from shingle at Blakeney Point by Watson. S. C. L.

Var. *aestiva* Hedw.

Rare. Damp walls.

East: 11, Drayton; 19, Tacolneston; 31, Woodbastwick. S. C.

T. marginata (B. & S.) Spruce

Rather rare. Brick walls.

West: 90, Scoulton and Woodrising.

East: 01, brick wall near river, Lyng, ACC; 13, Wolterton; 19, Tacolneston; 21, Frettenham; 31, Ranworth; 21, Crostwick. C. S.

Aloina rigida (Hedw.) Limpr.

A rare and somewhat critical species. At one time far more frequent on mud-capped walls in some of the north Norfolk coastal villages but the practice of mud-capping has now ceased and the species is now confined to chalk pits.

West: 74, Thornham and Titchwell; 78, old chalk pit, Weeting, 1958,!

East: 14, Sheringham. S. C. L.

A. ambigua (B. & S.) Limpr.

The commonest of the three species of *Aloina*, but like the last, no longer to be found on north Norfolk walls. Marl and chalk pits and bare calcareous soils of tracks, especially in Breckland.

West: 63, Heacham! 70, Cockley Cley! 73, Gt. Bircham, CPP; 74, Ringstead and Titchwell; 78, Weeting, CCT & ELS; 80, Gt. Cressingham! 82, Massingham Heath!

East: 11, Alderford and Drayton; 14, Sheringham; Weybourne! 20, Thorpe St Andrew; 24, Overstrand. S. C. L.

A. aloides (Schultz) Kindb.

Burrell expressed doubts about the validity of the records then available. The *Census Catalogue* credits Norfolk with all three species. Has been seen in the old chalk pit at Weeting, W78.

Pterygoneuron ovatum (Hedw.) Dix.

Very rare. On mud-capped walls formerly (under *Tortula pusilla*).

West: 74, Titchwell, WET; 98, roadside near Ringmere, 1967, BBS.
East: reputed to have occurred in v-c 27 but stations not specified. S. C. L.

P. lamellatum (Lindb.) Jur.

West: 98, roadside near Ringmere, 1967, BBS.   S. C. L.

Pottia lanceolata (Hedw.) C. Müll.

Frequent in chalk grassland.

West: 74, Ringstead Downs, BBS; 78, Weeting Heath, BBS; 98, Ring-
mere, BBS; o8, Garboldisham Heath, GHR & ELS.
East: 10, Ketteringham and Bawburgh; 11, Alderford and Drayton; 14,
Sheringham; 39, Broome.   S. C. L.

P. heimii (Hedw.) Fürnr.

Locally frequent. Restricted to maritime habitats.

West: 62, Wolferton salt marsh, 1959,! 74, Holme-next-the-Sea, Thorn-
ham and Titchwell, WET.
East: 14, Weybourne; 41, on piles in the R. Bure, Acle; 50, Gt. Yarmouth,
JP.   S. L.

P. intermedia (Turn.) Fürnr., var. *intermedia*

In view of the difficulty of determining this species satisfactorily, Burrell
rightly queried it. He had but few records and none from West Norfolk.
Most probably confused with *P. truncata*. Bare soil and occasionally in
stubble fields.

West: 62, Grimston Warren, 1965,! 78, Weeting Heath! 88, Santon! 98,
Ringmere, BBS, 99, Bragmere Pits near Stow Bedon, 1955, conf. CCT,!
99, stubble field, Cranberry Farm, Hockham, 1958, CCT & ELS.
East: 10, Bawburgh; 20, Norwich, DT; 32, Honing; 50, Gt. Yarmouth,
JP.   S. C. L.

P. truncata (Hedw.) Fürnr.

Widespread and abundant in stubble fields on acid soil and on sea-banks.

West: 61, E. Winch! 78, Weeting Heath! 83, Barmer! 84, sea-banks,
Burnham Overy Staithe and Burnham Deepdale, 1956,! 98, Ringmere,
BBS.   S. C. L.

P. wilsonii (Hook.) B. & S.

Very rare. On a sandy inland bank.

East: 14, Sheringham, Moss Exc., Club, 1907.
Not recorded for neighbouring counties.

P. davalliana (Sm.) C. Jens.

Recorded by Burrell as very common on bare earth (under *P. minutula*).

West: 63, detritus of old chalk pit, Heacham, 1956,! 78, Weeting, BFTD.
East: o1, stubble field, Elsing, EWJ; many stations listed by Burrell.
                                                                          S. C. L.

P. starkeana (Hedw.) C. Müll.

The East Norfolk record probably refers to that by Dawson Turner from
Acle, E40. No recent records.   S.

P. bryoides (Dicks.) Mitt.

Rare. Bare earth of chalk grassland.

West: 74, Ringstead Downs, BBS; 78, Weeting Heath, BBS; 98, Ring-mere, BBS.

East: 10, Bawburgh; 11, Ringland; 14, Sheringham; 20, Norwich.

S. C. L.

P. recta (Sm.) Mitt.

Locally frequent on chalk grassland.

West: 74, Ringstead Downs, BBS; 78, Weeting Heath, FR; 89, Grimes Graves near Weeting, FR & TVDS, 1965.

East: 12, Heydon; 14, Sheringham; 20, near Norwich, DT; 32, Stalham.

S. C. L.

Phascum curvicollum Hedw.

Occasional in marl pits and on fallow fields.

West: 64, Ringstead Downs, BBS; 78, Weeting Heath, 1958,! 89, Grimes Graves, FR & TVDS.

East: 14, Sheringham.

S. C. L.

P. cuspidatum Hedw., var. *cuspidatum*

Widespread and abundant on exposed base-rich soils, stubble fields and sea-banks.

S. C. L.

Var. *piliferum* (Hedw.) Hook. & Tayl.

West: 74, Titchwell, WET.

East: 50, Gt. Yarmouth, DT.

Var. *curvisetum* (Dicks.) Nees & Hornsch.

East: 30, Rockland St Mary; 31, Upton.

P. floerkeanum Web. & Mohr

Rare. No recent records. Bare earth.

East: 14, Sheringham; 23, Southrepps.

C.

Acaulon muticum (Brid.) C. Müll., var. *muticum*

Rare. Bare heathy soil.

West: 78, Weeting Heath, BBS; 98, Ringmere, BBS, 1967, new to West Norfolk.

East: 10, Bawburgh; 14, Beeston Regis and Sheringham; 23, Southrepps and N. Walsham, 1960, GHR.

S. C. L.

Barbula convoluta Hedw., var. *convoluta*

Very common throughout the county. Walls, stony ground of tracks, pits and exposed soils.

S. C. L.

Var. *commutata* (Jur.) Husn.

West: 99, Hockham, BBS, conf. ACC.

East: 11, Weston; 14, Weybourne; 30, Rockland St Mary; 31, Ranworth.

C.

B. unguiculata Hedw., var. *unguiculata*

Common. Colonises bare patches of soil in forest rides, chalk pits, sand of fixed dunes, stubble fields and walls. Fruit frequent.

S. C. L.

B. revoluta Brid.

Occasional on walls.

West: 64, Hunstanton; 70, near Swaffham; 73, Sedgeford, BBS; 74, Holme-next-the-Sea, c. fr., BBS; 84, Burnham Overy, BBS; 92, Brisley; 99, Hockham, BBS.

East: 13, Saxthorpe; 20, Thorpe St Andrew.   S. C. L.

B. hornschuchiana Schultz

Frequent. Sandy and stony ground.

West: 62, Sandringham Warren, BBS; 78, Weeting, CCT & ELS; 88, Santon Warren, ECW.

East: 01, Lyng; 10, Bawburgh; 11, Horsford and Drayton; 20, Norwich; 30, Rockland St Mary.   S. C. L.

B. fallax Hedw.

Widespread and abundant. Banks and waste land on calcareous soil.

B. reflexa (Brid.) Brid.

One record only.

East: 18, Diss, 1909, WRS (as *B. recurvifolia*).

B. rigidula (Hedw.) Milde

Rare. On walls. No recent records.

West: 08, S. Lopham; 62, Castle Rising; 64, Hunstanton.

East: 29, Newton Flotman; 31, Panxworth.   S. C. L.

B. trifaria (Hedw.) Mitt.

West: 73, Ringstead Downs, BBS, 1967, its only recent record; 74, on a wall, Titchwell, WET.   S. C. L.

B. tophacea (Brid.) Mitt.

Locally common in clay and marl pits.

West: 61, Bawsey! 62, Wolferton! 63, Snettisham! 73, Sedgeford, BBS; 74, Holme-next-the-Sea and Titchwell, WET; 99, Hockham, BBS.

East: 12, Booton Common, GHR & ELS; 14, Beeston Regis and Shering-ham; Runton; 33, Mundesley, EMH.   S. C. L.

B. cylindrica (Tayl.) Schimp.

Probably confused with *B. vinealis* by the early bryologists as there is no mention of this species in the 1914 *Flora*.

West: 70, bare soil of chalk grassland near Swaffham! ; 73, Ringstead Downs, BBS; 89, stonework of bridge over R. Wissey, Bodney, 1960, the first Norfolk record,! 99, Hockham, BBS.   S. C. L.

B. vinealis Brid.

Common on walls throughout the country.   S. C. L.

B. recurvirostra (Hedw.) Dix.

Occasional on walls and in hedgebanks.

West: buttress of railway arch, Swaffham-Fincham, 1966,! 78, Weeting Heath, FR.

East: 'Common', 1914, *Flora* (under *B. rubella*).   S. C. L.

Gyroweisia tenuis (Hedw.) Schimp.

Very rare although it may have been overlooked by reason of its small size.

West: 62, wall of Corn Exchange, King's Lynn, BBS, 1967; 82, Tatterford, JA; 90, in the village well, Woodrising, 1909, as *W. tenuis.*   C. L.

Eucladium verticillatum (With.) B., S. & G.

Very rare.

West: 82, base of bridge, S. Raynham, 1967, JAP.

East: 11, in village well, Costessey, 1909, as *W. verticillata.*   C. L.

Tortella inclinata (Hedw. f.) Limpr.

Only one record.

West: 64, Hunstanton, 1900, EMH, in Herb. Kew.   C. L.

T. flavovirens (Bruch) Broth., var. *flavovirens*

Restricted to maritime habitats. From high shingle to the lower zone of *Suaeda fruticosa.*

West: 63, Heacham; 64, Hunstanton, EMH; 74, abundant in the *Limonietum* at Holme-next-the-Sea, 1955, CCT; 84, Scolt Head Island, CVBM; under planted pines, Burnham Overy Staithe, BBS; 94, Wells-next-the-Sea.

East: 04, Blakeney Point, PWR; 14, Weybourne.   S. L.

Pleurochaete squarrosa (Brid.) Lindb.

Very rare. Chalk grassland in Breckland. Not recorded in the 1914 *Flora.*

West: 78, Weeting Heath, 1957, FR, its first record; 1966,!   S.

Trichostomum sinuosum (Mitt.) Lindb. ex Herzog

Apparently rare; no mention in *Flora.*

West: 89, on more or less submerged brickwork of bridge over R. Wissey, Bodney, 1960,!

East: 01, R. Wensum, Lyng, BBS; R. Wensum, Mill Street, ACC. C. S.

Weissia controversa Hedw.

Regarded as very common by Burrell on banks (under *W. viridula*).

West: 70, Gooderstone Fen! 74, Ringstead Downs, BBS; 78, Weeting Heath, BBS; 79, on ant-hills, Foulden Common, 1965,!

East: 03, east of Swanton Novers Great Wood, BBS.   S. C. L.

W. crispata (Nees & Hornsch.) Jur.

Found growing on an ant-hill on Foulden Common, W79, 1965,! and bearing the characteristic rudimentary peristome inserted below the orifice. Its only Norfolk record.

Not recorded for the neighbouring counties.

W. microstoma (Hedw.) C. Müll.

Occasional on exposed soil in basic grassland and thin turf of Breckland.

West: 70, Devil's Dike, Beechamwell! 71, E. Walton Common! 74, Thornham, WET; chalk detritus, Ringstead Downs! 78, Weeting Heath, CCT & ELS.

East: 03, Edgefield; 10, Wymondham; 12, Aylsham and Cawston; 14, Sheringham; 21, Newton St Faiths.   S. C. L.

W. crispa (Hedw.) Mitt.

Apparently rare. Chalk grassland.

West: 74, Ringstead Downs, BBS; 89, Grimes Graves, 1965, FR & TVDS; 90, Cranworth, ALB. S. C. L.

Leptodontium flexifolium (Sm.) Hampe

Very rare. No records in *Flora.*

West: 62, Sandringham Warren, 1956, MCFP. L.

GRIMMIACEAE

Grimmia apocarpa Hedw., var. *apocarpa*

Appears to be uncommon. Brick and stone work.

West: 64, Hunstanton; 74, Holme-next-the-Sea; 81, Castleacre; 88, Santon; 99, Hockham, BBS.

East: 01, Hockering Wood, BBS; 03, Holt; 20, Framingham Pigot; 21, Crostwick; 31, Upton and S. Walsham. S. C. L.

G. pulvinata (Hedw.) Sm.

One of the commonest mosses of walls, particularly in the mortar joints; also recorded from Blakeney Point growing on shell. S. C. L.

Rhacomitrium canescens (Hedw.) Brid.

Locally abundant on sandy heaths and dunes; conspicuous in autumn.

West: 60, Shouldham Warren! 62, Castle Rising! N. Wootton! Grimston Warren! 78, Weeting Heath! 84, Holkham; 88, Santon Warren! 89, Bodney; 98, Harling to Bridgham, GHR & ELS.

East: 21, Newton St Faiths; 50, Gt. Yarmouth, JP. S. C. L.

R. lanuginosum (Hedw.) Brid.

'On flat heaths in Norfolk', *Muscologia Britannica,* Hooker & Taylor, second edition, 1827. Burrell recorded it from Horsford, E11, and Newton St Faiths, E21. Now considered to be extinct. C.

FUNARIACEAE

Funaria hygrometrica Hedw.

Frequent on sites of fires in woods; exposed calcareous soil; mortar of damp walls; young dunes and shingle. S. C. L.

F. fascicularis (Hedw.) Schimp.

Rare. Bare earth. No recent records.

East: 14, E. Beckham and Sheringham; 30, Rockland St Mary and Lingwood. S.

F. obtusa (Hedw.) Lindb.

Very rare. No recent records.

East: 04, Kelling Heath, 1909, as *F. ericetorum.*

Not recorded for neighbouring counties.

Physcomitrium pyriforme (Hedw.) Brid.

Frequent in exposed calcareous soil.

West: 62, Grimston Warren! Sugar Fen, BBS; 71, Pentney! Marham Fen! 72, Congham, CPP; 80, Little Cressingham; 82, Tatterford, BBS; 88, Thetford, CCT; 89, West Tofts, CPP & ELS; 99, Rockland All Saints, GHR & ELS.

East: 11, Swannington, GHR & ELS. S. C. L.

P. eurystomum Sendtn.

New to the British Isles when discovered in 1961.

West: 99, in short turf and open vegetation, near the water-line, Lang-mere, 8th January 1961, BTFD det. EFW.

Not recorded for neighbouring counties.

Physcomitrella patens (Hedw.) B., S. & G.

Rare. Only once recorded in the 1914 *Flora* from Acle. One of the first records made by Sir James Paget (1814–99).

West: 70, dried-up pond, Foulden Common, 1955,! S. C. L.

SPLACHNACEAE

Splachnum ampullaceum Hedw.

No recent records. Believed to be extinct now although it still persists just outside the county boundary in Suffolk at Coney Weston.

EPHEMERACEAE

Ephemerum recurvifolium (Dicks.) Boul.

Reported by MCFP for West Norfolk but station not known. S. C. L.

E. serratum (Hedw.) var. *serratum*

Burrell states it to be rather rare on bare soil.

East: 03, east end of Swanton Novers Great Wood, 1967, BBS; 14, Sheringham; 31, Acle, DT. S. C. L.

Var. *minutissimum* (Lindb.) Grout

West: 78, Weeting Heath, 1967, BBS, its first Norfolk record.

East: 03, east end of Swanton Novers Great Wood, BBS.

SCHISTOSTEGACEAE

Schistostega pennata (Hedw.) Web. & Mohr

New to Norfolk. Found during the Brit. Bryol. Soc., excursion, 1967.

West: 62, very abundant and fruiting in Greensand cliff in sand-martins' burrows, Wolferton.

Norfolk only.

6 COLUMBINE   *Aquilegia vulgaris*   *Jarrolds*

7 BARBERRY   *Berberis vulgaris*   *Miss G. Tuck*

8 OREGON GRAPE *Mahonia aquifolium* agg. *Miss V. M. Leather*

9 YELLOW HORNED POPPY *Glaucium flavum* *R. Jones*

10 SEA ROCKET *Cakile maritima* *G. D. Watts*

11 TOWER MUSTARD *Turritis glabra* *Miss G. Tuck*

12 MARSH VIOLET *Viola palustris* *Miss G. Tuck*

13 BRECKLAND PANSY *Viola tricolor* ssp. *curtisii* *Miss D. M. Maxey*

14 SEA HEATH *Frankenia laevis* K. Wilson

15 SMALL MEDICK *Medicago minima* Miss D. M. Maxey

TETRAPHIDACEAE

Tetraphis pellucida Hedw.

Regarded by Burrell as very rare; only one record in the 1914 *Flora*.

West: 61, E. Winch Common, BBS; on decaying birch in carr, Roydon Common, 1956,! 92, Horningtoft! 03, abundant on old oaks, Swanton Novers Great Wood, FR & ELS.

East: 01, Hockering Wood! 30, Blofield, 1962, GHR.

BRYACEAE

Orthodontium lineare Schwaegr.

Has only appeared in the county during the past twenty-six years. Now widespread and abundant on the bases of trees in woodland and on sandy peat banks. 'I found it in both vice-counties in the early days of the War', 1941, E. F. Warburg *in lit.*

West: 62, Roydon Common, FR; N. Wootton, BBS; Sandringham Warren, BBS; 63, Dersingham Heath, CCT & ELS; 70, Toot Hill, Beechamwell! 78, Weeting! 82, Tatterford Common, BBS; 91, Horse Wood, Mileham! 92, Horningtoft, CPP & ELS; 98, Langmere and Ringmere, BBS.

East: 01, Hockering Wood, BBS; 03, Holt Lowes, FR; 30, Hassingham, GHR.   S. C. L.

Leptobryum pyriforme (Hedw.) Wils.

Rare in natural communities but frequent on soil in pots in greenhouses.

West: 62, Roydon Common, BBS; Sugar Fen, JA; 88, old gravel pit, Thetford, EFW; 09, dykeside on peaty soil, Swangey Fen!

East: 20, on wall in Norwich; 32, Sutton Marshes; 40, Acle, FR. S. C. L.

Pohlia nutans (Hedw.) Lindb.

Widespread and abundant. Conspicuous by the persistence of the setae after the capsules are shed. Heaths and woodland.   S. C. L.

P. rothii (Correns) Broth.

Although recorded in the *Census Catalogue* for both vice-counties, Burrell only gives it for East Norfolk under *Webera annotina* var. *erecta*.

East: 11, Horsford; 14, Beeston Regis.   S. L.

P. annotina (Hedw.) Loeske

Rather rare on sandy ground.

West: 62, Castle Rising; 91, Scarning, JES, 1780, with fruit, in his herbarium at Burlington House; 03, Swanton Novers Great Wood, 1967, BBS.

East: 14, Beeston Regis.   S. L.

P. proligera (Lindb.) Limpr., ex Arnell

Burrell regarded this as rather common on sandy ground. Not credited for the county in the *Census Catalogue* and, in view of probable confusion with forms of *P. annotina*, it is suggested that Burrell's records be omitted. Not recorded for neighbouring counties.

P. wahlenbergii (Web. & Mohr) Andr.
(*P. albicans* (Wahlenb.) Lindb.)

A new county record when found by CCT, August 1958, along a sandy track, Whin Hill Covert, Wolferton, W62.

East: 01, Hockering Wood, BBS. C. L.

P. delicatula (Hedw.) Grout

Occasional. Exposed clay banks.

West: 62, Sugar Fen, BBS; 70, Marham, JC; 80, Gt. Cressingham! 82, Tatterford, JA; 83, with *Sphaerocarpos* in mustard stubble field, S. Creake! 92, Horningtoft! Brisley; 98, Langmere, BBS; 99, banks of R. Thet, Rockland All Saints! 09, Swangey Fen, det. EVW,!

East: 10, Little Melton; 14, Sheringham; 39, Ditchingham. S. C. L.

Bryum mammillatum Lindb.

Very rare. No recent records.

West: 64, Hunstanton golf links, with *Juncus gerardi*, on damp sandy soil, 1902, WET. L.

B. pendulum (Hornsch.) Schimp.

Locally common along the coast on sand-dunes and shingle, occasionally fruiting.

West: 64, Hunstanton; 74, Holme-next-the-Sea and Titchwell, WET; 84, Holkham! Burnham Overy Staithe.

East: 04, Blakeney Point. S. L.

B. inclinatum (Brid.) Bland.

Rare. In similar habitats to the last.

West: 64, Hunstanton, EMH; 94, Wells-next-the-Sea.

East: 04, Blakeney Point S. L.

B. pallens Sw.

Neither recorded by Burrell nor given in the *Census Catalogue*.

West: 61, damp turf, E. Winch Common, BBS; 62, Sugar Fen, JAP; 78, Weeting Heath, FR; 82, Tatterford Common, JA; 84, 'frequent on the young dunes near the Ternery, Scolt Head', 1933, CVBM.

East: 04, Blakeney Point, 'fruiting on main shingle bank', 1927, PWR. S. C. L.

B. turbinatum (Hedw.) Turn.

Very rare. Wet sandy places.

West: 93, Fakenham, 1849, G. Fitt in *Bryologica Britannica*, 1855.

East: 04, Blakeney Point, high shingle of the Marrams, EVW, 'probably but no fruit'.

Norfolk only but occurrence suspect.

B. pseudotriquetrum (Hedw.) Schwaegr., var. *pseudotriquetrum*

Common in fens in which it is constant and associated with *Schoenus nigricans*, *Acrocladium cuspidatum* and *Ctenidium molluscum*.

Var. *bimum* (Brid.) Lilj.

Burrell regarded this as rather common. There appears to be little to distinguish it from the typical plant.

West: 08, S. Lopham.
East: 12, Gt. Witchingham; 19, Flordon; 32, Sutton. S. C. L.

B. intermedium (Brid.) Bland.
New to the county.
West: 88, wet ground in old gravel pit, Thetford, 1959, EFW. S. C. L.

B. caespiticium Hedw., var. *caespiticium*
Frequent on walls and occasionally in sand-dunes. S. C. L.

B. argenteum Hedw., var. *argenteum*
Widespread and abundant on roadsides, cinder tracks, sandy ground, walls, rotting fences and maritime habitats.

Var. *lanatum* (P. Beauv.) Hampe
In very dry, exposed habitats.
West: 64, Ringstead Downs! 82, Tatterford Common, JA; 84, Scolt Head.
East: 04, Blakeney Point. S. C. L.

B. bicolor Dicks.
Frequent. Chalk and marl pits, silt soil of salt marshes and in gravel pits.
West: localities too numerous to specify.
East: 01, Hockering, BBS; 14, Sheringham, 'rare', the only record by Burrell. S. C. L.

B. radiculosum Brid.
(*B. murale* Wils. ex Hunt)
Rare. No recent records.
West: 88, Croxton; 90, Wood Rising.
East: 20, Whitlingham. S. C. L.

B. erythrocarpum Schwaegr., agg.
Although Burrell regarded *B. erythrocarpum* as a rare plant of heaths, it is abundant in the autumn stubble fields but ploughing usually prevents fruiting. Consequently it has been confused with *B. bicolor* which grows in wetter places. Since the position has been clarified by Crundwell and Nyholm in *European Species of B. erythrocarpum Complex* in BBS, *Trans.*, IV, **4**, 1964, the following segregates have been found, some of them during the 1967 excursion of the BBS by A. C. Crundwell and H. L. K. Whitehouse.

B. ruderale Crund. & Nyh.
West: 74, chalk grassland, Ringstead Downs, BBS; 78, Weeting Heath, BBS; 98, bare ground of roadside, Ringmere, BBS.
East: 01, ride in Hockering Wood, BBS.

B. klinggraeffii Schimp.
West: 69, bank of new Cut, Wretton, F. G. Bell & J. M. Lock; 71, on ant-hills, E. Walton Common!
East: 01, earth by R. Wensum, Lyng, EWJ & HLKW, BBS.

75

B. micro-erythrocarpum C. Müll. & Kindb.

West: 62, Wolferton sea-bank! 70, chalk grassland, Devil's Dike, Beechamwell! 74, Ringstead Downs, BBS; 78, Weeting Heath, BBS; 91, *Callunetum* of Scarning Fen, BBS.

East: 03, east end of Swanton Novers Great Wood, BBS; 14, Beeston Regis, c. fr. 1904, leg. WHB in Herb. UKD (*op. cit.*).

B. bornholmense Winkelm. & Ruthe

West: 61, peaty ground, E. Winch Common, BBS; 62, sandy bank, Roydon Common! sandy ground, Dersingham Common, 1962, HLKW (*op. cit.*).

B. rubens Mitt.

West: 62, grass at edge of railway line, Sandringham Warren, BBS; 70, eroded chalk grassland and nearby arable land, Beechamwell! 71, ant-hills, E. Walton Common! 74, Ringstead Downs, BBS; 78, Weeting Heath, BBS; 91, Scarning Fen, BBS; 98, Langmere, BBS, ACC & HLKW; 99, Hockham, BBS.

East: 01, roadside bank, Hockering, BBS, ACC & HLKW; by R. Wensum at Lyng, ACC & HLKW; arable field west of Bylaugh Park, c. fr., EWJ & HLKW; gravel pit near Sparham, EWJ & HLKW.

B. donianum Grev.

Recorded once only from 'grey dune, top of Long Low', Blakeney Point, E04, EVW. Not in *Census Catalogue.*

B. capillare Hedw., var. *capillare*

Widespread and abundant. Walls, trees, thatched roofs, coastal dunes and shingle. S. C. L.

Var. *elegans* (Nees) Husn.

West: 74, Titchwell, WET.

B. obconicum Hornsch.

Reported to occur in West Norfolk (*Census Catalogue*) but locality not traced. L.

Rhodobryum roseum (Hedw.) Limpr.

Locally frequent on heaths and chalk grassland.

West: 62, Sugar Fen, JAP; 70, Foulden Common, GHR; 71, E. Walton Common, CCT; 72, Derby Fen! 78, Weeting Heath, 1957, FR, its first West Norfolk record; Pilgrims' Walk, Weeting! Gravel Pit Wood, Weeting! 08, Garboldisham Heath!

East: 03, Holt and Edgefield; 10, Bawburgh; 11, Attlebridge, Costessey and Horsford; 14, Beeston Regis. C. S.

MNIACEAE

Mnium hornum Hedw.

One of the most abundant species on the ground and about tree-trunks in woodland where it is often the dominant moss; peat banks of alder carrs. S. C. L.

M. cuspidatum Hedw.

Appears to be far less frequent than formerly. On damp sandy soil of woodland.

West: 62, Whin Hill Covert, Wolfteron, CCT & ELS; 78, Weeting, 1957, its first West Norfolk record!

East: chiefly in the area south-west of Norwich.   S. C. L.

M. longirostrum Brid.

There appears to have been some confusion here as Burrell regarded it as a rare plant of damp heaths. It is a frequent plant of woodland, both coniferous and deciduous, and chalk grassland.   S. C. L.

M. affine Bland.

Common. Base-rich fens, carrs and on ground in woodland rides.

M. rugicum Laur.

Rare. Bogs. Burrell gives only one station, Holt.

West: 99, Rockland All Saints Fen, GHR & ELS; 09, Swangey Fen, BBS, 1953.

East: 03, Holt, EMH; 40, Acle, FR.   S.

M. seligeri (Lindb.) Limpr.

Frequent. Growing in pale green mats in very wet base-rich fens and associated with *Juncus subnodulosus, Carex lepidocarpa* and *Schoenus nigricans.*

West: 62, Roydon Common, BBS; Sugar Fen, JA; 71, E. Walton Common, CCT; 91, Scarning and Potter's Fens, GHR & ELS; 99, Rockland All Saints! Thompson Common! 09, Swangey Fen, 'vf–a', FR.

East: 09, Old Buckenham Fen, FR; 11, Swannington, GHR & ELS; 12, Buxton Heath, GHR & ELS; 32, E. Ruston; 40, Acle, FR.   C. S.

M. undulatum Hedw.

Very common. Woodland banks particularly on base-rich or neutral soils; exceptionally on sand-hills at Scolt Head.   S. C. L.

M. punctatum Hedw.

Frequent and locally common in fens, bogs and woods.

M. pseudopunctatum B. & S.

Far less frequent than *M. punctatum* and in calcareous fens.

West: 62, Roydon Common, c. fr., FR; 71, E. Walton Common! 91, Scarning Fen, c. fr.,! 09, Swangey Fen!

East: 03, Holt Lowes! 12, Buxton Heath, JAP.   S.

Cinclidium stygium Sw.

Rare. Although stated to be found in only deep bogs, in Norfolk it occurs in fens. Burrell gave only one station in East Norfolk at Acle where it still persists and whence it was recorded in 1910.

West: 09, Swangey Fen, 1963, FR.

East: 03, Holt Lowes, 1959, FR; 12, Buxton Heath, 1959, very rare, FR; 30, Strumpshaw Marsh, 1959, GHR & ELS, so abundant there as to be sub-dominant; 40, Damgate near Acle, GHR, 1955.

Not recorded for neighbouring counties.

77

AULACOMNIACEAE

Aulacomnium palustre (Hedw.) Schwaegr., var. *palustre*

Widespread and abundant. Tolerating a wide range of soils, from base-rich fens to acid bog, but more frequent in the latter. Associated with *Molinia, Eriophorum* and *Sphagnum* spp.   S. L.

Var. *polycephalum* Hübn., occurred on sides of vehicle tracks in the wet fen area of Buxton Heath, E12, 1956, !

A. androgynum (Hedw.) Schwaegr.

Widespread and abundant with a preference for decaying tree stumps in carrs and damp woodland.   S. C. L.

MEESEACEAE

Amblyodon dealbatus (Hedw.) Brid.

Recorded by Crowe from St Faiths near Norwich and published in Smith's *Flora Britannica*, 1800–4. Considered now to be extinct. Not recorded for neighbouring counties.

BARTRAMIACEAE

Bartramia pomiformis Hedw.

Obviously far less common than formerly as such a striking moss is not likely to be overlooked. Shaded sandy banks where the sand is of low silica content.

West: 63, Ken Hill, Snettisham ! 92, Whissonsett, 1960, !

East: 11, Swannington, SAM; 22, Westwick ! Bryants Heath, FR. S. L.

Philonotis fontana (Hedw.) Brid.

Rare or perhaps overlooked owing to its resemblance to the common *P. calcarea*. Given as common by Burrell.

West: 62, Roydon Common, BBS.

East: 12, Buxton Heath, 1967, JAP; 19, Flordon, 1910.   S. L.

P. caespitosa Wils. ex Milde

Rare. In similar habitats to the last. No recent records.

West: 62, Grimston, EMH; 93, Kettlestone.

East: 03, Edgefield, EMH.

Not recorded for neighbouring counties.

P. calcarea (B. & S.) Schimp.

In base-rich fens where it is often abundant.

West: 61, Bawsey brick-yard ! 62, Roydon Common, FR; Grimston Warren ! 70, Gooderstone Fen, CPP & ELS; 91, Scarning Fen, FR & ELS; 93, Kettlestone Fen, FR; 09, Swangey Fen, FR.

East: 03, Holt Lowes ! 12, Buxton Heath, 'rich fen, pH 7·4', FR; 19, Flordon; 22, Bryants Heath, FR; 40, Acle, FR.   S. L.

ORTHOTRICHACEAE

Zygodon viridissimus (Dicks.) R. Br., var. *viridissimus*

Although regarded as a very common species throughout the British Isles, it appears to be far less so now in Norfolk.

West: 61, Cranberry Wood, E. Winch ! 70, Gooderstone Warren ! 71, Westacre ! 74, Ringstead Downs, BBS; 82, Eastfield Wood, Tittleshall ! 84, Burnham Overy, BBS; 89, Hilborough !

East: Burrell recorded it as very common.   S. C. L.

Var. *stirtonii* (Schimp. ex Stirt.) Hagen

West: 82, on elder, Tatterford Common, 1967, JAP, its first Norfolk record.

Not recorded for neighbouring counties.

Orthotrichum anomalum Hedw.

In the absence of natural rock in Norfolk, this species is restricted to walls of churches, priories and the like where it is occasional.

West: 62, S. Wootton ! 63, Ingoldisthorpe; 71, Pentney Abbey ! 74, Holme-next-the-Sea, WET; 81, Castleacre Priory ! 82, Tatterford, JA.

East: 14, Sheringham; 21, Horsham St Faiths; Coltishall; 24, Cromer; 31, Upton.   S. C. L.

O. cupulatum Brid.

Old records exist for East Norfolk but are not vouched for by Burrell.

West: 82, Helhoughton, 1967, JA.   S. C. L.

O. affine Brid.

Common and widespread on tree-trunks, irrespective of species. S. C. L.

O. striatum Hedw.

Rare on trees. Recorded by Burrell under *O. leiocarpum*.

West: not seen.

East: 13, Aylmerton; 14, Runton; 50, Gt. Yarmouth, JP.   S.

O. lyellii Hook. & Tayl.

Rare. On trees. Burrell gives several stations; only one recent record.

West: 78, on a chestnut tree, Weeting Castle, 1960, GHR & ELS.

O. schimperi Hammar

Included on the sole record by Burrell.

East: 13, Aylmerton, 'minute tufts on chestnut trees, c. fr.', 1907.   S.

O. tenellum Bruch ex Brid.

Rare. Trees. No recent records.

West: 80, Swaffham; 84, Holkham.

East: 13, Aylmerton.   S.

O. pulchellum Brunton

Rare. Trees.

West: 71, Narford; 82, Tatterford, JAP; 84, Holkham.

East: 13, Aylmerton; 30, on willow bushes, BBS, 1953; 31, S. Walsham Broad, 1966, GHR, 'on elder'.   S. L.

O. diaphanum Brid.

Common. Chiefly on elder but also on stonework. S. C. L.

Ulota phyllantha Brid.

Rare. No recent records. Trees.

West: 71, Narford.

East: 03, Holt, EMH; 14, Beeston Regis. S. L.

U. crispa (Hedw.) Brid.

Rather rare. Trees.

West: 62, on birch, Roydon Common ! 82, on ash, Eastfield Wood, Tittleshall ! Tatterford Common, JAP; 99, on elder, Cranberry Rough, Hockham ! 09, Swangey Fen, BBS, 1953.

East: 09, Attleborough and Hargham; 13, Aylmerton; 14, Beeston Regis and Sheringham; 21, Crostwick; 50, Yarmouth, JP. S. L.

U. bruchii Hornsch. ex Brid.

Found by Burrell growing with *U. ludwigii* at Aylmerton, E13.

No records for neighbouring counties.

U. ludwigii (Brid.) Brid.

East: 13, on a chestnut tree, Aylmerton, with *U. bruchii*; the only record. Norfolk only.

FONTINALACEAE

Fontinalis antipyretica Hedw.

Widespread and locally abundant. Either free-floating or submerged and attached to stones in streams. No fruiting plants seen but small caddis cases resembling capsules occur occasionally. S. C. L.

CLIMACIACEAE

Climacium dendroides (Hedw.) Web. & Mohr

Widespread and locally abundant. Fens, chalk grassland and most frequent on the margins of many of the Breckland meres. S. C. L.

CRYPHAEACEAE

Cryphaea heteromalla (Hedw.) Mohr

Rare. On trees.

West: 93, on elder, Thursford, 1956, CPP, the only recent record.

East: 09, Old Buckenham, DT; 11, Hellesdon; 14, Beeston Regis and Sheringham. S. C. L.

LEUCODONTACEAE

Leucodon sciuroides (Hedw.) Schwaegr.

Although recorded for both vice-counties in the *Census Catalogue*, it is scarcely as common as Burrell found it to be.

West: 72, mortar-filled joints of carstone wall, Hillington Hall, 1961, !

East: 'very common, WHB.' S. C. L.

Antitrichia curtipendula (Hedw.) Brid.

Not recorded since Dawson Turner (1775–1857) found it at Gt. Yarmouth and G. Munford (1794–1871) reported its occurrence. Most likely extinct now.   S.

NECKERACEAE

Neckera pumila Hedw.

Very rare. On trees. No recent records. The West Norfolk locality given in the *Census Catalogue*, is not known to the writer.

East: 14, Sheringham; 20, Arminghall; 28, Harleston.
Norfolk only.

N. complanata (Hedw.) Hüben.

Appears to be less frequent than formerly when it was stated to be very common on tree bases. Probably diminished by reason of felling.

West: 74, Ringstead Downs ! 80, Holme Hale ! 99, Wayland Wood near Watton ! 03, Swanton Novers Great Wood, FR & ELS.
East: 'very common', WHB.   S. C. L.

Omalia trichomanoides (Hedw.) B., S. & G.

One of the few instances where a species has become more frequent than formerly. Only one station given in the 1914 *Flora*. On tree bases in the boulder-clay woods of central Norfolk.

West: 82, Eastfield Wood, Tittleshall ! 91, Horse Wood, Mileham; 92, Elmham Gt. Wood, CPP & ELS; 92, Horningtoft; 99, Wayland Wood near Watton !
East: only one old record.

Thamnium alopecurum (Hedw.) B., S. & G.

Common and locally abundant particularly in woodland on the heavier soils of mid-Norfolk.   S. C. L.

HOOKERIACEAE

Hookeria lucens (Hedw.) Sm.

In 1808 the President of the Linnean Society, Sir James E. Smith, dedicated this genus to his friend, William Jackson Hooker, who found the plant on Holt Heath where it still persists.

East: 03, Holt Lowes, on the sides of drainage channels feeding the R. Glaven, in dense shade; in fruit, February 1967, !
Norfolk only.

LESKEACEAE

Leskea polycarpa Hedw.

Rare on tree roots near water. Burrell did not himself find this species but credits Norfolk with the record made by Bryant in Smith's *Flora Britannica*, 1800–4.

West: 80, Little Cressingham ! 99, Thompson Water, 1958, CCT & ELS, the first West Norfolk record.
East: 30, Wheatfen Broad, BBS, 1953.   S. C. L.

THUIDIACEAE

Anomodon viticulosus (Hedw.) Hook. & Tayl.

Occasional on trees, particularly ash, in woods on the boulder clay.

West: 90, W. Bradenham, GHR; 99, Wayland Wood near Watton !
East: 03, Seething, GHR.   S. C. L.

Thuidium abietinum (Brid.) B., S. & G.

Contrary to Burrell's observations, this species is considered frequent rather than rare. Often abundant in chalk grassland.

West: 70, railway cutting near Swaffham ! 78, Weeting Heath ! 80, Hilborough ! 88, Santon Warren ! 03, Swanton Novers, FR & ELS; 08, Garboldisham Heath, GHR.
East: 10, Bawburgh; 11, Swannington !   S. C. L.

T. tamarascinum (Hedw.) B., S. & G.

Widespread and abundant throughout the county on humus in woodland and in carrs.   S. C. L.

T. philibertii Limpr.

Rare. Restricted to calcareous soils.

West: 73, Ringstead Downs, (west end) BBS; 78, Weeting Heath, FR; 08, Garboldisham Heath, GHR.
East: 11, Alderford Common, 1914, WGC; 14, Sheringham.   S. C. L.

HYPNACEAE

Cratoneuron filicinum (Hedw.) Spruce var. *filicinum*

Common. In fens and along river margins. A variable species, the more typical plant, robust and pinnate, occurs occasionally, as along the bank of the R. Wissey above Hilgay on blue gault.

West: 62, Roydon Common, BBS; Castle Rising, CPP; 70, Beechamwell ! 71, water-cress beds, Westacre ! base of old mill, Pentney ! E. Winch, CPP & ELS; 72, Congham, CPP; 89, Bodney ! 93, Kettlestone Fen, FR.
East: 14, Beeston Common !   S. C. L.

Var. *fallax* (Brid.) Roth

West: 91, dried-up pond margin, Honeypot Wood, Wendling 1966! conf. JA. New to the county.   C. L.

C. commutatum (Hedw.) Roth var. *commutatum*

A frequent species in base-rich fens.

West: 93, Kettlestone Fen, FR; 99, Rockland All Saints, GHR & ELS.
East: 03, Holt Lowes ! 12, Buxton Heath, BBS, 1953; 19, Hapton and Flordon; 22, Swanton Abbot; Bryants Heath, FR.   S. C. L.

Var. *falcatum* (Brid.) Mönk.
West: 93, Kettlestone Fen, FR; 09, Swangey Fen, FR.
East: 03, Holt Lowes ! 12, Buxton Heath; 14, Beeston Bog, FR.   S. L.

Campylium stellatum (Hedw.) Lange & C. Jens.
Widespread and abundant. Bogs and base-rich fens in which it is constant.
S. C. L.

C. protensum (Brid.) Kindb.
Occasionally in base-rich fens with a high water-table. Not mentioned by Burrell.
West: 93, Kettlestone Fen, FR; 09, Swangey Fen, FR.
East: No records.   S. C. L.

C. chrysophyllum (Brid.) J. Lange
Not recorded by Burrell. Occasional in open basic grassland.
West: 74, Ringstead Down, BBS; 78, Devil's Dike, Weeting ! 80, detritus of lime kiln, near Swaffham ! 89, Grimes Graves, Weeting, FR; 98, E. Harling Heath, FR.
East: No records.   S. C. L.

C. polygamum (B., S. & G.) J. Lange & C. Jens.
Rare. Fens. Variable; vars. *stagnatum* Wils., and *minus* Schp., recorded by the early bryologists.
West: 08, S. Lopham (under *Hypnum polygamum*).
East: 19, Flordon, FR; 30, open fen, Wheatfen Broad, BBS, 1954; 41, Burgh Common, FR.   S. C. L.

C. elodes (Lindb.) Kindb.
Recorded as rather common in swamps, under *Hypnum*, in 1914 *Flora*. Appears to be frequent in base-rich fens.
West: 62, Derby Fen ! 70, Caldecote Fen, c. fr., 71, WHB & WGC; E. Walton Common ! 91, Scarning Fen, FR & ELS; 09, Swangey Fen, 'rare', FR.
East: 09, Old Buckenham Fen, FR; 19, Flordon; 30, Strumpshaw, GHR; 31, Ranworth; 32, Sutton.   S. C. L.

Leptodictyum riparium (Hedw.) Warnst.
Widespread and abundant throughout the county. Submerged wood and stone; rivers, dykes and gravel pits; marl pits subject to winter flooding and Breckland meres.

Amblystegium serpens (Hedw.) B., S. & G., var. *serpens*
Abundant on a variety of tree bases, particularly elder, on *Salix cinerea* in carrs, and in hedgebanks; occasionally on older coastal sand-dunes. Fruiting freely.

A. juratzkanum Schimp.
Neither this species nor the next two separated by Burrell.
West: 99, on a rotting branch of *Salix cinerea* in fen carr, Cranberry Rough, Hockham, October 1958, CCT & ELS, the first record. S. C. L.

A. kochii B., S. & G.

East: 30, carr at Wheatfen Broad, BBS, 1953, the only record.   L.

A. varium (Hedw.) Lindb.

West: 09, Swangey Fen, FR.

East: 03, Holt Lowes, on rotten branch in deep shade, 1956, ! 09, old rotten gate, Old Buckenham Fen, June 1956, ECW, the first record.

S. C. L.

Drepanocladus aduncus (Hedw.) Warnst.

Regarded by Burrell as common in swamps and pools (under *Hypnum*) with the vars. *polycarpon* Bland., *intermedium* Schimp., and *paternum* Sanio also being recorded. With such a variable species subject to habitat modification it is questionable if varietal distinctions are justified. Open fen, bog pools and abundant in some of the Breckland meres.

West: 62, Roydon Common ! 70, Caldecote Fen ! 71, E. Walton Common ! 88, Two Mile Bottom near Thetford ! 91, pond at Wendling ! 98, Ringmere, CCT & ELS; Langmere ! 99, marsh, E. Wretham, BBS.

East: 12, Gt. Witchingham Common ! 19, Flordon Common; 30, open fen, Wheatfen Broad, BBS, 1953.   S. C. L.

D. sendtneri (Schimp.) Warnst., var. *sendtneri*

Rare. Pools in fens and marshes.

West: 63, Ingoldisthorpe; 70, submerged in a pool, Foulden Common, 1956, conf. MCFP, !   L.

Var. *wilsonii* (Schimp. ex Lor.) Warnst.

West: 70, Foulden Common, 'in fairly deep pools with *Utricularia intermedia*', 1910, WHB & WGC.   S. L.

D. lycopodioides (Brid.) Warnst.

Rare. Very wet fens and bog pools.

West: 62, Roydon Common, FR; 70, Gooderstone Fen ! 91, Potter's Fen, Scarning ! 08, S. Lopham.

East: 11, Felthorpe; 21, Newton St Faiths.   S. L.

D. fluitans (Hedw.) Warnst., var. *fluitans*

Frequent in peat pools on heaths. The earlier bryologists recorded the vars., *gracile* Boul., *arnellii* Sanio and *falcatum* Schimp. Of these, only var. *falcatum* appears to be worthy of varietal distinction.

West: 61, E. Winch Common ! 62, Sandringham Warren ! Dersingham Fen ! Roydon Common ! 70, Foulden Common ! 82, Tatterford Common, JA; 98, Overa Heath, BBS, 1953.

East: 02, Bawdeswell, FR; 11, Alderford Common ! Swannington Common, GHR & ELS; 12, Buxton Heath, BBS, 1953.   S. L.

D. exannulatus (B., S. & G.) Warnst., var. *exannulatus*

In similar habitats to *D. fluitans* but far less common.

West: 61, E. Winch Common, BBS; 62, N. Wootton, EMH; Sandringham Warren !

East: 03, Holt Lowes ! 22, Bryants Heath, FR; 32, E. Ruston.   S. L.

Var. *rotae* (De Not.) Wynne
West: 62, Sugar Fen, 1967, JAP.

D. revolvens (Turn.) Warnst., var. *revolvens*
Common in bogs and fens. Burrell recorded *Hypnum revolvens* and *H. intermedium* as equally rather common in swamps, a distribution which applies today. There is a slight tendency for the typical plant to be associated with somewhat acid conditions but it will be seen that both it and the variety are common to three stations, of which Scarning Fen and Beeston Common are markedly calcareous. This suggests that the two are but habitat modifications and there appears to be little merit in according varietal status to the plant of more slender habit, greener colour, and shorter leaf-cells.

West: 62, Roydon Common ! In pools with *Myrica, Juncus subnodulosus, Phragmites, Vaccinium oxycoccus* and *Cirsium dissectum,* 'nice specimen of "type" *revolvens*', MCFP *in lit,* 1956; 70, Foulden Common ! Caldecote Fen ! 91, Scarning Fen ! 08, S. Lopham Fen !
East: 03, Holt Lowes ! 12, Buxton Heath ! 14, Beeston Common !
S. L.

Var. *intermedius* (Lindb.) Warnst.
West: 70, Gooderstone Fen ! 72, Derby Fen ! 91, Scarning Fen ! 99, Rockland All Saints Fen !
East: 03, Holt Lowes ! 12, Buxton Heath, CCT & ELS; 14, Beeston Common !
S. L.

D. vernicosus (Lindb.) Warnst.
Rare. Fens and marshes.
West: 09, Swangey Fen, 1952, ECW.
East: 02, Whitwell, GHR & ECW; 12, Buxton Heath, 1951, FR; 30, Strumpshaw Marsh, 1959, FR; 32, Brumstead and E. Ruston.
Norfolk only.

D. uncinatus (Hedw.) Warnst.
Sir James Paget recorded it as not uncommon near Yarmouth in East Norfolk. There are no recent records.
S. L.

Scorpidium scorpioides (Hedw.) Limpr.
Frequent in pools in fens and bogs.
West: 62, Roydon Common ! 70, Foulden Common ! 71, E. Walton Common ! c. fr., 08, S. Lopham ! 09, Swangey Fen, 'vl', FR.
East: 18, Roydon Fen near Diss, FR; 19, Flordon Common; 22, Bryants Heath, FR.
S. C. L.

Acrocladium stramineum (Brid.) Rich. & Wall.
Rare. In *Sphagnum* swamps.
West: 62, Roydon Common ! 71, Westacre, c. fr., WHB & WGC.
East: 03, Holt and Briston Common, EMH; 11, Felthorpe; 12, Buxton Heath, BBS, 1953; 22, Bryants Heath, FR; 32, E. Ruston.
S. L.

A. cordifolium (Hedw.) Rich. & Wall.

Very local. Wet peat in woods and carrs.

West: 61, E. Winch ! 62, Sugar Fen, JAP; N. Wootton, EMH & RC; Roydon Common, c. fr., ! 63, Ingoldisthorpe; 70, Foulden Common ! 71, Westacre; 82, Tatterford Common, JA; 99, Hockham !

East: 03, Holt Lowes, CPP; 31, Woodbastwick, GHR & ELS; 32, Sutton; 41, Thurne.   S. L.

A. giganteum (Schimp.) Rich. & Wall.

Variable in frequency; sometimes abundant in marshes; occasional in base-rich fens.

West: 61, E. Winch Common, BBS; 71, E. Walton Common ! 72, Hillington ! Derby Fen, CPP; 98, Overa Heath, BBS, 1953; 93, Kettlestone Fen, FR; 99, Rockland All Saints, GHR & ELS; 08, S. Lopham ! Garboldisham ! 09, Swangey Fen, FR.

East: 03, Holt Lowes ! 09, Old Buckenham Fen, FR; 12, Buxton Heath, pH 7·4, FR; 19, Flordon Common; 22, Bryants Heath, FR; 30, Wheatfen Broad, BBS, 1953; 32, E. Rushton, FR; 41, Burgh Common, FR.

                                                                        S. C. L.

A. cuspidatum (Hedw.) Lindb.

Very common and tolerating a wide range of habitats. Base-rich fens, bogs, pool margins, short turf on damp soils, chalk pits and chalk grassland.

                                                                        S. C. L.

Isothecium myurum Brid.

Apart from one station (Brisley) in West Norfolk, all Burrell's records are from East Norfolk (under *Eurhynchium*) but this is due to the concentration by the early bryologists on the latter vice-county, a trend which is similarly reflected in the distribution of the phanerogams. Although regarded by Burrell as rather rare, it is frequently seen covering the bases of trees in the woods and coppices on the heavier soils of central Norfolk.

West: 61, E. Winch Common ! 90, Potter's Carr, Cranworth ! 91, Horse Wood, Mileham, CPP & ELS; 99, Wayland Wood near Watton ! 92, Brisley; 03, Swanton Novers, FR & ELS.

East: 01, Hockering, BBS; 11, Swannington Common ! 14, Beeston Regis; 29, Brooke; 30, Little Melton.   S. C. L.

I. myosuroides Brid.

Rarer than the last species but in similar habitats.

West: 74, Ringstead; 82, Tatterford, JA; Eastfield Wood, Tittleshall ! 03, Swanton Novers Great Wood, FR & ELS.

East: 01, Hockering, BBS; 14, Sheringham; 31, Woodbastwick.

                                                                        S. C. L.

Camptothecium sericeum (Hedw.) Kindb.

Very common. One of the most abundant mosses of walls and trees.

                                                                        S. C. L.

C. lutescens (Hedw.) B., S. & G.

Common in chalk pits, on heaths, calcareous grassland, walls and occasionally in coastal sands.   S. C. L.

C. nitens (Hedw.) Schimp.

Rare. In base-rich fens with a high water-table. Has been found as 'subfossil remains . . . in some layers of the Norfolk Broads' (M. C. F. Proctor, BBS, *Trans.*, III, **1**, 1956). This glacial relict has become extinct in many counties but still persists in Norfolk where, as at Swangey Fen, it is locally abundant.

West: 62, Roydon Common, FR, rare in 1956 but has increased since; 09, Swangey Fen, 1952, FR, 'associated with *Cladium, Juncus subnodulosus, Peucedanum palustre, Phragmites, Schoenus nigricans, Aulacomnium palustre, Campylium stellatum, Ctenidium molluscum, Fissidens adianthoides* and *Mnium punctatum*', ELS field-notes.

East: 03, Holt Lowes, FR; 12, Buxton Heath, FR; 32, Smallburgh, 1911; 40, Acle Carr, 1955, GHR & ELS.

Norfolk only.

Brachythecium albicans (Hedw.) B., S. & G.

One of the common mosses of heathland and in open, dry places of sand-dunes and maritime shingle where it is frequently abundant.   S. C. L.

B. glareosum (Spruce) B., S. & G.

Uncommon. Marl and chalk pits but usually in small amounts. Recorded from grey dunes at Blakeney Point by Watson.

West: 60, Shouldham; 71, Narford, GHR; 72, Gayton chalk pit, CPP & ELS; 78, Weeting Heath, FR; 82 Massingham Heath ! 89, Ickburgh, GHR & ELS.

East: 10, Bawburgh; 14, Sheringham; 19, Flordon.   S. C. L.

B. salebrosum (Web. & Mohr) B., S. & G.

No mention by Burrell. Apparently very uncommon but probably over-looked by reason of its resemblance to *B. rutabulum*.

West: 88, Santon ! 91, Scarning Fen, FR & ELS, 1956; 99, Rockland All Saints, FR.

East: 19, Flordon, FR.   C. L.

B. rutabulum (Hedw.) B., S. & G.

This species and *Mnium hornum* are probably the two most abundant mosses. Widespread in a variety of habitats ranging from swamps to maritime sand-dunes but most characteristic of woodland.   S. C. L.

B. rivulare B., S. & G.

Not recorded by Burrell, and the *Census Catalogue* also omits it.

West: 62, Babingley, BBS; Roydon Common, 1960, ! 82, Tatterford Common, JA; 91, Scarning Fen, BBS; 93, Kettlestone Fen ! 98, Langmere, BBS; 99, Hockham, BBS; Rockland All Saints !

East: 03, Holt Lowes, 1956, ! Confd. JA; 09, Old Buckenham Fen, FR; 12, Buxton Heath, JAP.   S. C. L.

B. velutinum (Hedw.) B., S. & G.

Widespread and abundant on tree bases, floors of marl pits, carrs and hedgerows.   S. C. L.

B. populeum (Hedw.) B., S. & G.

Burrell reported it as rather common on banks and trees in north-east Norfolk. No recent records.

East: 10, Wymondham; 11, Drayton and Felthorpe; 14, Sheringham; 22, Scottow and Westwick; 39, Hedenham.   C. L.

Scleropodium tourretii (Brid.) L. F. Koch
(*S. illecebrum* auct.)

Very rare. Burrell had only one station and this record he considered doubtful.

West: 08, Garboldisham, 1956, GHR, its first West Norfolk record.

East: 12, Heydon; 31, on a bank at Woodbastwick, 1913 (under *Brachythecium*).   S. C.

Cirriphyllum piliferum (Hedw.) Grout

Not common. Woodland banks on the heavier soils.

West: 78, Emily's Wood near Brandon ! 82, Tatterford Common, JA; 92, Brisley; 99, Wayland Wood near Watton, GHR & ELS; Rockland All Saints, GHR & ELS; 03, Swanton Novers, FR & ELS.

East: 01, Hockering, BBS; 03, Holt; 10, Wymondham; 19, Flordon; 20, Framingham Pigot; 39, Chedgrave.   S. C. L.

Eurhynchium striatum (Hedw.) Schimp.

Frequent on woodland floors and occasionally in chalk grassland.

West: 78, Emily's Wood near Brandon ! 82, Tatterford Common, JA; 94, Danish Camp near Warham, CPP & ELS; 99, Wayland Wood near Watton, ! 03, Swanton Novers, FR & ELS.

East: 01, Hockering, BBS; 03, Holt Lowes, FR.   S. C. L.

E. praelongum (Hedw.) Hobk., var. *praelongum*

Widespread and abundant. Conspicuous on the ground of woods, about tree bases, shady hedgebanks and occasionally on old grey sand-dunes in shade.   S. C. L.

Var. *stokesii* (Turn.) Hobk.

No record in the 1914 *Flora*. Occurs in East Norfolk according to the *Census Catalogue*, but not found by author.

Not recorded for neighbouring counties.

E. swartzii (Turn.) Curn., var. *swartzii*

Common in chalk and marl pits and in stubble fields.   S. C. L.

Var. *rigidum* (Boul.) Dix.

West: 78, Weeting Heath, 1956, CCT, its first Norfolk record.   C.

E. schleicheri (Hedw. f.) Lor.

Apparently very rare.

East: 33, Mundesley, EMH, the only record (under *E. abbreviatum* Schp.). Norfolk only.

E. speciosum (Brid.) Milde

Burrell does not record this, which is somewhat surprising although it may have been confused with *E. swartzii* and overlooked.

West: 71, Marham Fen ! Narford, 1955, ! 78, Weeting Heath, FR; 99, Cranberry Rough, Hockham, 1955, CCT; 09, Swangey Fen, FR.
East: 31, Woodbastwick, BBS, 1953. S. C. L.

E. riparioides (Hedw.) Rich.
Frequent on brickwork and posts in the faster-flowing rivers.
West: 62, N. Wootton, CPP; Castle Rising, CPP; 71, Pentney ! Westacre ! 72, Hillington ! 80, Hilborough, CPP & ELS; 81, Castleacre; 82, Tatterford, JA; 88, Thetford. .
East: 03, Holt ! 19, Flordon; 31, Hemblington. S. C. L.

E. murale (Hedw.) Milde
Rather rare. Walls.
West: 62, Castle Rising, JWB; 78, Weeting Castle ! 09, S. Lopham.
East: 19, Flordon; 20, Whitlingham. S. C. L.

E. confertum (Dicks.) Milde
Very common. Walls, trees and occasionally on coastal sand-dunes.
S. C. L.

E. megapolitanum (Bland.) Milde
Only one record in the 1914 *Flora*, from Thornham. Bare, stony places in grassland.
West: 74, Holme-next-the-Sea, BBS; 78, Weeting Heath, 1967, det. ECW, !
East: 01, Lyng, ECW. S. C. L.

Rhynchostegiella pumila (Wils.) E. F. Warb.
(*R. pallidirostra* (Brid.) Loeske)
Burrell regarded this species, under *Eurhynchium pumilum*, as rare and gives two stations, both along the north Norfolk coast.
West: 74, Thornham, WET; Ringstead Downs, BBS, 1967; 99, Hockham, BBS.
East: 33, Mundesley, EMH. S. C. L.

R. tenella (Dicks.) Limpr., var. *tenella*
Although credited to East Norfolk in the *Census Catalogue*, Burrell and Clarke found it in both vice-counties, under *Eurhynchium tenellum*, on the mortar joints of walls but considered it to be rare.
West: 71, Narborough.
East: 11, Swannington; 21, Horstead, WGC. S. C. L.

Pseudoscleropodium purum (Hedw.) Fleisch.
Very common. Heaths, particularly amongst *Calluna*; floors of chalk pits, marl pits, edges of *Cladietum* in fens; occasionally on older sand-dunes.
S. C. L.

Pleurozium schreberi (Brid.) Mitt.
Very common and abundant around *Calluna* on the Greensand heaths and in acid grassland. S. C. L.

Isopterygium seligeri (Brid.) Dix. ex C. Jens.

Probably newly arrived in the county on imported conifers. Both records from dead wood of *Abies alba*.

East: 03, Hempstead near Holt; 14, The Dales, Upper Sheringham – both records by EFW and DGC, 1940.   S. C.

I. elegans (Hook.) Lindb.

Burrell gives it as locally common, under *Plagiothecium*, in woods in north-east Norfolk.

West: 03, Swanton Novers Great Wood, FR & ELS; 82, Tatterford, JA. East: 01, Hockering, BBS; 03, Stody; 14, Beeston Regis and Sheringham; 21, Newton St Faiths; 22, Westwick.   S. L.

Plagiothecium latebricola B., S. & G.

East: 01, Hockering Wood, BBS, 1967, its first Norfolk record, det. ECW.

P. denticulatum (Hedw.) B., S. & G., var. *denticulatum*

Frequent. Rotting tree stumps on acid, raw humus; sandy hedgebanks and alder carrs.

West: 61, Cranberry Wood, E. Winch ! 62, Sugar Fen, JAP; N. Wootton ! Roydon Common, BBS; 70, Beechamwell Fen ! Shingham Wood ! 72, Little Massingham ! 82, Eastfield Wood, Tittleshall ! Tatterford Common, JAP; 92, Whissonsett ! 03, Swanton Novers Great Wood, FR & ELS. East: 01, Hockering Wood, BBS; 03, Holt Lowes ! 12, Buxton Heath, JAP.   S. C.

P. ruthei Limpr.

Until the publication of 'The British Species of *Plagiothecium denticulatum-P. silvaticum* Group', by S. W. Greene (BBS, *Trans.*, III, **2**, 1957), the species in this complex had not been separated. The present species appears very distinct and has been found in carrs and damp woods on leaf-litter.

West: 62, Roydon Common ! 81, Litcham Common, EWJ; 91, Scarning Fen ! 99, Cranberry Rough, Hockham, CCT & ELS; Rockland All Saints ! 08, Swangey Fen, 1955, !   C.

P. curvifolium Schlieph. ex Limpr.

West: 82, Tatterford Common, JA confd. ACC, 1967; 03, Swanton Novers Great Wood, BBS, April 1967, its first Norfolk record. East: 01, Hockering Wood, BBS.

P. succulentum (Wils.) Lindb.

Damp sandy banks in woodland and mixed scrub.

West: 63, Ken Hill, Snettisham ! 89, Mundford Covert ! 03, Swanton Novers Great Wood ! East: 01, Hockering Wood, BBS; 11, Swannington, 1959, !   S.

P. sylvaticum (Brid.) B., S. & G.

Frequent on tree bases and earth banks in woods. Not given in *Census Catalogue*.

West: 62, Ling Common, N. Wootton, BBS; Hillington, BBS; 78, Weeting Fen ! 82, Eastfield Wood, Tittleshall ! Tatterford Common,

BBS; 91, Honeypot Wood, Wendling ! Horse Wood, Mileham ! 92, Brisley ! 93, Bulfer Grove ! 99, Wayland Wood near Watton ! 03, Swanton Novers Great Wood, FR & ELS.

East: 01, Hockering Wood, BBS; 03, east end of Swanton Novers Great Wood, BBS; Hunworth; 10, Wymondham; 11, Felthorpe; Swannington, GHR & ELS; 22, Bryants Heath, FR.   S. C.

P. undulatum (Hedw.) B., S. & G.

Rather common. On peat banks and litter in pine woods; carrs, particularly in north-east Norfolk where the rainfall is somewhat higher than in the rest of the county.

Hypnum cupressiforme Hedw.

Widespread and abundant. In the aggregate sense tolerating a wide range of habitats. Trees in the shelter of woodland, woodland rides, fens, chalk grassland, thatched roofs, heaths and coastal sand-dunes.   S. C. L.

Var. *resupinatum* (Wils.) Schimp.

Very common on tree-trunks particularly in woods on the heavier soils.

Var. *filiforme* Brid.

A very slender form on tree-trunks.

Var. *ericetorum* B., S. & G.

A slender form abundant on heaths, in woodland and fixed coastal sand-dunes.

Var. *tectorum* B., S. & G.

A robust and vigorous form on grassland on acid or neutral soils, thatched roofs and coastal sand-dunes.

Var. *lacunosum* Brid.

Again a robust form but the differences between this and var. *tectorum* appear to be merely relative and the epithet is applied to the vigorous plant found on chalk grassland.

H. imponens Hedw.

West: 61, wet heath, E. Winch Common, BBS, 1967, its first Norfolk record.

Not recorded for neighbouring counties.

H. lindbergii Mitt.

(*H. patientiae* Lindb., ex Milde)

Recorded by Burrell as rare in marshy pastures. No recent records.

West: 81, Litcham.

East: 14, Beeston Regis.   C. L.

Ctenidium molluscum (Hedw.) Mitt., var. *molluscum*

Locally abundant and luxuriant in base-rich fens in which it is constant; smaller forms occur in chalk grassland.   S. C. L.

Rhytidium rugosum (Hedw.) Kindb.

Although Burrell regarded this as very rare, it is to be found in most of the chalk grassland of the south-west Norfolk Breckland.

West: 78, near Snake Wood, Brandon ! Weeting Heath, abundant ! 79, Cranwich ! 88, Santon ! near Thetford ! 89, Grimes Graves, FR.   S. C.

Rhytidiadelphus triquetrus (Hedw.) Warnst.

Common. Woodland rides, mixed scrub, chalk grassland, pits in Breckland and locally abundant in coastal sand-dunes.   S. C. L.

R. squarrosus (Hedw.) Warnst.

Widespread and abundant in a variety of habitats. Woodland rides, poor pastures, maritime sand-dunes and frequently a pest in lawns. The Rev. W. E. Thompson found it fruiting at Thornham, W74. Fruiting at Sandringham Warren, W62, BBS.   S. C. L.

R. loreus (Hedw.) Warnst.

Very rare. Acid humus of woodland. Burrell gave only two stations, Blickling and Stratton Strawless, both in East Norfolk. It still persists in the former.

East: 11, Felthorpe, 1962, GHR; 12, Hercules Wood, Blickling, associated with *Plagiothecium undulatum* and *Campylopus flexuosus,* 1960, FR.
                    S. L.

Hylocomium splendens (Hedw.) B., S. & G.

Common. Calcicole. Chalk grassland, pits and older grey sand-dunes.
                    S. C. L.

PTERIDOPHYTA

LYCOPODIACEAE

Lycopodium inundatum L.   Marsh Club-moss

Native. Rare now. Wet heaths.

West: 62, N. Wootton Heath.

East: 03, Holt Lowes; 12, Buxton Heath; 22, Bryants Heath near Felming-ham, GHR.

EQUISETACEAE

Equisetum fluviatile L.   Water Horsetail

Native. Common. Ponds and ditches.

E. palustre L.   Marsh Horsetail

Native. Common. Wet places.

E. sylvaticum L.   Wood Horsetail

Native. Rare. Formerly known from a few damp woods in East Norfolk but now restricted to wet carr at Holt Lowes.

East: 03, Holt Lowes.

E. arvense L.   Common Horsetail

Native. Widespread and abundant in cultivated fields, hedgebanks and waste places.

E. telmateia Ehrh.   Great Horsetail

Native. Frequent. Wet shady places.

At Beeston Bog near Sheringham (E14), where it is abundant, a few plants bear both fertile and sterile stems = var. *serotinum* (A. Br.) Milde.

OSMUNDACEAE

Osmunda regalis L.   Royal Fern

Native. Rather rare now. Fens and bogs. More frequent in the Broads district.

West: 62, Roydon Common; 72, Leziate Fen; 90, Scoulton Mere.

East: 28, Denton, CF; 31, Horning; 32, Sutton Fen, RMB; E. Ruston, ACJ; 33, Edingthorpe Heath, RMB; 41, Martham, PER; Winterton Dunes, JHS; 42, Hickling Broad.

POLYPODIACEAE

Pteridium aquilinum (L.) Kuhn   Bracken; Brakes

Native. Very common on sand and gravel, and dominant over large areas of heath, where it replaces *Calluna* after burning; forms the ground flora of woods on sand; rarely seen on town walls.

93

Blechnum spicant (L.) Roth   Hard Fern
Native. Frequent on wet heaths on Greensand and, in the east, in damp woods on acid soils.
West: 62, 63, 72, 81.
East: 03, 04, 12, 20, 22, 23, 30, 33, 40–2, 49.

Phyllitis scolopendrium (L.) Newm.   Hart's-tongue Fern.
Native. Frequent on old walls and occasionally in woodland rides. In the absence of natural rock, many of our ferns are limited to old walls, church buttresses, wells and the like.
West: 41, 62, 71, 74, 78, 80–84, 88, 90–92, 99, 03.
East: 00, 02, 10, 12, 20, 22, 28–32, 39.

Asplenium adiantum-nigrum L.   Black Spleenwort
Native. Occasional. Old walls.
West: 62, 71–74, 79, 91, 99.
East: 01–03, 10, 13, 17, 19–21, 23, 30–33.

A. trichomanes L.   Maidenhair Spleenwort
Native. Occasional. Old walls.
West: 64, 72, 80, 81, 83, 88–90, 92, 94, 99.
East: 00, 04, 10, 17, 20, 21, 23, 29–31, 42.

A. ruta-muraria L.   Wall-rue
Native. Frequent on old walls throughout the county.

Ceterach officinarum DC.   Rusty-back Fern
Wall denizen. Very rare.
West: 92, on the ruins of the Saxon cathedral at N. Elmham until 1962.
East: 12, Heydon church from 1854; still there in 1942 teste Arthur Mee in *The King's England; Norfolk*; 19, six plants, Forncett St Peter Church, 1964, JHS; 39, old wall in Loddon, JHS, 1964.

Athyrium filix-femina (L.) Roth   Lady Fern
Native. Common. Woods.

Cystopteris fragilis (L.) Bernh.   Brittle Bladder-fern
Probably best regarded as a wall denizen. Very rare.
East: 21, beneath the platform of Salhouse station, 1958, RMB; 29, in the brickwork of drains of the runways of a disused airfield at Hardwick near Tasburgh, SA. Prior to these recent records, it was last seen *c.* 1874.

Dryopteris filix-mas (L.) Schott   Male Fern
Native. Common. Woods.

D. borreri Newm.   Golden-scaled Male Fern
Native. Uncommon in the west but, like many of the ferns, more frequent in the north-east of the county. In wet and shady woods.
West: 91, Beetley, EAE; near Spong Bridge, N. Elmham; 92, Horning-toft.
East: 03, 11, 14, 20, 23, 30–33.
First record: 1908, Edingthorpe (E33), F. Long, as '*D. filix-mas*', in Herb. Mus. Nor., det. J. P. Pugh.

D. cristata (L.) A. Gray   Crested Buckler-fern
Native. Rare and decreasing. Wet heaths and marshes.

West: 90, Scoulton Mere.

East: 11, Swannington, WGC, 1917/18; 30, Surlingham, EAE; 31, Hoveton Great Broad, GHR; Ranworth, EAE; 32, Irstead, EAE; 42, Horsey Mere, Horsey Warren, EAE; Hickling; Winterton, EAE; Potter Heigham, GHR, 1965.

There are many specimens in both the British Museum and Norwich Castle Museum, ranging from 1840 onwards.

D. carthusiana (Villar) H. P. Fuchs   Narrow Buckler Fern
Native. Frequent. Damp woods, marshes and wet heaths. Has increased.

West: 60–62, 70–72, 78, 81, 88, 91, 92, 99, 01, 03, 07.
East: 01, 08, 11, 12, 22, 29–32, 40–42, 51.

D. carthusiana × *cristata*
(*D.* × *uliginosa* (Newm.) O. Kuntze ex Druce)
Apparently a very rare hybrid.

West: 90, Scoulton Mere, 1965, det. ACJ.
East: 31, Hoveton Great Broad, GHR, 1958; 42, Hickling, 1934, JEL.

D. dilatata (Hoffm.) A. Gray   Broad Buckler Fern
Native. Common. Woods.

Polystichum setiferum (Forsk.) Woynar   Soft Shield-fern
Native. Uncommon. Damp, shady hedgebanks in north and East Norfolk. 'Polystichum angulare and *P. aculeatum* abound near Dereham, where more than twenty different varieties uniting the extreme forms of these two species may be collected in a single day.' (*The Fauna of Norfolk*, Lubbock, 1879.)

West: 79, Northwold Lodge, VML; 91, Northall Green near Dereham, DMM; 93, Bulfer Grove; 01, East Dereham; 03, Swanton Novers.
East: 12–14; 21, St Faith's airfield, ETD; 23, 30, 33, 41, 42.

P. aculeatum (L.) Roth   Hard Shield-fern
Native. More frequent than the last species, occurring in the woods of central Norfolk and dykesides on heavy soil.

West: 72, 80–82, 84, 90–93, 99–01, 03.
East: 01; 29, Stratton St Michael, JHS; 30; 31, Horning.

Thelypteris limbosperma (All.) H. P. Fuchs   Mountain Fern
Native. Very uncommon now. Woods.

West: 03, Swanton Novers, CPP, 1956.
East: 40; 49, Maps Scheme records.

T. palustris Schott   Marsh Fern
Native. Locally common. Confined to carr over peat, marshes and fens.

West: 62, 63, 71, 72, 83, 91, 94, 98, 99, 03.
East: 03, 12, 13, 21, 30–32, 40–42, 49.

T. dryopteris (L.) Slosson   Oak Fern
Apparently a newcomer to Norfolk as it is not recorded in the 1914 *Flora*.
East: 33, disused railway at Mundesley, KHB, 1959.

95

Polypodium vulgare agg.   Polypody
Native. The aggregate is frequent on hedgebanks, old walls, woodland and mature sand-dunes.

P. interjectum Shivas
Native. Distribution of this segregate not worked out but crops up on church and garden walls especially in north-east Norfolk; 'abundant on roadside banks at Salhouse (E21)', ACJ. Appears to be widespread throughout the county.
West: 84, 90, 92–94.
East: 02, 20–23, 30–33.

First record: 1841, Ewing in Herb. Norw. Mus., under *P. vulgare.*

MARSILEACEAE

Pilularia globulifera L.   Pillwort
Native. Very rare now. Shallow dykes on acid soil.
West: 72, Derby Fen until 1959; 61, E. Winch Common until 1929.
East: 11, Horsford Heath, EAE, 1951; 42, Eastfield near Hickling, NYS, 1926.

AZOLLACEAE

Azolla filiculoides Lam.   Water-fern
Nat. alien. Sometimes in extensive colonies locally. Dykes and Broads in East Norfolk.

E. A. Ellis has traced its history in Norfolk and the following is an abstract.

1903   First introduction in a pond in Chapel Field Gardens, Norwich, E20.
1908   Ditch, Woodbastwick side of Horning Ferry, F. H. Barclay, E31.
1913   Ludham, E31; S. Walsham Broad, E31; Stokesby, E41, and in dykes on the Acle marshes.
1916   River Yare and abundant in dykes connected with river at Surlingham.
1921   Incredible numbers at Denver Sluice, W50, having floated down the river Ouse.
1947   Decline set in.

Recent records:
East: 30, Wheatfen Broad, 1943; 31, Woodbastwick, JHS, 1963; Acle Marshes, ETD, 1964; 32, Catfield Dyke, 1950, ACJ; 40, Halvergate, ETD, 1965; 42, Potter Heigham, EAE, 1935; 49, Haddiscoe, DMM, 1965.

16 SPOTTED MEDICK *Medicago arabica* Miss D. M. Maxey

17 STRIATED CATCHFLY *Silene conica* Miss D. M. Maxey

18 SPANISH CATCHFLY   *Silene otites*   *Dr S. Clark*

19 BERRY CAMPION   *Cucubalus baccifer*   *J. E. Lousley*

20 MAIDEN PINK   *Dianthus deltoides*   *J. H. Fremlin*

21 SOAPWORT   *Saponaria officinalis*   *Mrs D. M. Dean*

22 SEA SANDWORT   *Honkenya peploides*   *A. H. Hems*

23 RUPTUREWORT   *Herniaria glabra*   *Miss G. Tuck*

OPHIOGLOSSACEAE

Botrychium lunaria (L.) Sw.   Moonwort

Native. Dry heath grassland. Much rarer now in curious contrast to its abundance in the Suffolk Breckland on Lakenheath Warren.

West: 98, one plant at West Harling, 1955; 88, Two Mile Bottom, DMM, 1968.

East: 04, High Kelling, 1948; 11, Ringland Hills, WGC, 1915; 28, Denton near Harleston, CF, 1956.

Plate 318 of *English Botany* was drawn from a plant collected by Mrs Kett from Seething (E39) 1st April 1796. First British record.

Ophioglossum vulgatum L., ssp. *vulgatum*   Adder's-tongue

Native. Occasional. Tolerant of widely differing habitats such as water-logged meadows, calcareous fens, chalk and marl pits, and sand-dunes.

West: 61–64, 70, 71, 73, 79, 84, 91, 92, 99.

East: 01–03, 10, 11, 19, 30–32, 39–42.

SPERMATOPHYTA

Gymnospermae

PINACEAE

We record only those species in which natural regeneration has been seen to occur in the extensive forests throughout the south-west Norfolk Breckland. We are grateful to Mr Chard of the Forestry Commission who, whilst stationed at Santon Downham, confirmed our observations and added to the list.

Abies grandis Lindl.   Giant Fir

Pseudotsuga menziesii (Mirb.) Franco   Douglas-fir

Picea abies (L.) Karst.   Norway Spruce

There are several large specimens of this species which is more at home in Norfolk than the Sitka spruce. The latter is very prone to damage by late frosts in Breckland.

Tsuga heterophylla (Raf.) Sarg.   Western Hemlock

Natural regeneration of this uncommon species also occurs at Weasenham, W82.

Larix decidua Mill.   European Larch

Naturalised in woodland throughout the county.

Pinus sylvestris L.   Scots Pine

Native. Common especially on the Greensand. Colonises heath in the absence of fires and rabbits.

P. nigra Arnold ssp. *laricio* (Poir.) Palabin   Corsican Pine

Now the most widespread and abundant species of conifer and extensively planted both in the Breckland and elsewhere. Much natural regeneration occurs and, like the Scots pine, colonises heath and sand-dunes, as at Holkham.

Ssp. *nigra*   Austrian Pine

Its timber, coarser and knottier than the Corsican pine, makes it a poor commercial species.

P. pinaster Ait.   Maritime Pine

Included in the extensive planting at Holkham Meols in 1850.

P. contorta Dougl. ex Loud.   Lodgepole Pine

Occurs on the Santon beat.

P. strobus L. Weymouth Pine

Regeneration occurs on the Mundford beat at Bunkers Hill.

CUPRESSACEAE

Thuja plicata D. Don   Western Red Cedar

Regeneration of this and the next species occurs at Congham, W72.

Chamaecyparis lawsoniana (A. Murr.) Parl.   Lawson's Cypress

Juniperus communis L., ssp. *communis*   Juniper

Nat. alien. Rare. Many old bushes still occur within the curtilage of the woodland near Widow's Hill Plantation, W.89, 1967, J. M. Schofield. Locality specified as Juniper Wood on the older Ordnance Survey maps (1882; 1889).

TAXACEAE

Taxus baccata L.   Yew

Naturalised throughout the county in plantations and churchyards.

Angiospermae

RANUNCULACEAE

Caltha palustris L.   Marsh Marigold; Kingcup

Native. Very common. Marshes, fens, riversides and wet woods.

Helleborus foetidus L.   Stinking Hellebore

Nat. alien. Rare. Woods and scrub on calcareous soil. At one time plentiful on the Castle Hill at Castleacre, W81, but uprooted by villagers for their gardens about 1913. Still in some quantity at Ditchingham, E39, on the Bath Hills whence it was recorded by the Bungay botanist, Daniel Stocks, over a hundred and fifty years ago.

West: 70, 71, 73, 79, 81, 94, 00.

East: 03, 04, 11, 12, 18, 39.

H. viridus L.   Bear's-foot

Nat. alien. Always very rare in Norfolk. Thickets.

West: 88, Thetford, HDH, until 1910; 90, Bradenham Wood, Miss Goddard, 1909, in Herb. Norw. Mus.; 01, Yaxham, 1918.

East: 03, Brinton Common, RPB-O, until 1953.

Eranthis hyemalis (L.) Salisb.   Winter Aconite

Est. alien. Locally abundant in a few deciduous woodlands.

West: 70, 80, 81, 88, 90, 91, 99.

East: 03, 28.

Aconitum napellus L.   Monkshood

Nat. alien. Rare. Usually a garden-escape. Has been known from damp woods near Hempstead ponds, E03, for eighty years.

West: 00, Whinburgh; 90, Scoulton.

East: 03, Hempstead; 23, Bradfield Common, since 1954, ETD.

Anemone nemorosa L.   Wood Anemone

Native. Common in deciduous woodland but very rare in the Thetford area. Occasionally, as in Wayland Wood near Watton, W99, and at Hockering Wood, E01, large plants with reddish-purple flowers occur= var. *purpurea* DC.

99

Clematis vitalba L.   Traveller's Joy, Old Man's Beard
Native. Locally frequent especially in East Norfolk. Hedgerows and wood margins and pits on chalk soil.
West: 63, 70, 72, 73, 79, 84, 88, 90, 91, 93, 94, 98, 99, 01, 08.
East: 02–04, 11, 12, 14, 18–21, 23, 24, 28–30, 32, 39, 40.

Ranunculus acris L.   Meadow Buttercup, Upright Crowfoot
Native. Common. Grassland.

R. repens L.   Creeping Buttercup
Native. Common. Wet meadows, arable land, grassland and riversides.

R. bulbosus L., ssp. *bulbosus*   Bulbous Buttercup
Native. Common. Drier grassland and especially meadows.

Var. *dunensis* Druce
East: 04, shingle bank, Cley, PDS.

R. arvensis L.   Corn Crowfoot, Bur Buttercup
Colonist. A rare cornfield weed now but found occasionally as a weed in gardens.
West: 62, S. Wootton; Bawsey Road, King's Lynn; 88, Thetford; 91, E. Dereham; 93, Gt. Walsingham, BT.
East: 18–20, 29, 31, 39.

R. sardous Crantz   Hairy Buttercup
Native. Characteristic of pastures near the sea, in which it may be frequent, and on sea-banks, but slow to appear in newly reclaimed land. Uncommon inland.
West: 62–64, 74, 84, 92, 00.
East: 02, Foxley, an inland station where it is frequent, ALB; 03, 04, 30, 31, 33, 40–42, 49, 51.

R. parviflorus L.   Small-flowered Buttercup
Native. The 1914 *Flora* listed eighteen stations but there are no recent records and it would appear reasonable to suggest that the severe winters of 1891–95 proved lethal to this species, the seeds of which germinate simultaneously, although there is a gathering in the Herb. Norw. Mus., dated 1894, from a stone wall at East Runton on the north coast.

R. auricomus L.   Goldilocks, Wood Crowfoot
Native. Not uncommon in deciduous woodland.
West: 61, 62, 71, 72, 89–92.
East: 03, 09, 10, 18, 19, 28, 29, 39.

R. lingua L.   Great Spearwort
Native. Marshes and fens where it has been frequent but is decreasing; still flourishing in some of the Broads particularly Calthorpe Broad. Dominant in a pond at Thompson (W99).
West: 62, 70–72, 89, 91, 98, 99, 08, 09.
East: 09, 30–32, 40–42.

R. flammula L., ssp. *flammula*   Lesser Spearwort
Native. Common. Wet places, tolerant of acid soils, and extending into bog.

R. sceleratus L.   Celery-leaved Crowfoot
Native. Common. Muddy ditches and ponds.

R. hederaceus L.   Ivy-leaved Water Crowfoot
Native. Common. On mud and in shallow water.

R. omiophyllus Ten.
(*R. lenormandii* F. W. Schultz)
Native. Rare. Shallow ponds or mud. Only one record in the 1914 *Flora* from West Runton, E14, by Babington. Although considerable time has been spent in field-work on the water buttercups, so far search has been unsuccessful in re-finding this rarity. In *The East Anglian Flora* (Sir E. J. Salisbury's Presidential Address to the Norf. & Norw. Nats. Soc., 1932) it is stated that its occurrence here is in conformity with the presence of other oceanic types.

R. fluitans Lam.   Water Crowfoot
Native. Frequent in the larger streams and abundant in the Breckland rivers.

R. circinatus Sibth.
Native. Frequent in ditches throughout the county.

R. trichophyllus Chaix ssp. *trichophyllus*   'Water Fennel'
Native. Common and abundant in ponds, ditches and woodland pools.

Ssp. *drouetii* (F. W. Schultz) Clapham
Forms approaching this subspecies with leaf-segments collapsing when taken from the water have been found, but the distinction is scarcely merited as this character depends on the amount of mineral incrustation.

R. aquatilis L.
(*R. heterophyllus* Weber)   Water Buttercup
Native. Common in ponds and ditches.

R. peltatus Schrank ssp. *peltatus*
Native. Frequent. Ponds, dykes and many of the Breckland meres as at Wretham and Breckles Heath.

Ssp. *pseudofluitans* (Syme) C. Cook
Native. Less frequent but characteristic of streams flowing from the chalk, especially the River Nar and its many tributaries.

R. baudotii Godr.
Native. Brackish ditches and ponds near the coast.

A variable species. Although usually with capillary leaves, drought results in the formation of laminate ones. Distribution imperfectly known; most of the plants seen have been in West Norfolk.
West: 62, 63, 74, 84, 94.
East: 04, 42.

R. ficaria L., ssp. *ficaria*   Lesser Celandine, Pilewort
Native. Common. Woodland, hedgebanks and streamsides.

Ssp. *bulbifera* (Marsden-Jones) Lawalrée

In response to an appeal by E. A. Ellis in 1936/37, it was found that this subspecies was less common.

A form with double flowers has persisted at South Runcton, W60, for fifty years. A tetraploid form, larger in all its parts, occurs in waste ground at Cringleford, E10, where it has been known for twenty-five years; also at Brundall, E30, EAE.

Myosurus minimus L.   Mousetail

Native. Very rare now. Cornfields. Plentiful in 1899 at Repps, E23.

West: 83, one plant at Quarles on the headland of an arable field, BT, 1962; also one plant, cabbage field, 1966, BT.

Aquilegia vulgaris L.   Columbine, Ladies' Bonnets

Native. Very local. Calcareous fens and thickets.

West: 70, Caldecote Fen; up to 1959 when drainage was carried out, this calcareous fen was coloured blue with the abundance of flowering plants, among which were a few purple and a very few chocolate-brown. Still present 1966; 90, Cranworth and Bradenham, ALB; Carbrooke Fen, 1914, FR; 91, Potter's Fen near Dereham, 1954, JSP; 99, Wretham Park, north side of Mickle Mere, 1963, PAW.

East: 00, Whinburgh, 1915, WGC; 03, Edgefield Heath, 1956, PHS; 10, Wymondham, 1924, WGC; 12, Booton Common, 1953, EAE.

Plate 297, *English Botany*, specimen from Newton St Faiths, 1793.

Thalictrum flavum L.   Common Meadow Rue

Native. Not uncommon on banks of streams and in fens.

West: 60, 61, 70, 71, 78, 79, 88–91, 98, 99, 07, 09.
East: 00, 04, 07, 09, 10, 18, 20, 28, 30, 32, 39, 49.

Thalictrum minus L., ssp. *minus*   Lesser Meadow Rue

Native. Locally frequent in suitable habitats such as dry banks and hedgerows on chalk; apparently restricted to West Norfolk, including Breckland. Has been given several varietal distinctions in the past, the one most favoured being *T. Babingtonii* Butcher, but these are scarcely merited.

West: 62, 68, 70–73, 78–80, 84, 88, 89, 91, 98, 01.

Ssp. *arenarium* (Butcher) Clapham

Native. Very rare. Exclusive to maritime sand-dunes.

West: 84, Scolt Head Island, 'flourishing in dense *Psamma* in the middle of the Massif', V. J. Chapman, 1930.

East: 50–51, sandhills between Yarmouth and Caister, EAE, 1961, where it has persisted for well over a hundred years as is shown by a specimen in Herb. Norw. Mus., from Caister, 1840; Caister-on-Sea, ETD, 1964.

BERBERIDACEAE

Berberis vulgaris L.   Barberry

Status doubtful. Occasional in hedges. Has recently been planted in Breckland mixed woodland. The relationship between this plant and the blight on wheat was known as long ago as 1805 when, in a letter, Joseph Banks observed: 'It has long been admitted by farmers, though scarcely credited by botanists, that wheat in the neighbourhood of barberry bushes, seldom escapes the blight.'

West: 63, 64, 70, 72, 74, 78, 79, 83, 84, 89–91, 93, 94, 98.
East: 03, 11, 20, 40, 42.

Mahonia aquifolium (Pursh) Nutt. agg.   Oregon Grape

Est. alien. Has been extensively planted for game cover and is abundant throughout Breckland where it reproduces both vegetatively and by seed.

West: 62, 70, 73, 79, 80, 83, 84, 88–93, 99, 09.
East: 01, 03, 19, 28, 30, 39.

NYMPHAEACEAE

Nymphaea alba L., ssp. *alba*   White Water-lily
Native. Occasional in ponds in fens but far more frequent in the Broads.
West: 61, 62, 70–72, 84, 91, 99.
East: 20, 22–24, 28–33, 39–42, 49.

Nuphar lutea (L.) Sm.   Yellow Water-lily, Brandy-bottle
Native. Common. In ponds and rivers.

CERATOPHYLLACEAE

Ceratophyllum demersum L.   Hornwort, loc. Ferret's-tail (Broads)
Native. Common. Ditches.

C. submersum L.

Native. Apparently rare but distribution imperfectly known. Ditches.

West: 51, Smeeth Lode, Tilney St Lawrence, 1953; 89, Tottington, CPP, 1967.
East: 04, Wiveton, FMD, 1956; 30, Wheatfen Broad, M. J. D. Cockle, 1938; 31, Upton Broad, EAE, 1938; 42, mouth of the R. Thurne, PDM, 1963.

PAPAVERACEAE

Papaver rhoeas L.   Corn Poppy, Red-weed, Canker-rose
Colonist. Very common and sometimes abundant in sandy, arable fields. As in other large plant populations, much variation occurs, including such examples as var. *hoffmanianum* O. Kuntze with black blotches at base of petal with a surrounding paler margin and, more frequently, var. *strigosum* Boenn., with appressed instead of spreading hairs.

P. dubium L. Smooth long-headed Poppy
Colonist. Often associated with the last but is less common.

P. lecoqii Lamotte   Babington's Poppy

Colonist. Very rare. Road-margins. Recorded for East Norfolk by both Babington and Linton towards the end of the nineteenth century.

West: 69, Methwold Hythe, ELS, 1965, by a roadside on calcareous peat. East: 28, Pulham St Mary, CF, 1967, garden weed for past twenty-one years.

P. hybridum L.   Round rough-headed Poppy

Colonist. Rare. Arable weed on calcareous soils of north-west Norfolk.

West: 62, N. Wootton; 63, Ingoldisthorpe; 70, Marham and Beechamwell; 72, Flitcham, RSC; 92, Wells; 93, Gt. Walsingham, GT; 04, Morston, JHC.
East: 04, Cley to Walsey Hills, FMD, 1954.

P. argemone L.   Long rough-headed Poppy

Colonist. Thinly but very widely distributed. Cornfields on light soil.

West: 62, 70–73, 79, 80, 83, 89–94, 98, 01.
East: 03, 08–10, 14, 20, 21, 29, 32, 33, 42.

P. somniferum L., ssp. *hortense* Hussenot   Opium Poppy

Est. alien. Rather rare. Formerly grown in the Fenland as a cure for ague. Occasionally grown these days as a source for poppy-seed oil. It also escapes from gardens.

West: 62, 79, 81, 88, 90, 00.
East: 11, 14, 20, 29, 51.

Glaucium flavum Crantz   Yellow Horned-poppy

Native. Confined to shingle beaches where it is often abundant, especially on the north-west coast.

Chelidonium majus L.   Greater Celandine

From the point of view of frequency, probably the best example of an old-established alien. Abundant throughout the county, particularly near habitations.

FUMARIACEAE

Corydalis claviculata (L.) DC.   White Climbing Fumitory

Native. Locally common, especially in recently cleared woods on acid soil.

C. lutea (L.) DC.   Yellow Fumitory

Est. alien. Locally common on old walls, especially in the Norwich area and the east of the county generally; apparently increasing. Rare in West Norfolk.

West: 62, 79, 83, 88, 93, 99, 00.
East: All but four of the twenty-six 10-km. squares.

Fumaria capreolata L.   Rampant Fumitory

Although this native is given as rather common in hedges in the 1914 *Flora* and frequent about Norwich, there are no recent records to support this.

F. bastardii Bor.

The only record to support its occurrence in Norfolk is that made by C. E. Salmon who found it growing on a roadside bank at Ranworth, E31, in 1915.

F. muralis Sond. ex Koch ssp. *boraei* (Jord.) Pugsl.

Native. Very rare. Hedgebanks.

West: 91, garden weed, Toftwood, 1955, JSP.
East: 41, between Caister and Hemsby, White and Salmon, 1915, in Herb. Norw. Mus.

F. micrantha Lag.

Native. Very rare. Arable land.
No records since the early years of the present century.

East: 30, Strumpshaw, F. Long; 32, Sutton, A. Bennett & C. E. Salmon.

F. officinalis L., ssp. *officinalis*   Common Fumitory

Native. Common. Arable land and waste places.
Forma *scandens* Pugsl.

East: 31, Ranworth, C. E. Salmon, 1915, 'festooning a row of peas . . . to a height of six feet'.

F. vaillantii Lois.

Native. Very rare. Calcareous arable land.

West: 60, Stow Bardolph, J. E. Little, 1917; 71, Narborough Field, ELS, 1958, conf. NYS; 89, Threxton, FR, 1917.

F. parviflora Lam.   Small-flowered Fumitory

Native. Very local. Calcareous arable land. A plant of the early phase of colonisation following the breaking-up of chalk grassland, hence the number of old records for Breckland. As soon as the vegetation becomes dense the plant disappears.

West: 70, Marham, B. Bray, *c.* 1880; Gooderstone, GT, 1966; 71, Narborough, B. Bray; 78, Weeting Heath, 1957; Pilgrims' Walk, Weeting, 1966, ELS; 79, Feltwell, WHB & WGC, *c.* 1914; 80, Gt. Cressingham, WCFN, *c.* 1912; 89, Bodney, WGC, *c.* 1914.

CRUCIFERAE

Brassica nigra (L.) Koch   Black Mustard

Colonist. Uncommon. Occasionally on banks in reclaimed land.

West: 51, 60, 62, 69, 90, 91, 00, 01.
East: 03, 04, 18, 20, 24, 28, 38, 41.

Sinapis arvensis L.   Charlock, loc. 'Garlick'

Colonist. Common as an arable weed, but fast disappearing since the introduction of differential spraying in 1940. Plants with stiffly hairy valve but glabrous beaks to the fruits=var. *orientalis* Koch & Ziz occur with the typical plants.

S. alba L.   White Mustard

Nat. alien. Often sown as a crop and persisting on field borders and in waste places.

Hirschfeldia incana (L.) Lagrèze-Fossat   Hoary Mustard

Est. alien. Locally abundant in the Norfolk Breckland in 'fire-breaks' and by the margins of conifer plantations.

West: 78, Weeting, 1957; 88, Santon, 1957.
East: 20, Household Heath near Norwich, RMB, 1959; Eaton Park, ETD, 1965; Harford tip, ETD, 1966.
First record: Santon; CPP, 1957.

Diplotaxis muralis (L.) DC.   Wall Rocket, Stinkweed

Est. alien. Common. Walls, railway tracks, rarely beaches, and frequently a feature of chalky stubble fields in autumn. Has increased in frequency as the 1914 *Flora* gives it as 'rather rare'.

Var. *caulescens* Kittel, a biennial or subperennial form with leafy stems, formerly under var. *babingtonii* Syme.

West: 70, Beechamwell, J. E. Little, 1919; 88, Thetford, FR, 1914.
East: 50, Yarmouth, FR, 1914.

D. tenuifolia (L.) DC.   Perennial Wall Rocket

Native. Not common. Old walls and waste places.

West: 61, Bawsey; King's Lynn Docks; 74, Ringstead chalk pit, PDM; 83, Quarles, BT; 84, Burnham Overy, BT; Burnham Thorpe, GT; 90, Watton, FR.
East: 20, city walls, Norwich, ETD; 21, Wroxham, KHB; 22; 23, N. Walsham, KHB; 50, Yarmouth, ETD; 50, 51.
First record: 1793 (Woodward, *English Botany*, t. 525, as *Sisymbrium tenuifolium*).

Raphanus raphanistrum L.   Wild Radish, White Charlock

Colonist. Very common as an arable weed especially on sandy, acid soils. Frequently abundant along wide roadside margins following the dumping of sugar beet. In dense populations it is possible to find two colour forms: one, forma *alba* F. Gér., having white flowers with lilac veins; and the other, forma *ochrocyanea* F. Gér., with pale yellow flowers but similarly veined. Another form, forma *hispidus* Lange with hispid fruits and white flowers with yellow veins, occurs rarely. The subspecies *landra* (Moretti) Bonnier was found at Gt. Yarmouth by J. H. Silverwood in 1964 and determined by J. E. Lousley.

Crambe maritima L.   Seakale

Native. Very rare now. Coastal shingle. Formerly described as abundant on the north Norfolk coast (*English Botany*). Now restricted to a single colony at Cley, E04, 1962, where seed was introduced by Oliver in 1912.

Cakile maritima Scop., ssp. *maritima*   Sea Rocket

Native. Locally common. Characteristic of coastal sands just above high-water mark, and often abundant in the early stages of colonisation here.
First record: *c.* 1744, Gt. Yarmouth, in Herb. Joseph Andrews, Herb. Mus. Brit.

Lepidium campestre (L.) R. Br.   Field Pepperwort
Colonist. Uncommon. Roadsides and arable land. A decreasing species.
West: 61, 62, 73, 83, 91, 93, 98, 01.
East: 04, 18, 19, 31, 39.

L. heterophyllum Benth. (*L. smithii* Hook.)   Smith's Cress
Colonist. Rare but formerly locally common. Habitat similar to the preceding species with which it is easily confused. Recent records only are given.
West: 91, E. Dereham, WGC, 1915; 01, Yaxham, WGC, 1915.
East: 00, Kimberley, WGC, 1915; 04; 11, Felthorpe, WGC, 1915; 20, 23; 30; 32, Dilham, WGC, 1915; 41, Martham, PER, 1930, in Herb. Norw. Mus.

L. ruderale L.   Narrow-leaved Dittander
Native. Occasional. Along the coast in scattered colonies but rare inland. First British record from 'Lynne in Norfolk' Ray, *Cat.*, 1670.
West: 62, King's Lynn; West Lynn; Gaywood; 74, Titchwell; Thornham; 84, Burnham Overy Staithe.
East: 20, 30, 40, 39, 49, 51.

L. latifolium L.   Dittander
Native. Rare. Waste places near the sea.
East: 04, Cley; 11, Drayton, WGC; 14, Beeston; 20, Thorpe St Andrew, WGC; 51, Ormesby.

Coronopus squamatus (Forsk.) Aschers.   Swine's Cress
Native. Common. Roadsides and arable land, particularly land reclaimed from the sea; about farm gates subjected to trampling by cattle.

C. didymus (L.) Sm.   Lesser Wart-cress
Colonist. Uncommon. Waste places, cultivated land; occasionally on manure heaps. The earliest records are from ports and harbours.
West: 60, Downham Market, CPP; 61, Shouldham Warren, ELS; 72, Massingham Heath, CPP; 84, Holkham, GT; 90, Watton, FR.
East: 11, Felthorpe, ELS; 12, Blickling, RCLH; 20, Harford, RMB; 30, Rockland St Mary, EAE; 31, Horning Ferry, JHS; 39, Broome, ETD; 40, Acle, ETD; 50, N. and S. Denes, Gt. Yarmouth, EAE.

Cardaria draba (L.) Desv., ssp. *draba*   Hoary Pepperwort.
Est. alien. Widespread and abundant in field borders, railway embankments, roadsides and sea-banks. Has a most aggressive vegetative spread in addition to occasional dispersal by seed. The 1914 *Flora* described it as rather rare. Reputed to have been introduced to this country in the straw used as bedding on the return of the military from the Walcheren expedition in 1809.
First record: near Ashwicken, W61, Dr John Lowe, 1865.

Thlaspi arvense L.   Field Penny-cress
Colonist. Locally common. Arable land, especially around stack-bottoms. Shows considerable increase.

Teesdalia nudicaulis (L.) R. Br.　　Shepherd's Cress

Native. Locally abundant. Ephemeral of sandy soils, especially on the dry Breckland heaths. Rare in coastal sands.

West: 60–63, 70–72, 78, 80, 88, 91, 98, 08.

East: 03, 04, 09, 11, 39, 40, 51.

Capsella bursa-pastoris (L.) Medic.　　Shepherd's Purse

Native. Common everywhere on cultivated soil and roadsides. In flower every month of the year.

Cochlearia officinalis L., ssp. *officinalis*　　Scurvy-grass

Native. Common. Characteristic of the late stages of muddy salt marshes, and along tidal river-banks.

First record: 1597, 'By the sea-side at Lynn', Gerard.

C. danica L.　　Ivy-leaved Scurvy-grass

Native. Rather rare. Shingle beaches and drier parts of salt marshes.

West: 63, Wolferton; 64, Hunstanton; Holme-next-the-Sea; 74, Thornham; Brancaster; 84, Burnham Overy Staithe; 94, Wells-next-the-Sea, ALB.

East: 04, Cley; 50, Gt. Yarmouth.

C. anglica L.　　Long-leaved Scurvy-grass

Native. In similar situations to *C. officinalis* but rarer.

West: 62, 63, 74, 94.

East: 04, 14, 30, 40–42, 49.

Alyssum alyssoides (L.) L.　　Small Alison

Est. alien. Very rare now. One of the typical Breckland specialities of sandy soils.

West: 71, Westacre, 1951; 78, near Brandon, H. S. Redgrave, 1934; between Hockwold and Weeting, HDH in Herb. Norw. Mus., 1922.

Berteroa incana (L.) DC.　　Hoary Alison

Est. alien. Rare. Roadsides and waste places.

West: 62, King's Lynn Docks; 79, Whittington, CPP; 88, Thetford, A. E. Ellis; 92, Hempton, RMB.

East: 20, Harford, JHS; 32, E. Ruston, WGC.

Erophila verna (L.) Chevall.　　Whitlow Grass

Native. Common spring ephemeral of walls, banks and fixed dunes. Many varietal names have been given to Norfolk gatherings in the past.

Armoracia rusticana Gaertn., Mey. & Scherb.　　Horse-radish

Est. alien. Common. Field-borders and waste places, spreading vegetatively.

Cardamine pratensis L.　　Cuckoo Flower, Lady's Smock

Native. Common. Wet meadows and fens. Forms with double flowers, sometimes accompanied by proliferation, are not infrequent.

C. amara L.　　Large Bitter-cress

Native. Less common than the preceding. Streamsides.

C. flexuosa With.   Wood Bitter-cress
Native. Frequent. Wet woods and streamsides.
West: 60, 62, 64, 80, 83, 88–93, 99, 08.
East: 00–03, 08, 10, 12, 14, 22, 28, 30–33, 39, 40–42.

C. hirsuta L.   Hairy Bitter-cress
Native. Abundant. Roadsides, railway tracks, cultivated land and waste places.

Barbarea vulgaris R. Br.   Yellow Rocket, Winter Cress
Native. Frequent. Streamsides.

B. stricta Andrz.   Small-flowered Yellow Rocket
Nat. alien. Appears to be increasing in frequency. Streamsides, especially on old wooden quay headings in Broadland.
West: 94, Warham and Stiffkey, PHS.
East: 20, Whitlingham, ETD; 30, Coldham Hall, ETD; Wheatfen, Surlingham, EAE; Buckenham Ferry, ETD.

B. intermedia Bor.   Intermediate Yellow-cress
Est. alien. Rare. Cultivated and waste ground.
West: 62, Castle Rising and N. Wootton; 80, Swaffham, ESE; 88, Thetford, HDH; 92, Mileham, ETD.
East: 02, Foulsham, ETD; 03; 04, Kelling Warren, ELS; 10, Hethel airfield, ETD; 23, Gimmingham and Trunch, JHC; 30, Surlingham, EAE; 33, Mundesley, JHC.
First record: 1903, Ranworth, E31, F. Long.

Arabis hirsuta (L.) Scop.   Hairy Rock-cress
Native. Frequent. Walls and dry banks on the chalk.
First record: 'Found growing on walls . . . in Lynn in Norfolk', Du Bois Herbarium, 1690–1723.

Turritis glabra L.   Tower Mustard
Native. Rare. Dry hedgebanks.
West: 80, Holme Hale; 89, Ickburgh, GT; 90, W. Bradeham, ALB; 99, Rockland All Saints and Stow Bedon, HDH; 02, Billingford, DMM.
East: 01, Bylaugh, ALB; 03, Holt; 11, Ringland; 13, Bodham, WGC; 20, Norwich, in Herb. Norw. Mus.; 28, Wortwell, CF.

Rorippa nasturtium-aquaticum (L.) Hayek   Watercress
Native. Widespread and abundant in wet places; luxuriant in chalk streams. Leaves and stems remain green in winter.

R. microphylla (Boenn.) Hyland.   One-rowed Watercress
Native. Distribution imperfectly known but records suggest it is frequent. Leaves and stems purplish brown in autumn and winter.
West: 60, 71, 72, 82, 83, 88, 91, 92.
East: 11, 14, 19, 24, 32, 39, 41, 42.
First record: 1833, Swaffham, Sir W. A. Trevelyan in Hancock Museum, Newcastle-on-Tyne, Howard and Lyon (*Distribution of the British Watercress species*, 1951, *Watsonia*, **2**, 91).

R. microphylla × *nasturtium-aquaticum*
(*R.* × *sterilis* Airy Shaw)

This hybrid appears to be rare.

West: 71, E. Walton Common; 72, Hillington, material collected by ELS in 1946 cited as the holotype of R. × *sterilis* Airy Shaw (see 1951, *Watsonia* **2**, 73); 81, Newton by Castleacre; 83, S. Creake; 91, E. Bilney, DMM. First record: 1890; Framingham, E20, R. S. Standen, in Herb. Mus. Brit., det. Howard and Manton.

R. sylvestris (L.) Besser   Creeping Yellow-cress

Native. Rare in West Norfolk. Streamsides and sometimes a persistent garden weed.

West: 61, Middleton railway-siding until 1945; 83, Syderstone, GT.

East: 00, 08, 12, 18, 19, 21, 30–33, 39, 49, 51

R. islandica (Oeder) Borbás   Marsh Yellow-cress

Native. Rather common. Ponds and ditches.

R. amphibia (L.) Besser   Great Yellow-cress

Native. Apparently increasing. Ditches and ponds. Tolerates shade. Early in the season with submerged leaves frequently pectinate has been noted as var. *variifolium* DC.

West: 60–62, 70, 71, 78–80, 82, 89–92, 98, 99, 01.

East: 09, 10, 18, 30.

Hesperis matronalis L.   Dame's Violet

Nat. alien. Rare. Waste places, frequently escaping from old cottage gardens.

West: 60, Shouldham and W. Dereham; 62, S. Wootton; 70, Boughton Fen, CPP; 72, Hillington; 91, E. Dereham, KD; 99, Knight's Fen, Hockham; Thompson.

East: 00, Hardingham, ETD; 01, Bawdeswell Heath; 29, Newton Flotman, ETD; 41, Clippesby, ETD.

Erysimum cheiranthoides L.   Treacle Mustard

Colonist. Common. Arable fields and waste places.

First record: 1744; Norwich, W. Holman in Herb. Joseph Andrews in Herb. Mus. Brit.

Cheiranthus cheiri L.   Wallflower

Nat. alien. Well established on many old walls and cliffs. Doubtless some of the old records refer to garden escapes but the plant described by J. E. Smith as *C. fruticulosus* with petals that 'became recurved and do not hang loosely flaccid like those of the true *C. cheiri*' and narrow, acute rigid leaves, still persists here and there on the walls of priories and abbeys.

West: 64, 71, 79–81, 93, 94.

East: 04, 10, 14, 20, 33, 41, 49, 51.

Alliaria petiolata (Bieb.) Cavara & Grande   Garlic Mustard, Jack-by-the-Hedge

Native. Abundant. Hedges and wood margins.

Sisymbrium officinale (L.) Scop.   Hedge Mustard
Native. Common. Hedgebanks, arable land and waste places.

Var. *leiocarpum* DC., easily distinguished at a glance by its yellowish-green coloration, strikingly glabrous, is not uncommon.

West: 61, Leziate and Ashwicken; 62, W. Newton, BSBI Exc., 1949; N. Wootton; 72, Hillington.

East: 04, Cley and Blakeney, FMD; 14, Beeston Regis, FMD; 31, Horning, Woodbastwick, Ranworth and S. Walsham, Salmon & White; 41, Thurne; 51, Scratby, plentiful, Salmon & White.

S. orientale L.   Eastern Rocket
Est. alien. Frequent on waste ground; bomb sites during the Second World War.

West: 62, 63, 71, 78–81, 88, 90, 94, 99.
East: 04, 14, 20, 21, 24, 41, 49, 51.

S. altissimum L.   Tall Rocket
Est. alien. Spreading rapidly in waste places.
'Numerous alien plants sprang up during the war of 1914–1918 on the sites of military camps at Thetford, Narborough and East Harling – at least fifty species have been recorded, but the only one that has persisted is *S. altissimum*' (R. R. Clarke in *In Breckland Wilds*, 1937, 2nd edit, p. 25).

West: 61, 62, 64, 70, 71, 73, 74, 78–80, 88, 89, 99.
East: 04, 11, 20, 28, 31, 51.

Arabidopsis thaliana (L.) Heynh.   Thale Cress
Native. Common. Walls, dry banks and sandy fields.

Descurainia sophia (L.) Webb ex Prantl   Flixweed
Colonist. Frequent. Arable fields, especially about stack sites.

RESEDACEAE

Reseda luteola L.   Dyer's Rocket, Weld
Native. Common. Waste places.

R. lutea L.   Wild Mignonette
Native. Locally common. More closely associated with the chalk.

VIOLACEAE

Viola odorata L.   Sweet Violet
Native. Common. Hedgebanks, scrub and woodland margins on basic soils. Plants with white petals suffused with some violet coloration and bearded lateral petals=var. *dumetorum* (Jord.) Rouy & Foucaud are frequent.

V. hirta L., ssp. *hirta*   Hairy Violet

Native. Curiously local in this county considering its frequency in Cambridgeshire on the chalk hills which are continued into Norfolk. Confined to calcareous banks and scrub.

West: o8, S. Lopham Fen: 70, Devil's Dike, Beechamwell; Narborough; 80, S. Pickenham; 90, Cranworth, ALB.
East: No records.

Var. *lactiflora* Rchb., with white flowers occurred in a meadow at Reffley, W62, until 1948.

V. riviniana Rchb., ssp. *riviniana*   Common Wood Violet

Native. Common. Woods and hedgebanks. Forms with white flowers, perhaps forma *luxurians* Becker, occur at Kettlestone, W93, EAE.

Var. *pseudo-mirabilis* Coste, either 'very rare or overlooked on account of its evident affinity with' the normal plant (Mrs E. S. Gregory in *British Violets*, 1912).
West: 99, Wayland Wood near Watton, FR, 1919.

Forma *nemorosa* Neuman, with dark spur and shorter calyx appendages.
West: 91, Rawhall, BSBI Exc., 1954; 92, Horningtoft, BSBI Exc., 1954; Beeston, ESE, 1937.

Ssp. *minor* (Gregory) Valentine

The distribution of this subspecies is imperfectly known but it appears to favour exposed situations around gorse bushes on fens.
West: 72, Derby Fen; Leziate Fen; 79, Foulden Common; 91, Scarning Fen.
East: o1, Hockering, NNNS Exc., 1962.

V. reichenbachiana Jord. ex Bor.   Pale Wood Violet

Native. Less widely distributed than *V. riviniana* with which it occasionally hybridises where the two grow together as at Horningtoft, W92. Abundant at times in woods. Puzzling forms with pale or dark and thick furrowed spurs occur in large populations (?=var. *punctata* Greg.). A white flowered form appears to be very rare (*forma leucantha* (Beck.) Airy Shaw) and has been found once at Fincham, W60.

V. canina L., ssp. *canina*   Dog Violet

Native. Frequent. Heaths and dunes. A very variable species. The normal plant occurs on established dunes.

Var. *ericetorum* Rchb.   Heath and Hill Dog Violet

'Differs from type in its lowly habit and small leaves on long petioles' (Gregory).
West: 70, Shingham, ESE; 71, Westacre, ESE; 81, Litcham, ESE det. P. M. Hall.

Var. *pusilla* Bab.

'The variation from type consists in a much-branched *woody* stem; flowers of a deeper, more *hirta*-like blue, intensified by contrast with the large yellowish eye and yellow spur' (Gregory). Recorded for both vice-counties by Mrs Gregory.

West: 84, Scolt Head Island.

East: 51, N. Denes, Gt. Yarmouth, FR, 1914.

Var. *lanceolata* Mart.-Don.

Plant ascending diffuse; leaves longer than petiole, lanceolate, sub-cordate. 'Miss Pallis's Violet' (Gregory, p. 82) from fixed dune between Warham (*sic*, read Waxham) and Palling, E42, probably belongs here.

West: 91, E. Dereham, Mrs Russurim det. Gregory.

V. canina × *stagnina* (*V.* × *ritschliana* W. Becker)

West: 60, W. Dereham Fen, A. Templeman, 1921.

V. palustris L., ssp. *palustris*   Marsh Violet

Native. Locally frequent in carr developed over peat.

West: 60, 62, 72.

East: 03, 12, 13, 22, 32, 33, 40, 42, 49.

First record: 1797; Dawson Turner in *English Botany*, t. 444.

ssp. *juressi* (Neves) P. Coutinho

Differing from preceding by its peduncles, petioles and veins of the underleaf surfaces being usually hairy and in larger flowers.

West: 60, Stow Bardolph, J. E. Little, 1919.

V. tricolor L., ssp. *tricolor*   Wild Pansy, Heart's-ease

Native. Widespread but not abundant in cultivated land and waste places.

West: 62, 71–73, 89, 90.

East: 03, 04, 08, 13, 14, 18, 19, 39, 41, 42.

Ssp. *curtisii* (E. Forst.) Syme   Sea Pansy

Although a maritime plant elsewhere in Britain, it is with us restricted to Breckland where it is locally abundant along the railway margins and fire-breaks at Santon.

West: 88, Santon, 1919 onwards; Croxton, FR, 1918; 89, Lynford, 1958.

East: 14, Sheringham, F. Long, 1917, the only seaside record.

V. arvensis Murr.   Field Pansy, Heart's-ease

Native. Common. Cultivated and waste ground. Although modern treatment has shown that the many older segregates are not specifically separable, it would appear that a good case can be made out for the retention of *V. segetalis* Jord., and *V. obtusifolia* Jord., since they remain constant in cultivation. Although they become more luxuriant in the garden, plants lose none of their characters. They occur in the wild in light soil especially in root crops.

POLYGALACEAE

Polygala vulgaris L.   Milkwort

Native. Frequent. Base-rich grassland.

West: 62, 64, 70, 71, 74, 79, 80, 83, 90, 91, 93, 98, 99, 09.

East: 00, 03, 04, 09, 11–14, 19, 29, 30, 40–42.

P. serpyllifolia Hose   Heath Milkwort
Native. Common. Heaths and acid grassland.

HYPERICACEAE

Hypericum androsaemum L.   Tutsan
Native. Very rare now. Damp woods.
West: 79, Didlington, RMB, 1960.
East: 14, Roman Camp Woods, W. Runton, EAE, 1958; 24, Overstrand, KHB, 1959; 11; 30.

H. calycinum L.   Rose of Sharon
Nat. alien. Frequent in plantations, spreading vegetatively.
West: 79, 84, 88, 89, 94, 98.
East: 01, 12, 19, 23, 30, 39.

H. perforatum L.   Common St John's Wort
Native. Abundant. Extensive colonies are found along hedgebanks and in open woods on dry soils. A variety with narrow leaves, perhaps the var. *angustifolium* DC., reported to be abundant in the Swaffham district by E. S. Edees.

H. maculatum Crantz ssp. *obtusiusculum* (Tourlet) Hayek   Imperforate St John's Wort
Native. Far less common than *H. perforatum*. Moist ditches and damp hedgebanks. Sepals lanceolate with eroded margins and petals with dark linear glands, see Robson, 1957, *Proc., B.S.B.I.*, **2**, 237.
West: 60–63, 69, 71, 72, 80, 82, 91.
East: 01, 14, 18–20, 33, 39.

H. tetrapterum Fries   Square-stemmed St John's Wort
Native. Common. Ditches and fens.

H. humifusum L.   Trailing St John's Wort
Native. Common. Gravelly heaths and open woodland.

H. pulchrum L.   Slender St John's Wort
Native. As this species usually occurs singly, one gets the impression that it is rare, which its widespread distribution contradicts.

H. hirsutum L.   Hairy St John's Wort
Native. Rare. Characteristic of woods on the boulder clay.
West: 62, 80, 82, 90–92, 99.
East: 02, 10, 18, 19, 28, 29, 39.

H. montanum L.   Mountain St John's Wort
Native. Very rare. Copses on gravelly or chalky soil.
East: 03, Hunworth Common, WGC, 1917; 14, Weybourne Springs, where the water seeps out from a gravel hill glacial in origin; 20, 'near Norwich', Dr F. Moor.

H. elodes L.   Marsh St John's Wort
Native. Confined to bogs in which it is often abundant.

CISTACEAE

Helianthemum chamaecistus Mill.   Common Rockrose

Native. Practically confined to chalk grassland in which it is abundant. The rare cream-coloured form occurs in chalk pits at Gayton, W72, and Ringstead, W74.

West: 61, 62, 64, 70–74, 79–81, 89, 94, 98, 99, 03

East: 03, 04.

FRANKENIACEAE

Frankenia laevis L.   Sea Heath

Native. Frequent at the landward side of sandy salt marshes, reaching its British northern limit in Norfolk.

West: 63, Heacham; 64, Holme-next-the-Sea; 74, Brancaster; 84, Burnham Overy Staithe.

East: 04, Blakeney.

First record: 1746; on the coast of Norfolk near Lynn, Blackstone's *Spec. Bot.*

CARYOPHYLLACEAE

Silene dioica (L.) Clairv.   Red Campion

Native. Woods and hedgebanks. Probably less common than formerly owing to the considerable felling of woodland but still abundant in the woods of the north-east on richer soils.

S. alba (Mill.) E. H. L. Krause   White Campion

Native. Cultivated land and waste places. Appears to be far more frequent than formerly owing to the increase in acreage of arable land.

S. alba × *dioica*

This hybrid is seen occasionally, especially where woodland and arable land adjoin.

West: 62, 71, 72, 89, 91.

East: 04, 33, 42.

S. noctiflora L.   Night-flowering Campion

Colonist. Common weed of arable land on light soils.

S. vulgaris (Moench) Garcke   Bladder Campion

Native. Common. Arable land, roadsides and waste places. The hairy variety (var. *hirsuta* S. F. Gray) is widespread, especially where there is an abundance of chalk.

S. maritima With.   Sea Campion

Native. Exclusive to maritime shingle in which it is often abundant. In large populations, plants lacking the anthocyanin pigmentation are frequent; very rarely, double forms occur. At Wolferton, W63, two or three plants were observed with normal leaves but considerably reduced calyces, 13 mm. × 3 mm., compared with the average size of 15 mm. × 10 mm., and correspondingly narrow petals.

S. conica L.　　Striated Catchfly

Native. Much rarer than formerly. Dry chalky fields of the south-west but also on cliff-top at Sheringham.

West: 70–72, 78, 79, 88, 89, 98.

East: 04, Cley, FR, 1914; 11, Drayton Brecks, G. J. Cooke, 1936; 14, Sheringham.

S. gallica L., agg.　　Small-flowered Catchfly

Native. Frequent. Sandy or gravelly arable land.

Var. *anglica* (L.) Clapham with small petals, usually dingy white or yellowish, is the commonest of the varieties.

West: 60–62, 70, 72, 73, 83, 84, 90, 91, 93.

East: 03, 04, 11, 13, 14, 23, 32, 33, 42.

Var. *sylvestris* (Schott) Asch., & Graebn., petals pale pink or rose.

A rare casual.

West: 71, Westacre, in a field of lucerne, 1953. Material sent to BSBI Exchange Club.

East: 50, Gt. Yarmouth, weed in garden, 1967, JHS.

Var. *quinquevulnera* (L.) Mert., & Koch, petals spotted with reddish blotches at base. Has persisted on the railway embankment at Sheringham for many years.

East: 14, Sheringham; 17, Billingford, DMM, 1952; 23, Northrepps, CG, 1957.

S. otites (L.) Wibel　　Spanish Catchfly

Native. Much rarer than formerly when it occurred in several localities 'so abundant as to appear a hay-crop growing to a height of nearly two and a half feet' (*In Breckland Wilds*, 1937). One of the 'steppe' species confined to Breckland heaths of south-west Norfolk but reaching its northerly limit at Gayton, W72, in a chalk pit.

West: 70, Gooderstone; Cockley Cley to Shingham, ESE; Beechamwell, FR, 1918; 72, Gayton; 78, Weeting, GHR, 1960; 79, Cranwich; 88, Thetford, PAW; 98, Bridgham; W. Harling; E. Harling; 08, Garboldisham, HDH, 1933.

Lychnis flos-cuculi L.　　Ragged Robin

Native. Common. Ditches and fens.

Occasionally white-flowered forms occur; a double-flowered form seen at Martham Broad, L. R. Lloyd, 1929.

Agrostemma githago L.　　Corn Cockle

Colonist. Formerly a frequent weed of cornfields, now almost disappeared.

West: 71, Pentney, 1945; 99, Merton, 1949.

East: 01, Yaxham, 1937; 03, Holt, 1928, CPP; 04, Cley, 1928, CPP; High Kelling, 1938; 20, Keswick, about 200 plants in 1965, ETD; 24, Cromer.

Cucubalus baccifer L.　　Berry Catchfly

Although native in woods in central and southern Europe, we regard this great rarity as an established alien where it grows in a plantation at

Merton, W99. It was first discovered in 1914 by Fred Robinson who considered it indigenous and confined to this one station. In 1961, E. J. Campbell recorded it elsewhere in woodland rides in Knight's Fen, Great Hockham, and at Hills and Holes in the same parish about 1950, and in 'several places in remote woodland' (J. E. Lousley, 1961, *Proc., B.S.B.I.* **4**, 262). It would appear to owe its dispersal to bird-carriage. In 1965, Miss E. R. Noble found it at Broad Flash, Merton, near the classic site and also at Thompson; a further station is at Tottington where it has been known since 1933. As it has not been seen on the Isle of Dogs in the Thames for 100 years, the Norfolk localities are the only known British stations.

Dianthus armeria L.   Deptford Pink

Native. Very rare. Hedgerows on sandy soil. Reappeared in a hedgerow following clearing operations in 1964 in the Beck Meadows at Pulham St Mary where Miss C. Forrest has known it for several years. Has been recorded for the Harleston area for over 100 years.

West: 80, Saham Toney, FR, 1912; 90, Watton, FR, 1925.
East: 04, Bayfield, CPP, 1929; 19, Wacton, WGC, 1920; 28, Pulham St Mary, CF, 1964.

D. deltoides L.   Maiden Pink

Native. Rare. Chalk grassland in south-west Norfolk.

West: 70, Beechamwell; 72, Gayton; 78, Weeting Heath; 80, Swaffham Heath, FR; 81, near Castleacre, ESE; 88, Thetford, PAW; 89, Ickburgh; 98, Brettenham, HDH.
East: 04, Gravel Pit Hill near Cley, PHS.

Saponaria officinalis L.   Soapwort, Bouncing Bet

Est. alien. Frequent. Roadsides, waste places and sand-dunes. Becomes so invasive in gardens that it is frequently thrown out and establishes itself on roadsides. The double-flowered form is frequent; var. *hirsuta* Wierzb., recorded by J. P. M. Brenan from sand-dunes at Hunstanton and Holme-next-the-Sea.

West: 62–64, 70–72, 79, 83, 88, 91, 94, 98, 99.
East: 00, 04, 09, 11, 14, 19, 20, 22, 23, 28, 30–32, 39–41.

Cerastium arvense L.   Field Mouse-ear Chickweed

Native. Common. Heaths, dry banks and road margins.

Var. *latifolium* Fenzl, a very luxuriant plant with the habit of *C. holosteoides* found by Fred Robinson in hedgebanks at Rockland, W99, persisted for many years up to 1945.

C. tomentosum L.   Dusty Miller, Snow-in-Summer

Est. alien. Like the soapwort, becoming increasingly recorded as a garden escape; recorded also from maritime shingle.

West: 62, Castle Rising; 63, Snettisham beach; 88, Thetford; 90, Cranworth, ALB; 99, Snetterton Heath, JHS.
East: 00, Hackford, ALB; 20, Harford; 31, 33.

C. holosteoides Fr.   Common Mouse-ear Chickweed

Native. Common. Grass- and arable land, dunes and waste places.

C. glomeratum Thuill.   Sticky Mouse-ear Chickweed

Native. Very common. Arable land, wall-tops and maritime shingle.

C. atrovirens Bab.   Dark-green Mouse-ear Chickweed

Native. Locally frequent but restricted to coastal shingle and stable dunes.

West: 63, Wolferton and Snettisham; 64, Old Hunstanton; 74, Holme-next-the-Sea, Thornham and Brancaster; 84, Burnham Overy Staithe.

East: 04, Cley; 33, Mundesley; 51, Gt. Yarmouth, FR; 42, Winterton.

C. semidecandrum L.   Little Mouse-ear Chickweed

Native. Common as a spring ephemeral of heaths, sand-dunes and dry places.

Myosoton aquaticum (L.) Moench   Water Chickweed

Native. Frequent in wet places and sometimes abundant on the peat in the Fenland.

West: 60–62, 70–73, 78–80, 82, 83, 88–92, 98, 99.

East: 00, 02, 03, 10, 13, 18–21, 23, 29–31, 33, 39–41.

First record: 1794, found by J. Crowe near Lynn.

Stellaria media (L.) Vill.   Chickweed

Native. Abundant in arable and waste land. Apetalous forms occur.

S. pallida (Dum.) Piré   Lesser Chickweed

Native. Locally common, especially on sandy soils of Breckland, and on sand-dunes. The scarcity of records in the 1914 *Flora*, where this plant has but two, each from the Norfolk coast, suggests it may have been over-looked or confused with petal-less forms of *S. media*. In 1952 it was sug-gested that the Breckland plants growing in the shade of pines were apetalous *S. media* which became normal in exposed situations. Seeds of the yellowish apetalous plants from Santon were sown and produced dark green plants in garden soil with larger leaves with the appearance of apetalous *S. media*. A chromosome count, kindly carried out by Dr J. K. Morton, showed $2n = 22$, the number given for *S. pallida* and we consider the Breckland plants are undoubtedly this species. The number of stamens ('90 per cent have 2 stamens', I. K. Gibson, *ined.*) is a far more reliable character than the shape and colour of the seeds.

S. neglecta Weihe   Greater Chickweed

Native. Rare but probably overlooked on account of its close affinity to luxurious forms of *S. media*. Hedgerows in damp, sandy and shady places. No mention in the 1914 *Flora*.

West: 60, Downham Market, A Webster, 1913; 62, N. Wootton, CPP, 1964.

East: 00, damp woodland, Hardingham, ELS, 1960; 02, carr at Guist, FRo, 1959; 03, woods near Holt Lowes, BFD; 04, abundant in shady hedgebank, Bard Hill, Salthouse Heath, ELS, 1962; 12, Booton Common, FRo, 1960.

S. holostea L.   Greater Stitchwort

Native. Common. Hedgerows and woods. Curiously rare in Breckland.

S. palustris Retz.   Marsh Stitchwort
Native. Frequent in marshy places on alkaline soils.

S. graminea L.   Lesser Stitchwort
Native. Common. Heaths and acid grassland.

S. alsine Grimm   Bog Stitchwort
Native. Common. Ditches and pond margins.

Moenchia erecta (L.) Gaertn., Mey., and Scherb.   Upright Chickweed
Native. Rare. Gravelly heaths.
West: 80, Swaffham, ESE.
East: 04, Blakeney, ELS; Salthouse Heath, CPP; 11, Swannington Upgate Common, EAE; 22, Bryants Heath near Felmingham, EAE; 29, Stratton St Mary, WGC; 39, Ditchingham.

Sagina apetala Ard.   Common Pearlwort
Native. Common. Walls and sandy places.

S. ciliata Fr.   Ciliate Pearlwort
Native. Frequent. Bare ground on heaths, tracks, sand-pits and disused railway tracks. Often with *S. procumbens*; stems become purple as in *S. maritima*.
West: 50, 60–62, 70, 73, 74, 78, 79, 83, 84, 88, 92, 94, 99.
East: 04, 09, 10, 14, 22, 28, 31, 32.
First added to the British flora by W. W. Newbould in 1847 from Thetford.

Var. *filicaulis* (Jord.) Corbière
Plants with very slender stems and more or less completely glandular occur in West Norfolk.
West: 62, Whin Hill Covert, Wolferton; 78, Weeting Heath; 79, Devil's Dyke, Cranwich.

S. maritima Don   Sea Pearlwort
Native. Locally frequent. Compacted soil of older dunes and landward side of sandy salt marshes. At Wolferton the plant is abundant around bushes of *Suaeda fruticosa* with *Plantago coronopus* and *Parapholis incurva*. West: 62, Wolferton; 63, Snettisham, Heacham; 64, Holme-next-the-Sea; 74, Thornham, Titchwell, Brancaster; 84, Burnham Overy Staithe, Holkham.
East: 04, Wiveton Bank, FMD; Blakeney; 42, Waxham.

S. procumbens L.   Procumbent Pearlwort
Native. Common. Roadsides, streamsides and a persistent weed in lawns.

S. nodosa (L.) Fenzl   Knotted Pearlwort
Native. Frequent. Tolerates a wide range of habitats. Fens, maritime sand and shingle, heaths and margins of meres in Breckland.

Var. *moniliformis* (G. F. W. Meyer) Lange
A prostrate form of maritime sand and shingle bearing few flowers but, when mature, readily detachable fascicles of leaves which propagate the plant vegetatively. Seen at Wolferton, W62.

Var. *glandulosa* Bess.

Markedly glandular-hairy plants occur which have been placed under this variety.

West: 70, Beechamwell, ESE; 80, Swaffham, ESE; 89, Tottington, FR.

Minuartia hybrida (Vill.) Schischk   Fine-leaved Sandwort

Native. Occasional. Old walls, sandy arable fields, railway embankments and, rarely, on dry banks. Variable in indumentum; glabrous plants more frequent than those with glands. Form with double flowers seen at Hillington, W72. Small plants with five stamens = var. *laxa* (Jord.) occur in Breckland.

West: 62–64, 70, 72, 74, 78–83, 88, 89, 94, 98, 99.
East: 02, 11, 19, 39.

Honkenya peploides (L.) Ehrh.   Sea Sandwort

Native. Confined to maritime sand and shingle, where it is characteristic and abundant.

Moehringia trinervia (L.) Clairv.   Three-nerved Sandwort
Native. Common. Woods and plantations.

Arenaria serpyllifolia L.   Thyme-leaved Sandwort
Native. Common. Dry places including walls.

Var. *macrocarpa* Lloyd

Plants with stout, compact rosettes, large capsules and seeds up to 1·0 mm. × 0·8 mm. occur in maritime shingle.

West: 62, Wolferton; 63, Snettisham; 74, Titchwell.
East: 04, Blakeney.

A. leptoclados (Rchb.) Guss.   Lesser Thyme-leaved Sandwort

Native. Probably less common than *A. serpyllifolia*, in identical habitats. Has been a source of much confusion. It is noteworthy that Fernald in Gray's *Manual of Botany*, 8th edition, 1950, reduces this plant to a variety as var. *tenuior* M., & K., under *A. serpyllifolia* whilst Oostrom in *Flora van Nederland* (1957) reduces it to subspecific rank.

Spergula arvensis L.   Corn Spurrey

Native. Common and often abundant in arable land on acid soils, both var. *arvensis* and var. *sativa* being represented.

Spergularia rubra (L.) J. & C. Presl   Sand Spurrey
Native. Common. Sandy and gravelly heaths.

S. media (L.) C. Presl   Greater Sea Spurrey
Native. Frequent in salt marshes.

West: 62–64, 74, 84, 94.
East: 03, 14, 40–42.

Under *S. marginata*, Druce described a variety *glandulosa* reduced to forma *glabrescens* by Pugsley (*J. of Bot.*, 1921, 130), a plant of sea-cliffs and rocky shores rather than salt marshes.

West: 64, Hunstanton, T. B. Blow in Herb. Mus. Brit.

24 SHRUBBY SEABLITE   *Suaeda fruticosa   J. C. E. Hubbard*

25 MEADOW CRANESBILL   *Geranium pratense   Jarrolds*

26 ROUND-LEAVED CRANESBILL   *Geranium rotundifolium*   Miss G. Tuck

27 PETTY WHIN   *Genista anglica*   Miss G. Tuck

28 HYBRID MEDICK   *Medicago × varia*   Miss V. M. Leather

29 PURPLE MILK-VETCH   *Astragalus danicus*   *Miss D. M. Maxey*

30 YELLOW VETCH   *Vicia lutea*   *Miss D. M. Maxey*

31 EVERLASTING PEA   *Lathyrus latifolia*   *J. E. Lousley*

32 MARSH PEA   *Lathyrus palustris*   R. *Jones*

33 HYBRID GEUM   *Geum × intermedium*   *Miss D. M. Maxey*

34 WALL-PEPPER   *Sedum acre*   *A. H. Hems*

S. marina (L.) Griseb.   Lesser Sea Spurrey

Native. Characteristic of salt marsh and often abundant at a higher level than *S. media*.

Herniaria glabra L.   Glabrous Rupture-wort

Native. Very rare. One of the so-called 'steppe' species formerly flourishing in dry sandy soil in the south-west Norfolk portion of Breckland. Now appearing only sporadically.

West: 70, Beechamwell, ESE, 1943; between Cockley, Cley and Shingham, ESE, 1943; Langwade Cross, GT, 1966, about a dozen plants; 71, Narford, GHR, 1961, one plant; 78, Weeting Heath, ELS, 1966, 2/3 plants; 79, Feltwell, WGC, 1922, abundant; 80, Gt. Cressingham, RPL, 1954; Swaffham, FR, 1918.

East: 21, Frettenham gravel pits, EAE, 1945, det. Kew as a very hairy form; Horstead, J. F. Buxton, 1917, again a hairy form.

Scleranthus annuus L.   Annual Knawel

Native. Common. Dry sandy arable land and waste places.

S. perennis L.   Perennial Knawel

Native. Very rare now. Dry sandy fields. One of the Breckland specialities. All the material in the Castle Museum herbarium, Norwich, from both Thetford and Snettisham, has been determined by P. D. Sell as ssp. *prostratus* Sell.

West: 63, Snettisham beach until 1954; 79, Weeting, EL, 1951.

East: 04, Walsey Hills, Cley, RSRF, 1954.

First record: *c.* 1745, Thetford, in Herb. J. Andrews in Herb. Mus. Brit.

PORTULACACEAE

Montia fontana L., agg.   Blinks

Native. Frequent. Usually in damp acid soils but occasionally on dry commons. With the exception of one record of ssp. *intermedia* (Beeby) S. M. Walters for East Norfolk, all our plants belong to the ssp. *chondrosperma* (Fenzl) S. M. Walters (see *Montia fontana* by Walters in *Watsonia*, **3**, 1–6, 1953).

M. perfoliata (Willd.) Howell   loc. 'Buttonhole Flower'

Est. alien. Has spread rapidly since 1914 and is now locally abundant in sandy soils, especially in the Norfolk Breckland. 'Quite recently has spread along the pine heaths where the undergrowth is sparse' (1937), *In Breckland Wilds*.

West: 61–63, 74, 78, 79, 84, 88, 90, 94.

East: 03, 12, 20, 23, 30, 39.

First record: 1851, Norwich, O. Corder.

CHENOPODIACEAE

Chenopodium bonus-henricus L.   Good King Henry, Mercury

Est. alien. Long established near villages as a relic of cultivation.

C. polyspermum L.   Many-seeded Goosefoot, All-seed
Colonist. Uncommon. Arable land and waste ground.
West: 60, 62, 69, 79, 82, 90.
East: 04, 18–21, 29, 38, 39, 49.

Var. *cymosum* Moq. (var. *obtusifolium* Gaud.)
East: 12, Blickling, FR, 1921.

C. vulvaria L.   Stinking Goosefoot
Colonist. Very rare now; formerly 'rather rare' and usually found in 'dry places near houses', particularly about ports and harbours; also, a weed of cultivation.
East: 50, Gt. Yarmouth, FR, 1918, the last record.

C. album L.   Fat Hen
Colonist. An abundant weed in arable land.
Many varietal names have been applied to Norfolk material in the past but these are of doubtful value in such a variable species.

C. ficifolium Sm.   Fig-leaved Goosefoot
Colonist. Frequent. Manure heaps, rubbish tips and arable land. An under-recorded species which is the probable reason for it being assessed as rare in the 1914 *Flora*.
West: 61, 62, 69–71, 79, 99.
East: 20, 29, 42.

C. rubrum L.   Red Goosefoot
Colonist. Common. Arable land, manure heaps, rubbish tips and occasionally maritime shingle. A variable species, dwarf forms being designated var. *pseudobotryodes* Wats., in the past.

C. glaucum L.   Oak-leaved Goosefoot
Colonist. Very rare. Weed of cultivation and waste places. Formerly more frequent in West Norfolk.
East: 23, N. Walsham cattle-market, 1959, KHB, abundant; 42, Winterton Ness, RMH, 1962.

Beta vulgaris L., ssp. *maritima* (L.) Thell.   Sea Beet
Native. Characteristic of sea-banks, where it is often abundant, but not appearing until the banks are many years old.
West: 62–64, 74, 94, 04.
East: 04, 14, 20, 24, 31–33, 40, 42, 49.

Atriplex littoralis L.   Shore Orache
Native. Common. Muddy shores above salt-marsh level. The first of the maritime plants to colonise newly made sea-banks. Extended inland to the limit of the sea-floods in 1953, but did not persist. The var. *serrata* (Huds.) Grey scarcely merits varietal distinction as both serrated and entire leaves may be seen on one and the same plant.
First record: 'Found at Lynn in Norfolk by Mr. J. Sherard' (British Plants in the Du Bois Herbarium, 1690–1723).

A. hastata L., sensu lato.   Hastate Orache

Native. Common and very variable. One of the plants of the foreshore community, persisting in arable land reclaimed from the sea and occasionally inland.

Var. *genuina* Gren. & Godr., forma *salina* Moss & Wilmott, a dwarf prostrate form with small leaves.
West: 52, Terrington Marsh, ELS det. A. J. Wilmott.

Var. *oppositifolia* (DC.) Moq.
West: 62, Wolferton beach, ELS det. AJW.

Var. *deltoidea* (Bab.) Moq.
A frequent inland taxon.
West: 78, Hockwold, ELS.

A. glabriuscula Edmondst.   Babington's Orache

Native. Rare. On maritime shingle and sea-banks above high-tide mark. Distribution imperfectly known but possibly overlooked.
West: 52, Terrington Marsh; 63, Wolferton beach; Snettisham beach; 84, Holkham.
East: 04, Cley; 14, Sheringham; 33, Happisburgh, ALB; 51, Caister on Sea, ETD.

A. glabriuscula × *A. hastata*
West: 62, Wolferton beach, det. AJW.

A. patula L.   Common Orache

Native. Common. Occasionally maritime but more frequent as an arable weed.

Var. *bracteata* Westerl., frequent by the sea.
East: 42, Horsey, det. AJW.

A. laciniata L.   Frosted Orache

Native. Rare. Sandy beaches just above high-water mark, but much less constant than other plants of this habitat, e.g., *Cakile, Salsola.*
West: 64, Old Hunstanton; 84, Holkham.
East: 04, Blakeney Point, RSRF; Morston ETD; 42, Winterton Ness, RMB.

Halimione portulacoides (L.) Aell.   Sea Purslane

Native. Abundant in salt marshes, especially on the sides of creeks. Although large bushy forms with wide leaves predominate, occasionally dwarf forms with much narrower leaves are found which would come under the var. *parvifolia* (Rouy). These, according to Chapman in *Ann. Bot.*, 1937, remain constant in cultivation.

Suaeda maritima (L.) Dum.   Herbaceous Seablite

Native. Abundant in salt marshes and shows considerable variation.

Var. *macrocarpa* (Desv.) Moq., with short, blunt leaves and large seeds 2·5 to 4 mm. broad.

West: 62, Wolferton, ELS; 64, Holme-next-the-Sea, ELS; 84, Burnham Overy Staithe, ELS; Scolt Head Is., V. J. Chapman.
East: 51, Gt. Yarmouth, 1883, E. S. Marshall.

Var. *flexilis* (Focke) Rouy with erect stems, short branches ascending and flowering later than var. *macrocarpa*.

West: 84, Scolt Head Is., V. J. Chapman.

Var. *procumbens* Syme with stems procumbent. This is often in continuous societies but of doubtful varietal distinction as erect, decumbent and prostrate forms occur.

West: 62, Wolferton; 63, Snettisham; 94, Wells-next-the-Sea, F. Long.
East: 04, Salthouse, A. R. Horwood.

S. fruticosa Forsk.   Shrubby Seablite

Native. Characteristic of maritime shingle and neighbouring salt marsh. Now reaches its British northern limit in West Norfolk and the adjoining Lincolnshire coast.

West: 63, Wolferton; Snettisham; Heacham; 64, Holme-next-the-Sea; 74, Thornham; Titchwell; Brancaster; 84, Burnham Deepdale; Burnham Overy Staithe; Holkham; 94, Warham; Stiffkey; Morston.
East: 04, Cley; Salthouse; 14, Weybourne.
First record: 1690, '*in litore Norfolciae*', Sir Thos. Browne in Ray, *Cat.*

Salsola kali L.   Prickly Saltwort

Native. Characteristic of sandy beaches just above high-water mark, where it is common.

SALICORNIA   Glasswort, Marsh Samphire (loc. 'Samphire' in the absence of *Crithmum*)

In the course of 100 years the number of species has increased from four in J. E. Smith's *English Flora* (1828) to nine in Druce's *British Plant List* (1928) at which figure they remain in both editions of the *Flora of the British Isles* (Clapham, Tutin & Warburg), although it has been suggested that further segregation may be necessary. One can readily sympathise with the treatment in Bentham & Hooker's *Handbook of the British Flora* (7th edition, 1930) where all the British forms, including *S. perennis*, were reduced to a single species!

There is no question that the species present considerable difficulties in determination. This arises in great part from their reduced floral structure and the response to habitat factors such as salt-marsh succession, periods of immersion by tides and nature of the substratum. Apart from giving a general outline of habit, itself a doubtful character, herbarium material is useless. Photographs of uniform representatives are of some assistance. It is essential to study populations in the field, making precise notes of habitat, habit, branching, colour, shape and numbers of internodes, and floral characters. In 1945, the late A. J. Wilmott stressed the importance of population study in no uncertain terms. 'Singletons will not do. If you could send me 50 or 100 which were so alike that you felt at any rate this was a *kind of Salicornia*, then send me 3–6 according to size'. (AJW to ELS *in lit.*)

Recent work has shown the value of Sir E. J. Salisbury's observation in his paper, 'Ecological Aspects of Plant Taxonomy', in *The New Systematics* (edit. Julian Huxley, 1940): 'The annual British species of *Salicornia herbacea* L., agg., afford a striking example of an aggregate species of which component critical segregates are not only ecologically specialized as shown by their spatial distribution, but exhibit also a serial range of morphological characters that suggests a definite sequence of evolution.'

There is no doubt that birds, on their autumn migration, assist in dispersal, as twites, linnets and shore-larks have been observed feeding on the readily disarticulating spikes. Seeds have also been seen floating, attached to their floral 'envelopes', in water-filled depressions on the landward side of N. Wootton salt marsh where some species occur. The seeds which bear an indumentum of non-hygroscopic but mucilaginous hairs, many of them hooked, sink at once and it is assumed that these hairs assist in anchoring the seed to the substratum. In the following spring when germination begins, the embryo plants rise and float away with the help of the expanding cotyledons until the roots secure a foothold.

The following account is based on the treatment in the first edition of the *Flora of the British Isles* as it has not been found possible to correlate the Norfolk representatives with the names used in the second edition. Field-study has been confined to West Norfolk.

Salicornia perennis (Gouan) Mill.
Native. Locally common. Drier parts of salt marsh, where sand or gravel combine.
West: 74, Holme-next-the-Sea, Thornham, Titchwell, Brancaster; 84, Burnham Overy Staithe; 04, Blakeney Point.

S. dolichostachya Moss
Native. Frequent. One of the phanerogamic pioneers of the lowest zones of salt marsh, colonising the bare mud, though it is gradually being displaced from this role by the so-called *Spartina* × *townsendii*. Also occurs on the banks of creeks.
West: 62, N. Wootton, Wolferton; 84, Burnham Overy Staithe; 04, Blakeney Point, C. E. Moss.

S. dolichostachya × *ramosissima*
West: 62, Wolferton, det. AJW.

S. stricta Dum. (*S. europaea* auct., *S. herbacea* auct.)
Native. Common and variable. Frequently dominant on the upper levels of salt marsh. The earlier botanists referred most of their records to *S. herbacea* L.

S. ramosissima Woods
Native. Locally frequent. Salt marsh, particularly where sand occurs.

125

Very variable as on most salt marshes there are complete intergradations from extreme simple to fully branched forms.

West: 62, N. Wootton, Wolferton; 74, Holme-next-the-Sea; 84, Burnham Overy Staithe; Holkham. RMB conf. Dalby; 04, Blakeney, HAS.

East: 50, Bure marshes.

S. appressa (Dum.) Dum.

Native. Distribution not worked out. Our plants were growing on the landward side of a salt marsh and were strikingly prostrate. A doubtful species, probably best regarded as a prostrate variety of *S. ramosissima*.

West: 62, N. Wootton, det. AJW; 74, Titchwell.

East: 04, Salthouse.

S. smithiana Moss

Native. On the highest part of the salt marsh.

West: 62, Wolferton.

S. disarticulata Moss (*S. pusilla* Woods)

Native. Locally frequent. Landward margins of salt marsh. The most distinct of the annual species by reason of its one-flowered cymes.

West: 64, Holme-next-the-Sea; 74, Titchwell; 84, Burnham Overy Staithe.

TILIACEAE

Tilia platyphyllos Scop.   Large-leaved Lime

Apparently the least common of the introductions. Possibly more frequent than the records suggest.

West: 78, Weeting; 79, Didlington; 90, Cranworth, ALB; 99, Illington, ALB.

East: 04, 20, 49.

T. cordata Mill.   Small-leaved Lime

Questionably native in Norfolk.

West: 78, Weeting; 81, Castleacre, ALB; 90, Cranworth, ALB; 99, Illington, ALB; 03, Swanton Novers Great Wood, FRo.

East: 01, Hockering, ELS; 20, 21.

T. × europaea L.   Common Lime

Naturalised. Widespread and frequent.

West: 70, 79, 82, 89, 92, 98, 99.

East: 00–04, 10, 12–14, 18–20, 29–32, 40, 41, 51.

MALVACEAE

Malva moschata L.   Musk Mallow

Native. Frequent. Roadsides and dry banks, not confined to chalk. White-flowered forms occasional. Var. *integrifolia* Lej. & Court., with lower leaves entire and upper faintly lobed is very rare but is constant in cultivation.

West: 61, 62, 70–72, 78–80, 88–90, 91, 98, 99.
East: 00, 01, 03, 04, 10, 11, 18, 19, 23, 29, 30, 39, 41.

Var. integrifolia Lej. & Court.
West: 62, Vincent Hills, near W. Newton.

M. sylvestris L.   Common Mallow, 'Pick-cheese'
Native. Common. Roadsides and arable land.

Var. *angustiloba* Celak, small leaves with narrow lobes and deep sinuses; appressed pubescence.
West: 88, Thetford, Miss A. B. Cobbe, 1920.
East: 20, Harford tip, ELS, 1961.

M. neglecta Wallr.   Dwarf Mallow
Native. Common but less so than *M. sylvestris*. Roadsides and waste places.

Althaea officinalis L.   Marsh Mallow
Native. Very rare in the west but locally abundant in Broadland on silt soils.
West: 51, Tilney St Lawrence, persisting in ditches from which the sea has been excluded for many years.
East: 30; 39, 40, Heckingham to Reedham Ferry, ETD; Halvergate, ETD; 41, near Martham Broad, JEL; Stokesby, ETD; Thurne, ETD; 42, Horsey, RMB; 49, near the R. Waveney at Haddiscoe, EAE.

LINACEAE

Linum bienne Mill. (*L. angustifolium* Huds.)   Pale Flax
Native. Much rarer than formerly when some fourteen stations were listed in the 1914 *Flora*.
West: 94, sea-bank at Wells-next-the-Sea, 1963, the only recent record.
East: 20, Norwich, ETD, until 1965.

L. anglicum Mill. (*L. perenne* auct. angl.)   Perennial Flax
Native. Very rare. Has become almost extinct. Some eight records from chalky pastures appeared in the 1914 *Flora*.
East: 14, Aylmerton, WGC, 1920, the last record.

L. catharticum L.   Purging Flax
Native. Common on chalk grassland, less so on heaths, but widely distributed.

Radiola linoides Roth   All-seed
Native. Rare. On sand which is kept moist by seepage from springs, when taller plants are kept down by trampling.
West: 61, E. Winch Common; 62, S. Wootton; Roydon Common; Whin Hill Covert, Wolferton; 81, Litcham Common, ALB.
East: 22, Bryants Heath near Felmingham, BSBI Exc., 1938; 42, Winterton-Horsey dunes, EAE, 1933.

GERANIACEAE

Geranium pratense L.   Meadow Cranesbill

We do not consider this native as it occurs in such a variety of artificial habitats and there is but one record from a meadow. Frequent as a garden-escape and in churchyards.

West: 80, S. Pickenham, JSP, 1953, hedgebank near church; 91, E. Bilney and Beetley, DMM, 1960; 94, Warham churchyard, 1949; 98, Overa, HDH, 1943, 'an obvious escape'.

East: 01–04, 11, 13, 18, 19, Ashwellthorpe, RMB; 20, 28, Denton, 'fine sight from the vicarage to the church', CF; 29, Tasburgh, SA; 30, 33, 39, Seething, JHS; 40, 41.

G. endressii Gay

Est. alien. Apparently increasing as a garden-escape.

East: 29, Morningthorpe churchyard, SA & ELS, 1961; 31, Woodbast-wick churchyard, KHB, 1959; N. Burlingham.

G. endressii × *versicolor*

East: 29, Morningtoft; 31, Woodbastwick (see previous species).

G. versicolor L.   Pencilled Cranesbill

Est. alien. The most frequent of these garden-escapes.

West: 89, Hollow Heath near Hilborough, E. I. Newman.

East: 13; 19, Tacolneston, ETD, 1964; 20, Thorpe, G. J. Cooke, 1933; 29, Brooke, A. C. Armes, 1943; Tasburgh, SA, 1960; 31, between S. Walsham and Upton, Salmon & White, 1915, 'fairly established for 50 yards'; N. Burlingham; 32, Tunstead, G. J. Cooke, 1933; 40; 42, near Horsey Hall, EAE, 1933, 'abundant'; Hickling, ETD, 1963.

G. phaeum L.   Dusky Cranesbill

Est. alien. Occasional on hedgebanks. An old garden plant.

West: 82, Houghton; 90, Saham Hills; Watton, FR; Wood Rising, ALB; 93, Gt. Walsingham, G. Scott; 98, Brettenham, HDH.

East: 12, Brandiston, WGC; 13, W. Beckham, PHS; Bodham; 19, Forncett St Peter, MBA; 20, Earlham, ETD; 29, Morningthorpe, SA & ELS; Tasburgh, SA; 39, Ditchingham, MIB.

G. pyrenaicum Burm. f.   Mountain Cranesbill

Est. alien. Has spread widely since the 1914 *Flora*. Now frequent in hedgebanks and waste places.

Var. *pallida* Gilmour & Stearn in *J. Bot.*, 1932, *Supp*.

West: 98, rail bridge, Roudham Heath, ELS, 1966.

G. columbinum L.   Long-stalked Cranesbill

Native. Occasional, Dry hedgebanks on the chalk.

West: 63, 64, 70–74, 80, 83, 84, 98.

East: 01, 03, 04, 09, 14, 18, 19.

G. dissectum L.   Cut-leaved Cranesbill

Native. Common on heavy soil in grassland and arable.

G. rotundifolium L.   Round-leaved Cranesbill

Denizen. Widespread but not common. Dry hedgebanks. Apparently not known in the county until 1882 but has now spread widely. Observations kept on the Thetford plants by H. Dixon Hewitt from 1909 who reported '1916, still persists; 1917, much reduced by late spring frosts; 1921–22, increased range; 1928, again reduced; 1936, range again increased; 1940–46, abundant'. Its native habitat is rocks, hence its status with us.

West: 88, 91, 94, 98, 99.
East: 03, 09, 10, 14, 20–22, 28, 29, 39, 49, 51.
First record: 1882, Thetford, E. F. Linton.

G. molle L.   Dove's-foot Cranesbill

Native. Common. Grassland, cultivated ground and dunes. Rarely with white flowers.

Var. *grandiflorum* Lange

East: 31, Ranworth, C. E. Salmon, 1915, 'with flowers 11–12 mm. in diam.'.

G. pusillum L.   Small-flowered Cranesbill

Native. Common. Roadsides and waste places. Continues flowering long after *G. molle*.

G. lucidum L.   Shining Cranesbill

Native. Rare in the west. Old walls and hedgebanks.

West: 71, Pentney, Narford and Westacre; 88, Green Lane, Thetford; 00, Reymerston, Garveston and Thuxton, ALB.
East: 03, 04; 20, Postwick, JHS; 23; 29, Tasburgh, SA; 30, 39.

G. robertianum L., ssp. *robertianum*   Herb Robert

Native. Common. Mainly a woodland plant.

Ssp. *maritimum* (Bab.) H. G. Baker

Confined to maritime shingle.

West: 62, Wolferton beach, whence it was first recorded in 1939, ELS; 63, Snettisham beach.

Erodium moschatum (L.) L'Hérit.   Musk Storksbill

Native. Very rare. The only record in the 1914 *Flora* for Cromer (E24) taken from the *Botanists' Guide* (1805) was considered doubtful by H. D. Geldart.

West: 84, dunes at Scolt Head, 1932/33, V. J. Chapman.
East: 39, Ditchingham, MIB conf. Miss C. M. Rob, 1945.

E. cicutarium (L.) L'Hérit., ssp. *cicutarium*   Common Storksbill

Native. Common. Heaths and sandy places. White-flowered forms frequent in Breckland.

Ssp. *dunense* Andreas occurs frequently on the sand-dunes of the coast from Gt. Yarmouth in the east to Burnham Overy Staithe in the west. Recorded also from Breckland (BSBI Exhib., 1962).

E. glutinosum Dum.

Reported as found by E. F. Warburg in 1938 at Blakeney Point in *Biol. Medd. Dan. Vid. Selsk.*, **23**, no. 6. This is probably identical with *E. cicutarium* B *glandulosum* Van den Bosch (*Prod. Fl. Batav.*, 1850) given in the 1914 *Flora* for the same station, E. J. Salisbury, 1912.

OXALIDACEAE

Oxalis acetosella L.   Wood-sorrel

Native. Common. Woods but apparently absent from Breckland.

O. corniculata L., var. *corniculata*   Procumbent Yellow Sorrel

Est. alien. Occasionally by roadsides and a persistent garden-weed.

West: 62, W. Newton, 1945; King's Lynn, 1948; S. Wootton; 79, Northwold, VML; 84, Burnham Thorpe, BT; 94, Stiffkey, BT.
East: 20, Norwich; Harford; Gt. Plumstead, JHS.
For other species see list of casuals.

BALSAMINACEAE

Impatiens capensis Meerburgh   Orange Balsam, Jewel Weed

Est. alien. A remarkable example of a plant that has shown explosive spread in a short time. Unrecorded in the 1914 *Flora* and first introduced 'on one of the tributaries of the river Bure near Aylsham in 1927', EAE. 'It has now colonised alder carr, swamps and river banks throughout Broadland.' EAE.

East: 12, Aylsham, RMB, 1963; 20, Postwick, ETD, 1964; 21, Horstead Mill and Coltishall, ETD, 1963; Wroxham, RMB, 1952; 30, Surlingham and Brundall, EAE; 31, Woodbastwick, JHS, 1963; Horning, RMB, 1956; Ranworth, Salhouse and S. Walsham; 32, Barton Turf, BFTD, 1962; 41, Thurne.

I. parviflora DC.   Small Balsam

Est. alien. Occasional but increasing. Waste sandy places and woodland. Both the King's Lynn and Norwich records are from timber-yards suggesting introduction from its native country by way of imports of Russian timber. Apparently not recorded before 1914.

West: 62, Sandringham, 1954; King's Lynn, 1949; W. Newton, RSC, 1965; 63, Dersingham, 1951; 94, Wells-next-the-Sea, ALB, 1964, one plant; 99, Shropham, FR, 1916; Gt. Hockham, ALB, 1966; 09, Hargham, FR, 1916.
East: 10, Wymondham, C. Smith, 1956; 12, Aylsham, AEE, 1938; Blickling, RCLH, 1961; 20, Norwich, ETD, 1965; 22, Brampton, EAE, 1938; 23, N. Walsham, PJB; 49, near Beccles, SA, 1962.

I. glandulifera Royle   Policeman's Helmet

Nat. alien. Locally abundant. Riversides and waste places. Well established in Broadland. Not recorded in the 1914 *Flora*.

West: 61, Gayton Road station yard, 1943; 69, Hilgay; 90, Cranworth, ALB, 1963; 99, Stow Bedon, 1963; 00, Reymerston, ALB.

East: 01, Lyng, ETD, 1964; 11, Lenwade and Ringland, ETD, 1964, 14, W. Runton brick-yard, R. Creed, 1916; 20, Norwich; 30, near Coldham Hall, Surlingham, ETD, 1964; 49.
First record: 1916, W. Runton, R. Creed.

ACERACEAE

Acer pseudoplatanus L.   Sycamore
Denizen. Common. Woodland and hedges. Reproduces freely.

A. platanoides L.   Norway Maple
Denizen. Introduced throughout Thetford Forest in Breckland and conspicuous in late autumn by reason of leaf colour. Natural regeneration occurs.

A. campestre L.   Field Maple
Native. Common. Hedges and woods

HIPPOCASTANACEAE

Aesculus hippocastanum L.   Horse-chestnut
Denizen. Common. Roadsides and woodland. Natural regeneration occurs.

AQUIFOLIACEAE

Ilex aquifolium L.   Holly
Native. Common. Woodland and hedges.

CELASTRACEAE

Euonymus europaeus L.   Spindle-tree
Native. Frequent. Hedges, woodland and scrub on chalk; less frequent in East Norfolk.

BUXACEAE

Buxus sempervirens L.   Box
Denizen. Locally abundant. Planted extensively for game cover and luxuriant in chalk soils as in Weeting beech-woods and throughout Breckland. Much less common in East Norfolk.

RHAMNACEAE

Rhamnus catharticus L.   Buckthorn
Native. Frequent. Characteristic of fen carr and widespread also on chalk.

Frangula alnus Mill.   Alder Buckthorn, 'Berry Alder'
Native. Frequent in fen carr, with *Rhamnus catharticus*, and characteristic of carr developing on acid bog.

131

PAPILIONACEAE

Lupinus arboreus Sims   Tree Lupin

Est. alien. Locally abundant on the cliffs and waste places along the north coast. Introduced at Blakeney Point by the bird-watcher before the Second World War. Recently planted for game cover at East Wretham in Breckland.

West: 94, Blakeney Point; 99, E. Wretham, SA.

East: 14, Beeston Hill; W. Runton and Sheringham, KHB; 20, Keswick, ETD; 22, Felmingham, ETD; 23, Sidestrand, ETD; 24, Overstrand, ETD; 33, Mundesley, KHB.

Genista tinctoria L.   Dyer's Greenweed

Native but formerly cultivated as a dye-plant. Rare now. Dry commons and rough pastures.

West: 71, Lamb's Common, Gayton Thorpe.

East: 01, Yaxham, 1937, JBE; 08, Shelfanger, 1952, ALB; 18, Dickleburgh Fen, 1933, PER; 19, Wacton Common, WGC; 1966, still plentiful, ETD; 29, Hardwick airfield, locally abundant, 1963.

Genista anglica L.   Petty Whin, Needle Furze

Native. Frequent. Wet heaths and bogs, often forming a transition zone between the two.

Ulex europaeus L.   Gorse, Furze, Whin

Native. Common. Heaths and rough grassy places.

U. europaeus × *gallii*

Only three stations are known but may be widespread as the two species often grow together and are completely interfertile.

West: 62, Dersingham Common, 1967, JHS.

East: 20, Mousehold Heath, 1967, JHS; 50, Gt. Yarmouth, 1967, JHS confd. P. M. Benoit.

U. gallii Planch.   Dwarf Furze

Native. Locally common. Heaths.

West: 62/63, Dersingham Common; 73, Docking, Gt. Bircham; 83, Syderstone, Coxford; 91, Beetley, DMM; 92, N. Elmham, DMM.

East: 01, Sparham Hole, DMM; 03, Holt Lowes; 04, Salthouse Heath, Kelling Heath; 11, St Faith's Common, Felthorpe; 12, Buxton Heath; 20, Mousehold Heath, ETD; 22, Bryants Heath near Felmingham; 32, E. Ruston.

U. minor Roth   Dwarf Furze

We are much indebted to Mr J. H. Silverwood for pointing out errors of distribution and identification of this species hitherto recorded as rare in the 1914 *Flora*, rare in *West Norfolk Plants Today*, 1962, and for which *The Atlas of the British Isles* showed several stations, particularly in East Norfolk. Both *U. gallii* and *U. minor* can be very variable vegetatively and the flowers, normally the best means of separation, can show characters intermediate both in size and colour although they do not hybridise. Mr

35 MOSSY TILLAEA *Crassula tillaea* *Miss D. M. Maxey*

36 GRASS OF PARNASSUS *Parnassia palustris* *A. H. Hems*

37 SPURGE LAUREL *Daphne laureola* *Miss G. Tuck*

38 BASTARD TOADFLAX *Thesium humifusum* *Miss G. Tuck*

39 SEA HOLLY *Eryngium maritimum* *A. H. Hems*

40 WILD ANGELICA *Angelica sylvestris* *R. Jones*

41 CYPRESS SPURGE   *Euphorbia cyparissias*   *A. C. Jermy*

42 WOOD SPURGE   *Euphorbia amygdaloides*   *Miss D. M. Maxey*

43 SEASIDE CURLED DOCK   *Rumex crispus* var.   *J. C. E. Hubbard*

Silverwood suggests it is these intermediate forms which have caused errors in identification in the past.

As a result of considerable field-work and checking of records he has shown that those published in the past for *U. minor* are in fact all referable to *U. gallii*. On the evidence at present available, the conclusion must be reached that, except for a very small, isolated and almost certainly adventitious colony of *U. minor* found by him at Snetterton in 1964, this species does not grow in Norfolk. This confirms the view of Babington who, in 1877, questioned the existence of this plant in Norfolk. In recent years, *U. minor* and *U. gallii* have been the subject of detailed study, notably by Dr M. C. F. Proctor (1959 *Proc., B.S.B.I.*, **3**, 332 and 1965 *Watsonia*, **6**, 177.), and these investigations have shown that even where colonies of these plants exhibit intermediate or overlapping characters, by examining a number of individual plants and establishing the mean measurements of their flowers, the correct species can be accurately determined.

West: 99, Snetterton, 1964, JHS.

Sarothamnus scoparius (L.) Wimmer ex Koch   Broom

Native. Common. Heaths and waste ground on sandy soil.

Ononis repens L., ssp. *repens*   Restharrow

Native. Common. Dry grassland, banks and sand-dunes.
First record: 1725, Gt. Yarmouth, in Herb. Joseph Andrews in Herb. Mus. Brit.

O. spinosa L.   Restharrow

Native. Rarer than *O. repens*. Grassland on heavy soil.

West: 60, 62, 70–72, 79, 82, 90–92, 94, 99, 08.
East: 00, 03, 08–10, 18, 19, 28, 29, 39, 51.

Medicago falcata L.   Sickle Medick

Native. Fairly frequent on dry roadsides and chalk grassland in the south-west.

West: 61, 70–72, 78–80, 89, 90, 98, 99, 08.
East: 11, Ringland and Swannington, WGC; 14, Sheringham cliff-top, FMD; 24, Cromer, JBE; 31.

M. falcata × sativa
(*M. × varia Martyn*)

Appears sometimes in abundance where the two parents grow together. Flowers variously coloured; yellow, green, blue, purple and black.
First record: 1686, near Norwich, Ray, *Hist. Plant.*

M. sativa L.   Lucerne, Alfalfa

Est. alien. Common. Waste places and field borders.
First record: *c.* 1700, 'gathered wild near Norwich', Jas. Sherard in Du Bois Herbarium.

M. lupulina L.   Black Medick

Native and alien. Common. Grassland and waste places.
A variable plant. In its typical form it is an undoubted native but the var.

willdenowiana Koch with large leaves and glandular-hairy pods grows in fodder crops, escapes and becomes an established alien; var. *eriocarpa* Rouy occurs in places which are predominantly calcareous. For a similar phenomenon, see *Trifolium pratense*.

M. minima (L.) Bartal.   Small Medick
Native. Frequent throughout the Breckland. Sandy fields and heaths. Flowers continuously to late autumn.
West: 62, 70, 78–80, 88, 89, 98, 99.
East: 11, Honingham, ALB.

M. polymorpha L.   Hairy Medick
(*M. hispida* Gaertn., incl. *M. denticulata* Willd.)
Native. Apparently very rare. Sea-banks.
West: 62, N. Wootton sea-bank, in some quantity, CPP, 1966, associated with *Bellis perennis*, *Torilis japonica*, *T. nodosa* and *Agropyron pungens*; 94, Wells rail embankment, F. Long, 1888, in Herb. Norw. Mus., as *M. denticulata* Willd.; 1961, GT.
East: 04, Blakeney, Cley; 28, Harleston area.

M. arabica (L.) Huds.   Spotted Medick, 'Calvary Clover'
Native. Frequent. Peculiarly characteristic of the coastal belt; roadsides and arable land.
West: 61, 62, 64, 74, 78, 83, 84, 91, 92, 94, 04.
East: 02–04, 14, 18, 20, 28–30, 33, 39, 40, 42.

Melilotus altissima Thuill.   Tall Melilot
Status doubtful, probably denizen. Locally frequent. Sandy fields and waste places.
West: 59, 61, 69–72, 80, 91, 92, 98.
East: 11, 14, 19, 20, 29.

M. officinalis (L.) Pall.   Common Melilot
Colonist. Frequent. Fields and waste places.
West: 61–64, 72, 74, 79, 88, 94, 98
East: 00, 29, 33.

M. alba Medic.   White Melilot
Colonist. Frequent on roadsides and in cultivated fields, especially in Breckland where it is occasionally grown as a fodder-crop for sheep.
West: 61, 62, 72, 73, 79, 80, 82, 88, 89, 94.
East: 04, 11.

Trifolium ornithopodioides L.   Birdsfoot Fenugreek
Native. Rare. Dry banks and paths especially near the sea.
West: 62–64, 72–74, 84.
East: 04, 11, 14, 28, 33.

T. micranthum Viv.   Slender Trefoil
Native. Widely distributed but less common than the two following species, and typical of poorer soil.

T. dubium Sibth.   Lesser Yellow Trefoil.
Native. Common. Grassland and roadsides.

T. campestre Schreb.   Hop Trefoil
Native. Common. Grassland and roadsides.

T. hybridum L.   Alsike Clover
Colonist. Regarded as rather rare in 1914, it is now a common plant as a result of its increased use in agriculture.

T. repens L.   White or Dutch Clover
Native. Common. Grassland and a feature of newly made paths and road-sides. Forms with foliaceous calyces = var. *phyllanthum* Seringe are met with occasionally.

T. glomeratum L. Clustered Clover
Native. Rare. Sandy and gravelly places. East Norfolk only.
East: 04, Kelling, CPP; 11, Hellesdon, WGC, 1915; 14, Sheringham, MVD, 1962; 24, Overstrand, 1967, R. Pankhurst; 39, Ditchingham, 1954, MIB; 49, Haddiscoe, PER, 1928.

T. suffocatum L.   Suffocated Clover
Native. Very rare. Sandy and gravelly places near the sea.
East: 04, Cley Channel near Blakeney, EAE, 1963.
First record: 1792, Gt. Yarmouth Denes, L. Wigg, its first British record.

T. fragiferum L.   Strawberry Clover
Native. Frequent in wet pastures, especially near the sea.
West: 51, 61, 62, 64, 71, 72, 74, 78, 80, 82–84, 88, 92, 98, 99.
East: 04, Cley Marshes, FMD; 22, Bryants Heath near Felmingham, BSBI Exc., 1938; 40, 41.

T. medium L.   Zigzag Clover
Native. Has increased considerably and is now locally frequent.

T. arvense L.   Hare's-foot
Native. Common. Sandy arable land, Breckland heaths and maritime sand and shingle.

T. scabrum L.   Rough Trefoil
Native. Occasional. Dry places, especially sea-banks.
West: 62–64, 70–72, 74, 78–80, 89, 98.
East: 03, 04, 11, 14, 24, 40, 50, 51.

T. striatum L.   Soft Trefoil, Knotted Clover
Native. Frequent. Dry banks and pastures
First record: 1807, Gt. Yarmouth Denes, Dawson Turner, *English Botany*, t. 1843.

T. pratense L.   Red Clover
Mixed status. Common. Grassland and roadsides.
Distribution of the small wild plant with bright red flowers is augmented by escapes from cultivation of the var. *sativa* Schreb., a much larger plant with dull pink flowers. Now that considerable reseeding of pastures has

taken place the cultivated form has become more common than the wild plant.

T. ochroleucon Huds.   Sulphur Clover

Native. Rare in West Norfolk but locally abundant on the boulder clay deposits of the south-east.

West: 61, between Blackboro' End and E. Winch, 1965, one plant; 91, roadsides at Beeston and Mileham, DMM; 90, Cranworth, ALB.

East: 02, 08–10, 18, 19, 28, 29, 31, 39.

T. subterraneum L.   Subterranean Clover

Native. Rare. Dry gravelly places and sea banks.

West: 62, N. Wootton Gongs, CPP; 63, Snettisham; 84, Burnham Overy Staithe.

East: 04, Blakeney sea-bank, Cley and Holt; 14, Beeston Bog, PHS; Beeston Hill, GT; 39, Ditchingham, MIB.

Anthyllis vulneraria L.   Ladies' Fingers, Kidney Vetch

Native. Frequent in chalk grassland.

Lotus corniculatus L.   Birdsfoot-trefoil, Bacon and Eggs

Native. Common. Grassland.

First record: 1725, Gt. Yarmouth, in Herb. Joseph Andrews in Herb. Mus. Brit.

Var. *crassifolius* Pers.

East: 04, shingle bank, Cley, PDS.

L. tenuis Waldst., & Kit., ex Willd.   Slender Birdsfoot-trefoil

Native. Very rare. Dry hedgebanks.

West: 63, Snettisham; 70, Oxborough.

East: 50, Gt. Yarmouth.

L. pedunculatus Cav.   Greater Birdsfoot-trefoil

Native. Common. Wet meadows and fens. Ranging in indumentum from pubescent to glabrous.

Robinia pseudoacacia L.   Acacia

Nat. alien. Occasional. Woods. Natural regeneration occurs.

West: 72, Hillington; 78, Weeting; 88, near Thetford.

East: 03, 14, and undoubtedly elsewhere.

Astragalus danicus Retz.   Purple Milk-vetch

Native. Locally frequent in chalk grassland in south-west Norfolk.

West: 60, 70, 71, 74, 78–80, 88, 90, 98, 99.

East: no records.

A. glycyphyllos L.   Milk Vetch

Native. Very rare. Bushy places.

West: 80, Holme Hale, FR, 1914; 98, Harling road, Thetford, HDH, up to 1939.

East: 14, Dead Man's Hill track to Weybourne cliff, PHS, 1950; Weybourne cliff, RPB-O, 1953, in Herb. Norw. Mus., 24, Cromer, HDH, 1916.

Ornithopus perpusillus L.   Birdsfoot
Native. Common. Dry sandy and gravelly places.

Hippocrepis comosa L.   Horse-shoe Vetch
Native. Exclusive to chalk grassland in which it is not infrequent.
West: 64, 70, 71, 72, 74, 79, 94.
East: 14, Beeston Hill.
First record: *c.* 1725, Castle-yard, Norwich, Herb. Joseph Andrews in
Herb. Mus. Brit.

Onobrychis viciifolia Scop.   Sainfoin
Status doubtful but probably a naturalised alien. Well established as a
relic of cultivation. All the erect form.

Vicia hirsuta (L.) S. F. Gray   Hairy Tare
Native. Common. Arable weed and in grassy places.

V. tetrapserma (L.) Schreb.   Smooth Tare, Slender Vetch
Native. As *V. hirsuta* but much rarer.
West: 61–63, 72, 79, 82, 92.
East: 00, 08, 20, 22, 28–33, 39.

V. cracca L.   Tufted Vetch
Native. Common. Hedges and fens.
The well-known botanist, Fred Robinson, found a plant at the top of the
railway cutting on clay soil, Saham Toney (W90), in 1915 'with flowers the
colour of *Wistaria* and nearly as large. It is very distinct in habit and
appearance from the common form. Moreover, it flowers three weeks to a
month before the type'. Dr Thellung placed it between the subspecies
incana and *tenuifolia* (BEC, 1915, 336). A similar plant was sent to us by
Mrs B. C. Hibberd from a hedge at Coltishall in 1965.

V. sepium L.   Bush Vetch
Native. Rather common. Woodland margins and hedges.

Var. *ochroleuca* Bast., with creamy-white flowers.
East: 10, Wymondham, Mrs Pomeroy, 1923.

V. lutea L.   Yellow Vetch
Est. alien. Very rare. Roadsides.
West: 92, Brisley, DMM, 1958–64.
East: 11, 21, colony in the lane leading from Horsford Crown Inn to St
Faiths, A. C. Armes, 1927,

V. sativa L., ssp. *sativa*   Common Vetch
Although this species is strictly a casual, being a relic of cultivation which
does not persist, it is listed here as it is so frequently and erroneously
recorded. It is reported for all squares except two in the Maps Scheme
and is confused with *V. angustifolia* var. *segetalis*. The true plant, in
addition to being larger in all its parts, always has brown pods and is
rare.
West: 61, E. Winch; 70, Beechamwell; 84, Burnham Thorpe.

V. angustifolia L.   Common Vetch

Native. Common. Hedgerows and grassland. Very variable; we discard var. *bobartii* Koch on the grounds that this slender-leaved plant with solitary flowers is a habitat-form only. Var. *segetalis* Koch is common and, although resembling *V. sativa*, always has black pods.

V. lathyroides L.   Spring Vetch

Native. Frequent. Heaths, dry grassy places and sand-dunes.
First record: 1796, 'sent from Norwich by Mr Pitchford', *English Botany*, t. 30.

Lathyrus nissolia L.   Grass Vetchling

Native. Rare. Grassy places and sea-banks. Probably overlooked when not in flower by reason of its grass-like leaves.
West: 62–63, sea-banks from N. Wootton to Snettisham; 84, Burnham Deepdale; 94, Wells-next-the-Sea.
East: 19, Tharston, SA; 29, Hempnall, ELS.

L. pratensis L.   Meadow Vetchling

Native. Common. Grassland.

L. tuberosus L.   Earth-nut Pea

Nat. alien. Rare. Grassland and hedgebanks.
West: 78, large colony in a forest ride, near Grimes Graves, AMS, 1950.
East: 04, High Kelling, 1937; 21, Mousehold Heath, ETD, 1964; 42, Horsey, TBR, 1964, 'well-established in roadside hedge'.

L. sylvestris L.   Narrow-leaved Everlasting Pea

Denizen. Rare. Woodland margins and sand-dunes.
West: 84, dunes, Burnham Overy Staithe; 88, margins of conifer plantation, Santon; Thetford, PAW; 94, Morston, ETD.
East: 31, 33 (Maps Scheme).
Plate number 805 made from a plant sent by Mr W. Humphrey for *English Botany*, from Brundall near Norwich, 1800.

L. heterophyllus L., var. *unijugus* Koch

Two large colonies have been established for many years in damp hollows of sand-dunes at Burnham Overy Staithe. They are introduced plants probably owing their origin to garden-escapes from hut sites some seventy years ago. The variety differs from the normal form in having all leaves of one pair of leaflets instead of two. Material determined by Dr Melderis, Brit. Mus. 1957 *Proc., B.S.B.I.*, **2**, 238–40. Apparently new to the British Isles.
First record: 1949, the above, ELS.

L. latifolius L.   Everlasting Pea

Est. alien. Occurs with the preceding plant. Has large flowers, 15–30 mm. long, of a rich magenta colour.
West: 84, Burnham Overy Staithe, 1956, ELS.
East: 10, Bawburgh, ETD; 23, Trimingham, KHB; 24, Overstrand, KHB; 32, Dilham, ETD.
First record: 1956, ELS.

L. palustris L. Marsh Pea

Native. Much rarer now but locally abundant on the Strumpshaw marshes in 1959. Fens and marshes.

West: 61, Cranberry Fen near Blackboro'; 09, Swangey Fen, GHR.
East: 30, Wheatfen Broad, EAE; Strumpshaw marshes, GHR & ELS; 31, Hoveton, G. C. Druce, 1928; Ranworth marshes, RMB; Upton Broad, BSBI Exc., 1938; Woodbastwick, JHS; 41, Burgh St Margaret, WGC, 1917; Filby, G. C. Druce, 1928; 42, Calthorpe Broad, GG; 49, Aldeby marshes, GG.

L. japonicus Willd. Sea Pea
(*L. maritimus* Bigel.)

Although abundant and a conspicuous feature of the shingle-banks along the adjoining Suffolk coast, this plant does not form part of the indigenous flora of Norfolk. Seeds from the Chesil Bank were introduced at Blakeney in 1912 by F. W. Oliver. Following the 1953 sea-floods, seeds were sown at Cley by E. A. Ellis.

West: 84, Holkham, E. N. Mennell, 1935, 'shingly dell behind the dunes', EAE.
East: 04, Cley, 1962, one small colony only seen; Salthouse, A. R. Norwood, 1921.

ROSACEAE

Filipendula vulgaris Moench Dropwort

Native. Frequent. Exclusive to chalk grassland.

West: 63, 64, 70–74, 79, 80, 82, 83, 88, 89, 91, 93, 94, 98, 99, 08.
East: 08, 19.

F. ulmaria (L.) Maxim. Meadow-sweet

Native. Common. Fens and rough pastures.

Rubus idaeus L. Raspberry

Native. Common. Hedges and woods.

Cultivated on a large scale in the county, the cultivar 'Norfolk Giant' appearing as a chance seedling in a Norfolk garden. Cultural experiments with seed from varying soil types in West Norfolk showed late flower-bud development, segregation for male plants, large fruits suggesting escapes from cultivation and the least die-back. See 'The Raspberry Wild in Britain', Dr Gordon Haskell, 1960, *Watsonia*, **4**, 238–55.

R. caesius L. Dewberry

Native. Common on chalk, rare elsewhere. Hedgerows and scrub.

Rubus fruticosus agg. Blackberry

In 1878, the Rev. E. F. Linton accepted the living of Sprowston near Norwich and, until he removed to Bournemouth ten years later, devoted much of his leisure time to the study of two difficult genera, *Rubus* and *Salix*. Later, in conjunction with the Revs. W. Moyle Rogers, R. P. Murray, and his brother, W. R. Linton, he edited a *Set of British Rubi* (1892–6) which were readily taken up and are to be found in many private and public herbaria. The frequency with which 'Sprowston' and 'Mousehold' appear in the following account testify to his work whilst at Norwich. The account of the *Rubi* in the 1914 *Flora* was largely drawn from the results of his studies. His name is commemorated by *R. lintoni* Focke. His work is so important that we have built upon the foundation he laid, adding additional records from the literature of the B.E.C., B.S.B.I., Norfolk references in W. C. R. Watson's *Handbook of the British Rubi*, notes by E. S. Edees during his visits to the county and our own somewhat limited work.

Although endless diversity of opinion still prevails as to the number of species and much uncertainty hangs over their determination, so far as we are able we have attempted to bring the nomenclature in line with the present-day treatment.

Undated references refer, in the main, to Linton's records.

NCR = New County Record.

Section I   Suberecti

Rubus nessensis W. Hall
West: 62, King's Lynn, 1930, Miss Roper in BEC.
East: 21, Sprowston (as R. *suberectus* Anders.).

R. plicatus Weihe & Nees
West: 09, Hargham, HJR, 1925, 'in great quantity'; 'flowers frequently pink', in BEC.
East: 21, Sprowston, Beeston St Andrew.

R. fissus Lindl.
East: 22, Westwick (as *R. Rogersii* Linton).

R. affinis Weihe & Nees
West: 62, Wolferton.
East: 20, Mousehold; 24, Cromer; 51; Ormesby.

Section II   Triviales

R. conjungens (Bab.) W. Wats.
Frequent in hedgerows and margins of woodland.
West: 70, Foulden Common, 1964, BSBI Exc., KM confd. B. Miles; 80, Swaffham district, ESE, 1944; 84, Burnham Overy Staithe, KM, 1964, BSBI Exc.; 90, Cranworth, ALB, 1963.
East: 51, Caister, Miss C. G. Trower, *c.* 1913, in *British Brambles*, BEC, 1928, with plate.

R. sublustris Ed. Lees
West: 62, Sugar Fen, Leziate, Miss I. M. Roper, 1930, in WBEC; 80, Swaffham district, 'widespread', ESE, 1953; 99, Thompson Common, ELS.
East: 02, Guist, ELS, 1952; 14, Sheringham, CCT det. B. Miles; 20, Mousehold Heath, ESE, 1953; 21, Sprowston, 'the common form about here', EFL.

R. halsteadensis W. Wats.
Recorded for West Norfolk in the *Handbook*.

R. tuberculatus Bab.
West: 90, Wood Rising, ALB, 1965.
East: recorded in the *Handbook*.

R. scabrosus P. J. Muell.
  (*R. dumetorum* var. *ferox* sensu Rogers)
West: 71, E. Walton Common, 1964 BSBI Exc., KM conf. B. Miles.

Section III   Sylvatici

R. gratus Focke
East: 03, Holt Lowes, CCT det. B. Miles, 1959; 20, Mousehold, ('still there, 1953'), ESE; 21, Sprowston.

R. carpinifolius Weihe
West: 09, Hargham Heath, HJR, 1925, in BEC as NCR; 71, E. Walton Common, Kew Exc., 1964.

R. nemoralis P. J. Muell.
(*R. selmeri* Lindeb.)
West: 61, Ashwicken, ELS, 1953; 62, Wolferton, ELS, 1952; 71, E. Walton Common, BSBI Exc., 1964, KM conf. B. Miles; 72, Stone Pit Hills, Grimston, ESE & ELS, 1953; 80, 'rare in the Swaffham district', ESE, 1953.
East: 03, Holt Lowes, CCT conf. B. Miles, 1959; 12, Cawston; 14–24, Runton to Cromer; 20, Mousehold; 21, Sprowston.

R. laciniatus Willd.
A garden-escape.
East: 20, Mousehold Heath, ESE, 1953.

R. lindleianus Ed. Lees
West: 62, Wolferton; N. Wootton, ELS, 1954; 63, Dersingham; 80, Swaffham, ESE, 1953; 91, Beetley and Gressenhall; 92, N. Elmham.
East: 20, Mousehold; 23, Roughton; 28, Redenhall; 41, Burgh St Margaret, 1929, in WBEC.

R. egregius Focke
Recorded for East Norfolk in the *Handbook*.

R. macrophyllus Weihe & Nees
West: 80, Swaffham; 91, Gressenhall; 92, N. Elmham.
East: 12, Cawston to Aylsham; 20, Thorpe; 21, Beeston St Andrew and Sprowston; 24, Cromer.

R. schlechtendalii Weihe
East: 21, Beeston St Andrew; 51, Ormesby (both as var. *macrophyllus*).

R. amplificatus Ed. Lees
East: 21, Sprowston, in WBEC.

R. pyramidalis Kalt.
West: 61, Blackborough End, ELS, 1954.
East: 20, Mousehold and Thorpe; 21, Sprowston (Church Wood), HJR, 1925; Beeston St Andrew.

R. mollissimus Rogers
West: 63, Dersingham, A. B. Jackson, 1902, in BEC as NCR.

R. hirtifolius Muell. & Wirtg.
East: 20, Mousehold Heath, W. Watson, 1931; 33, Happisburgh.

R. salteri Bab.
East: 21, Sprowston.

R. poliodes W. Wats.
East: 11, Ringland Heath; 20, Mousehold Heath, W. Watson, 1956.

R. insularis Aresch.
Recorded for East Norfolk in the *Handbook*.

R. macroacanthus Weihe & Nees
The only British records.
East: 20, Mousehold Heath; 21, Sprowston, W. Watson, 1928.

R. incurvatus Bab.
East: 21, Sprowston (and in 1925, HJR in BEC); 28, Brockdish.

R. iricus Rogers
Record for both vice-counties in the *Handbook*.

R. polyanthemus Lindeb.
Under *R. pulcherrimus* Neum. in Linton's records.
West: 81, Bartholomew Hills near Castleacre, ESE, 1953; 83, East Common, N. Creake, ELS, 1953; 92, N. Elmham.
East: 03, Holt Lowes, CCT det. B. Miles, 1959; 12, Cawston; 20, Thorpe and Mousehold; 21, Sprowston; 24, Cromer; 28, Redenhall; 42, between Hickling and Sea Palling, CCT det. B. Miles, 1959.

R. rhombifolius Weihe
West: 62, Roydon Common, BSBI Exc., 1964, KM conf. B. Miles.

R. cardiophyllus Lef. & Muell.
East: 20, Mousehold Heath, ESE, 1953.

R. imbricatus Hort.
West: 09, Hargham; 99, Merton (both 1925, HJR as NCR in BEC).
East: 20, about Norwich, 1925 (as NCR in BEC, HJR).

Section IV Discolores

R. ulmfolius Schott fil.
Our most widespread and abundant species. Common in both vice-counties.

Section V   Sprengeliana

R. sprengelii Weihe
East: 03, Holt Lowes, CCT det. B. Miles, 1959; 21, Sprowston.

R. lentiginosus Ed. Lees
East: 20, Mousehold Heath.

Section VI   Appendiculati

R. vestitus Weihe
(*R. leucostachys* sensu Rogers)
West: 72, Stone Pit Hills, Grimston, ESE & ELS, 1953; 80, N. Pickenham
Warren, ESE, 1953; 83, East Common, N. Creake, ELS, 1953.
East: 20, Mousehold Heath, ESE, 1953; 22, Westwick, ELS, 1953.

R. adscitus Genev.
East: 21, Sprowston (as *R. hypoleucus* Lefv. & Muell.).

R. leucostachys Smith
West: 91, Beetley.
East: 12, Cawston and Blickling; 20, Thorpe; 21, Sprowston; 24, Cromer;
28, Redenhall; 33, Happisburgh.

R. boreanus Genev.
West: 91, Beetley.
East: 20, Mousehold; 23, N. Walsham; 24, Cromer.

R. criniger (E. F. Linton) Rogers
(*R. gelertii* var. *criniger* E. F. Linton)
West: 71, Pentney, ESE, 1953; 81, Lexham; 87, Newton by Castleacre,
ESE, 1953; 99, Merton, ELS, 1953.
East: 20, Mousehold Heath, ESE, 1953, 'and many other places about
Norwich', HJR in BEC, 1925; 21, Beeston St Andrew and Sprowston.

R. mucronifer Sudre
(*R. mucronatus* Blox.)
East: 22, Westwick.

R. gelertii Frid.
East: 21, Beeston St Andrew and Sprowston.

R. radula Weihe
East: 12, Cawston; 20, Mousehold ('still there, 1953', ESE); 21, Sprowston;
28, Redenhall; 30, 39, Buckenham to Loddon.

R. uncinatiformis Sudre
The only British record.
East: 11, Ringland Heath, W. Watson.

R. discerptus P. J. Muell.
East: 20, Mousehold Heath, ESE, 1953. Linton regarded this, under
R. rusticanus Merc., as common.

R. echinatoides (Rogers) Druce
East: 20, Mousehold.

R. rudis Weihe
East: 20, Mousehold.

R. granulatus Lef. & Muell.
Recorded for East Norfolk in the *Handbook*.

R. foliosus Weihe & Nees
East: 20, Mousehold Heath, HJR in BEC as NCR, 1925.

R. fuscus Weihe
East: 22, Westwick.

R. pallidus Weihe
East: 21, Sprowston.

R. menkei Weihe
East: 21, Sprowston.

R. obcuneatus Lef. & Muell.
Recorded for East Norfolk in the *Handbook*.

R. scaber Weihe
East: 20, Mousehold Heath; 21, Sprowston, W. Watson, 1932.

R. rufescens Lef. & Muell.
West: 99, Stow Bedon, Druce, 1918 (as *R. infecundus* Rogers).

R. apiculatus Weihe
East: 21, Sprowston (as *R. anglosaxonicus* Gelert.).

R. raduloides Rogers
East: 20, Mousehold.

R. phaeocarpus W. Wats.
Recorded for East Norfolk in the *Handbook*.

R. rotundifolius (Bab.) Bloxam
East: 20, Mousehold (as *R. hirtus* Waldst & Kit., B *rotundifolius* Bab.).

R. rosaceus Weihe
East: 21, Sprowston.

R. hystrix Weihe
East: 22, Westwick.

R. infestus Weihe
East: 20, Mousehold Heath, HJR, 1925; 21, Church Wood, Sprowston, HJR, 1925, in BEC as NCR (under *R. colemanii* Bloxam).

R. dasyphyllus (Rogers) Druce
West: 72, Stone Pit Hills near Grimston, ESE & ELS, 1953.
East: 23, Roughton; 24, Cromer.

R. oegocladus Muell. & Lef.
East: 20, Mousehold; 21, Sprowston (as *R. oigoclados*).

44 COMMON SEA LAVENDER *Limonium vulgare* *W. H. Palmer*

45 MATTED SEA LAVENDER *Limonium bellidifolium* *Miss G. Tuck*

46 ROCK SEA LAVENDER *Limonium binervosum* *Dr Lodge*

47 THRIFT *Armeria maritima* A. H. Hems

48 COWSLIP *Primula veris* *Jarrolds*

49 OXLIP *Primula elatior* R. *Jones*

50 WATER VIOLET   *Hottonia palustris*   *R. Jones*

51 YELLOW LOOSESTRIFE   *Lysimachia vulgaris*   *R. Jones*

52 BOG PIMPERNEL   *Anagallis tenella*   *A. H. Hems*

53 MARSH GENTIAN   *Gentiana pneumonanthe*   *P. Banham*

54 BOGBEAN   *Menyanthes trifoliata*   *A. H. Hems*

55 EASTERN COMFREY   *Symphytum orientale*   *Miss G. Tuck*

R. bellardii Weihe
East: 21, Sprowston.

R. lintonii Focke ex Bab.

East: 20, 'woods, lanes and thickets on a heath, near Norwich', HJR in BEC, 1925; 21, Sprowston (not v-c 28 as in *Handbook*) 'still at Sprowston between church and inn', ESE, 1946.

Potentilla palustris (L.) Scop.   Marsh Cinquefoil
Native. Frequent. Bogs and wet meadows with acid soil water. Not typical of fen.

P. sterilis (L.) Garcke   Barren Strawberry
Native. Common. Hedges and woods.

P. anserina L.   Silverweed
Native. Common. Waste places, paths, roadsides and damp meadows.

P. argentea L.   Hoary Cinquefoil
Native. Frequent. Sandy heaths, especially on the Greensand of West Norfolk.

P. recta L.   Sulphur Cinquefoil
Est. alien. Rare. Heathy ground and waste places.
All our plants belong to the var. *sulfurea* DC., with pale primrose-coloured flowers.
West: 61, Bawsey; 71, Westacre; 79, Feltwell, J. A. Ward det. EAE; 83, N. Creake, BT; Dunton, GT; 93, Sculthorpe; 98, Bridgham, HDH, 1942.
East: 08, Quidenham, JHS; 11, Taverham, ETD; 14, Weybourne; 20, Norwich, ETD; 24, Cromer, J. L. Fielding; 28, Denton, JHS.
First record: 1942, Bridgham, HDH.

P. tabernaemontani Aschers.   Spring Cinquefoil
(*P. verna* L., p.p.)
Native. Very rare. Chalk grassland.
West: 99, Garboldisham, '70/80 plants in a grass-grown chalk pit', ALB, 1952; not seen in 1957.

P. erecta (L.) Rausch.   Common Tormentil
Native. Common. Heaths.

P. erecta × *reptans* (*P.* × *italica* Lehm.)
West: 62, S. Wootton Heath, CPP det. D. E. Allen; 72, Roydon Common; formerly regarded as *P. anglica* Laichard.

P. anglica Laicharding   Trailing Tormentil
Native. Rare. Heaths.
East: 00, Reymerston, ALB; 01, Bawdeswell Heath, PH; 04, Kelling Warren, ELS.

P. reptans L.   Creeping Tormentil
Native. Common. Roadsides and waste places.

145

Fragaria vesca L.   Wild Strawberry

Native. Common. Woods and scrub on chalky soil and along railway embankments.

Var. *berscheriensis* (Druce) Druce, larger in all its parts with leaves whitish below and darker above.

West: 99, Thompson, Druce and FR, 1915.

F. × ananassa Duchesne   Garden Strawberry

Est. alien. Frequent as a garden-escape by roadsides and in gravel pits.

West: 61, Blackborough End; 79, Whittington, GT; 83, Stanhoe, GT; 84, Burnham Market; 91, Beetley, DMM; oo, Reymerston, ALB.

East: oo, o3, 12, 20, 21, 30, 32.

Geum urbanum L.   Herb Bennet, Wood Avens

Native. Common. Woods.

G. rivale L.   Water Avens

Native. Frequent. Wet meadows, fens and damp woods.

G. rivale × urbanum (*G. × intermedium* Ehrh.)

Occurs in woodland where both parents are found. Is very variable in both form and colour of flowers.

West: 60, Brickyard Wood, Wallington, J. E. Little; 70, Shingham; 71, E. Walton; Gayton; 72, Congham; Hillington; 80, Saham Toney, ETD; 90, Cranworth, ALB; 91, Rush Meadow, Dereham, PH; Honeypot Wood, Wendling, DMM; 99, Wayland Wood near Watton; Hockham.

East: oo, Kimberley; o2, Reepham, DMM.

Agrimonia eupatoria L.   Common Agrimony

Native. Common. Hedges and woods.

A. odorata (Gouan) Mill.   Fragrant Agrimony

Native. In similar situations to the last, but less common.

West: 61, 62, 84, 89, 90–92, 98, 99.

East: 11, 12, 30, 31, 39.

Alchemilla vulgaris agg.   Lady's Mantle

Under the aggregate name, some seven records appear in the 1914 *Flora* together with 'c. *filicaulis* (Buser)'; from the known present-day distribution the latter is considered to be an error.

    A. vestita (Buser) Raunk.

    Native, Very rare. Damp grassland.

    West: 91, Beetley Common, Miss Goddard, 1901, in Herb. Norw. Mus.

    East: o1, woodland path, Hockering Wood, ALB and ELS, 1966, two plants.

    A. glabra Neygenfind

    Native. Very rare.

    West: o3, Swanton Novers, CPP det. S. M. Walters, 1956, who suggested it was possibly a recent introduction.

Aphanes arvensis L.   Parsley Piert
Native. Common. Sandy arable land, on both neutral and acid soils.

A. microcarpa (Boiss. & Reut.) Rothm.
Native. Common. Heaths.

Sanguisorba officinalis L.   Great Burnet, Bloodwort
Native. Very rare now. Damp grassland and fens.
West: 07, Blo' Norton Fen.
East: 14, Beeston Regis.

Poterium sanguisorba L.   Salad Burnet
Native. Exclusive to chalk grassland in which it is abundant.
West: 64, 70–74, 79, 80, 84, 88, 90, 94, 98, 99.
East: 00, 02–04, 08, 09, 18–20, 29, 30, 40.

P. polygamum Waldst. & Kit.
Est. alien. An escape from cultivation which persists.
West: 61, Gayton; 71, Leziate; 73, Choseley; 78, Hockwold.
East: 10, Hethersett; 49, Haddiscoe, RMB.

Acaena anserinifolia (J. R. & G. Forst.) Druce
Est. alien. Locally frequent. Heaths of north-east Norfolk. Although this native of Australia and New Zealand is frequently recorded as an alien owing its presence in Britain to introductions by way of wool imports, the extensive colonies on Kelling Heath appear to have been the result of dumping garden rubbish, including *Solidago* and *Aster* spp., from Sheringham. It has been known there for forty-five years.
East: 04, Kelling Heath and Warren; 13, Selbrigg Pond; Bodham; 14, Sheringham; 24, sand-pits near Newman's Hill, Cromer, J. L. Fielding.

ROSA

Whilst holding the living of Sprowston, near Norwich, from 1885 to 1888, the Rev. E. F. Linton included this difficult genus in his critical studies of British plants, and the pages of the 1914 *Flora* are enriched with many of his local records. Some 50 years elapsed before much further progress was made when collecting was carried out in the Thetford area by E. B. Bishop and his sister, Mrs C. L. Wilde and their records appeared in 'Rosa Notes for 1934' (*Bot. Exc. Club Reports*, 1932–4). All the records of these workers have been incorporated in the following account, allowance being made for name changes and doubtful finds, together with the results of recent field-work. Many thanks are due to Dr R. Melville of Kew for his generous help in naming or confirming finds. Considerable use has been made of A. H. Wolley-Dod's 'A Revision of the British Roses' (*Journal of Botany*, 1930–1, *Supplts.*).

Rosa arvensis Huds.   Field Rose
Native. Frequent in hedges on the heavier soils.

R. pimpinellifolia L. (*R. spinosissima* L., p.p.)   Burnet Rose

Native. Frequent locally. Although recorded for Fakenham, W92, Coxford, W83, in the 1914 *Flora* and in 1942 by H. Dixon Hewitt for Rushford, W98, it has not been seen in West Norfolk and appears to be restricted to the north-east coastal region.

East: 04, Salthouse Heath; Kelling Heath; 14, Weybourne Heath; 33, Mundesley, JHC.

Var. *ripartii* Déségl.

East: 21, Sprowston, EFL.

R. × involuta (Sm.) W.-Dod. (*R. pimpinellifolia* × subsect. *Villosae*)
East: 21, Sprowston, EFL.

R. rugosa Thunb.   Ramanas Rose

Est. alien. Considerable numbers have been planted in Breckland mixed woodland for game-cover. They have rapidly become established, forming, at times, impenetrable thickets. Does not appear in the 1914 *Flora*.

West: 78, Weeting, ELS; 98, Kilverstone, ELS; 99, mixed scrub at Snetterton Heath, JHS; Thompson Hall Plantation. ELS.

East: 23, cliffs at Sidestrand, KHB; 33, walls of churchyard and on cliffs at Mundesley, KHB.

R. stylosa Desv. (*R. systyla* Bast.)

This 'columnar-styled Dog Rose' was recorded for both vice-counties in the last *Flora* and described as rather common in hedges. In spite of much searching this species has not been seen and its distribution cannot be confirmed; *The Flora of the British Isles* queries its occurrence in Norfolk.

Var. *congesta* (Rip.) R. Kell. 'The first definitely accepted record in Britain was made from specimens collected by my sister, Mrs C. L. Wilde, near Thetford, West Norfolk, 9th September 1934' (EB in BEC, 1937, 448).

R. canina L.   Dog Rose

The most variable species. Widespread and abundant in hedges, scrub and woodland.

Var. *lutetiana* (Lem.) Baker

A common variety.

West: 80 Swaffham, EFL; 81, Castleacre, EFL; 91, Beetley, EFL; 92, N. Elmham, EFL; 98, E. Harling, EFL; 99, Wretham and Rockland All Saints, ELS.

East: 19, Flordon; 20, Thorpe; 21, Sprowston; 28 Harleston (*Fl. Pl. H.*); 39, Ellingham; 41, Ormesby – all but one by EFL.

Forma *lasiostylis* Borb.

West: 99, Rockland All Saints, ELS.

Var. *sphaerica* (Gren.) Dum.

West: 81, Castleacre; 91, Dereham; Beetley to Gressenhall; 92, N. Elmham.

East: 12, Cawston; 21, Sprowston; 39, Ellingham to Geldeston – all by EFL.

Var. *senticosa* (Ach.) Baker

West: 99, Rockland All Saints, ELS.

East: 39, Loddon, EFL.

    Forma *oxyphylla* (Rip.) W.-Dod

    West: 88, near Thetford, EB.

Var. *spuria* (Pug.) W.-Dod

West: 88, Croxton, EB; 99, E. Wretham and Rockland All Saints, ELS.

Var. *globularis* (Franch.) Dum.

West: 88, Thetford, EB.

Var. *dumalis* (Bechst.) Dum.

West: 80, Swaffham, EFL; 81, Newton, EFL; 88, Croxton, EB; 91, Beetley to Gressenhall, EFL; 98, Larling, EFL; 99, Rockland All Saints, ELS.

East: 21, Sprowston, Hainford and Frettenham, EFL; 28, Harleston (*Fl. Pl. H.*); 39, Loddon, EFL.

Var. *stenocarpa* (Déségl.) Rouy

West: 78, Weeting, EB.

Var. *fraxinoides* H. Br., forma *recognita* Rouy

West: 99, Wretham, EB.

Var. *surculosa* (Woods) Hook.

East: 18, Tivetshall, AB; 21, Frettenham, EFL; 39, Loddon, EFL.

Var. *verticillacantha* (Mér.) Baker

West: 81, Lexham, EFL; 88, near Thetford, EB.

East: 21, Sprowston, Beeston St Andrew and Frettenham, EFL.

    Forma *clivicola* Rouy

    West: 88, Croxton, EB.

R. dumetorum Thuill.

Native. Included under *R. canina* agg., in *Flora Brit. Is.*, but is kept separate here to conform with Wolley-Dod's *Revision*.

Var. *typica*

West: 81, Castleacre, EFL.

East: 21, Sprowston, EFL; 23, Trunch, EFL.

    Forma *urbica* (Lem.) W.-Dod

    West: 88, Thetford and Croxton, EB; 99, Wretham, EB. Linton retained varietal status for the following records: West: 60, Denver (1837, in Herb. Salmon); 91, Scarning and Beetley; 98, Larling.

    East: 19, Flordon; 20, Thorpe and Postwick; 21, Sprowston, Hainford and Old Catton; 28, Harleston (*Fl. Pl. H.*); 39, Loddon and Ellingham.

Var. *ramealis* (Pug.) W.-Dod
West: 88, near Thetford, EB.

Var. *gabrealis* (F. Gér.) R. Kell.
West: 78, Weeting, EB; 88, near Thetford, EB.

Var. *sphaerocarpa* (Pug.) W.-Dod
West: 98, Kilverstone, EB.

Var. *hemitricha* (Rip.) W.-Dod
West: 98, Kilverstone, EB.

R. dumalis Bechst.
West: 99, Thompson Common, ALB.
East: 21, Sprowston, EFL, as *R. coriifolia* Fr.

Var. *watsonii* (Baker) W.-Dod
East: 10, 20, between Swardeston and Swainsthorpe, AB.

R. obtusifolia Desv.
Native. Apparently rare but probably overlooked. Hedges.
West: 71, E. Walton Common, JEL; 73, Ringstead Downs, CPP; 80, Swaffham, EFL; 98, Larling, EFL.
East: 11, Horsford, EFL; 21, Sprowston, EFL.

Var. *tomentella* (Lem.) Baker
West: 91, Scarning and Beetley, EFL.
East: 21, Frettenham; 39, Loddon, EFL.

Var. *borreri* (Woods) W.-Dod
East: 21, Sprowston; 41, Ormesby, EFL, as the species.

R. tomentosa Sm.   Downy-leaved Rose
Native. Hedges, particularly on the chalk. Although given as rather rare in the 1914 *Flora*, the following recent records suggest it is not uncommon in both vice-counties.
West: 70, Beechamwell, ELS; 78, Weeting, ELS; 88, Fowlmere, WGC; 90, Cranworth, ALB; 91, Rawhall, CPP; 99, Breckles, ALB.
East: 10, Hethel, WGC; 28, Pulham St Mary, ELS; 29, Tasburgh, SA; 32, Crostwight-Ridlington district, JHS.

Var. *pseudo-cuspidata* (Crép.) Rouy
West: 99, Rockland All Saints, ELS.

Var. *scabriuscula* Sm.
West: 91, Dereham, EFL.
East: 21, Rackheath, EFL; 39, Ellingham, EFL.

R. sherardii Davies (*R. omissa* Déségl.)
Native. Apparently very uncommon and not recorded in the 1914 *Flora*. Hedges on chalky soil.
West: 73, Sedgeford, 1962, ELS–determined by Dr Melville as the forma *resinosoides* (Crép.) W.-Dod; 88, near Thetford, 1934, EB, under var. *omissa* (Déségl.); 90, Bradenham, ALB.

R. villosa L. (*R. mollis* Sm.; *R. molissima* Willd.)
Native, Apparently rare. Hedges.
West: 62, lane near Roydon Hall, ELS.
East: Recorded for the Maps Scheme in squares 03, 04 and 09 but not seen by the writer; also given for East Norfolk by AB.

R. rubiginosa L.   Sweet Briar
Native. Common, equally on heath, chalk grassland, in chalk pits and hedges on calcareous soil in both vice-counties.

R. micrantha Borrer ex Sm.   Small Sweet Briar
Native. Rare or perhaps overlooked. Chalk scrub.
West: 70, Beechamwell, CPP; 73, Ringstead, CPP; 99, Thompson Common, ALB.
East: 11, Ringland Hills, WGC; 21, Rackheath, Sprowston and Frettenham, EFL; 39, Loddon, EFL; 41, Between Ormesby and Filby, AB.

Prunus spinosa L.   Blackthorn, Sloe
Native. Common. Hedges and scrub.

Var. *macrocarpa* Wallr., flowers appearing with the leaves and large fruits, to 16 mm. diam.
West: 92, Horningtoft, BSBI Exc., 1954.

P. domestica L., ssp. *domestica*   Wild Plum
Denizen. Frequent in hedges usually on or near the sites of old cottage gardens. Natural regeneration occurs.

ssp. *insititia* (L.) C. K. Schneid.   Bullace
Native. Frequent in hedges and similar to the last but occasional bushes are found remote from dwellings.

P. cerasifera Ehrh.   Cherry Plum
Est. alien. Common. Often used for hedging and shelter belts near orchards but isolated fruiting trees occur.

P. avium (L.) L.   Gean, Wild Cherry
Native. Frequent. Woods and hedges on the heavier soils in mid- and East Norfolk.
West: 61, 62, 70, 72, 83, 84, 89–94, 99, 01
East: 02–04, 09, 13, 18–20, 29–32, 39.

P. cerasus L.   Sour Cherry
Nat. alien. Occasional. More frequent than the last in mid-Norfolk.
West: 73, 81–83, 88, 90–93, 99, 03.
East: 03, abundant in a hedge near Field Dalling, ELS, 1963. The 1914 *Flora* lists many more stations.

P. padus L.   Bird Cherry
Native. Widely distributed and reproducing freely in damp woods, particularly characteristic of fen carr.

P. laurocerasus L.   Cherry Laurel

Est. alien. Planted occasionally in woods where occasionally natural regeneration occurs.

West: 99, Merton; Thompson Hall Plantation. 08, Snetterton Heath, JHS.

Cotoneaster microphyllus Wall. ex Lindl.   Rockspray

Est. alien. Large colonies established for many years on the sandhills at Burnham Overy Staiths.

West: 84, Burnham Overy Staithe.
East: 20, 23, as garden-escapes.

Crataegus oxyacanthoides Thuill.   Midland Hawthorn

Native. Rare. In hedges as isolated specimens.

West: 80, Swaffham, ESE; 88, Croxton; 90, Watton, Druce, 1918; Cranworth, ALB; 91, Broom Hill near E. Bilney, DMM; E. Dereham.
East: 02, Foxley Wood; 03, 09, 20, Arminghall, JHS; 31, 32.

C. monogyna Jacq.   Hawthorn, May

Native. Common and very variable. Hedges and scrub.

C. monogyna × *oxyacanthoides* is found occasionally on the margins of woods and, as *C. oxyacanthoides* is generally absent, points to the possibility of the last having been formerly planted in the wood itself and dying out.

West: 62, Reffley; 07, S. Lopham, GHR.
East: 20, Mousehold Heath, JHS.

Sorbus aucuparia L.   Mountain Ash

Native. Common in woods on sandy soil.

S. aria (L.) Crantz   White Beam

Nat. alien. Rare. Woods but usually planted.

West: 78, Weeting, HDH.
East: 00, Hardingham, ALB; 03, Selbrigg Pond, PHS.

S. torminalis (L.) Crantz   Wild Service Tree

Native. Very rare. Woods on boulder clay.

West: 91, Rawhall, CPP; 03, Swanton Novers Great Wood, FRo.
East: 02 Foxley Wood, ALB & ELS.

Pyrus communis L.   Pear

Denizen. Rare. Hedges usually near old cottages. Rarely fruiting.

West: 62, Congham; 90, Cranworth, 'one tree in a remote hedgerow with fruits 6–8 cms.', ALB; 98, Larling; 99, Wretham.
East: 00, 04, 20.

Malus sylvestris Mill., ssp. *sylvestris*   Crab Apple

Native. Common. Hedgerows and woods.

The many puzzling forms make it difficult to classify our crab apples. This subspecies is the wild native and appears to be less frequent than the next. A count of trees along the old coaching road at Great Hockham showed 80 per cent to be the pubescent ssp. *mitis* and 20 per cent the ssp. *sylvestris* but here, as elsewhere, intermediates occur. Crab apples are conspicuous features of hedgerows along the Peddars Way.

Ssp. *mitis* (Wallr.) Mansf.
Native. More common than the preceding. In similar habitats.

CRASSULACEAE

Sedum telephium L.　　Orpine, Livelong, 'Orphan John'
Native. Frequent. All our plants are ssp. *purpurascens* (Koch) Syme with carpels grooved on the back. 'The country people in Norfolk hang it up in their cottages, judging by its vigour of the health of some absent friend' (*English Botany*).
West: 60, 61, 71, 73, 83, 91–93, 99.
East: 00–03, 13, 22, 24, 31, 33, 39.

S. anglicum Huds.　　English Stonecrop
Native. Rare. Sandy places, usually near the sea.
East only: 04, 13, 14, 20, 41, 51.
First record: 1793, Gt. Yarmouth Denes, Dawson Turner in *English Botany*, t. 171.

S. album L.　　White Stonecrop
Nat. alien. Uncommon. Old walls and hedgebanks in the north-east.
West: 64, Old Hunstanton dunes; 83, Stanhoe, GT.
East: 10, Bawburgh, ETD; 13; 14, Sheringham and Beeston; 23; 24, Cromer, JBE; 32; 33, Mundesley, JC; Happisburgh, ACD.
Ssp. *micranthum* (DC) Syme, much smaller than the typical plant, with flattened leaves and more acute petals.
East: 14, Beeston Regis, ELS, 1961.

S. acre L.　　Wall-pepper, loc, 'Creeping Sarah'
Native. Common. Dry banks, heaths, walls, maritime shingle and railway tracks.

S. forsteranum Sm.　　Rock Stonecrop
Nat. alien. Locally common in hedgerows in East Norfolk. Has been established, as *S. rupestre*, since 1838.
East: 03, Saxlingham, CPP; 21, Coltishall, WGC; 22, Felmingham, FR; 23, Southrepps, WGC; 39, Ditchingham.
On the Southrepps plants, sent to the Wats. BEC in 1901, as *S. rupestre* L., a. *majus* Syme, H. D. Geldart noted: 'This has taken possession of a tract of land some miles square in north-east Norfolk, but it very rarely flowers, and when it does so, is much injured by insects, which bite through the flower stem just before the flowers expand. It is not of recent introduction, for I have a specimen dated 1838.'
Var. *minus* (Syme), stems more slender, branches of the cyme not exceeding 2·5 cm.
East: 31, between S. Walsham and Upton, Salmon and White, 1915, 'extraordinarily abundant and a feature of the vegetation by the roadside'.

S. reflexum L.

Est. alien. Increasing. Old walls and banks; garden-escape.

West: 90, Cranworth, ALB; 93, Fakenham, CPP; 99, Rockland St Peter, FR; Swangey Farm road, HDH.

East: 03, 09, 10, 12, 14, 21, 29, 30, 39, 41, 49.

Var. *albescens* (Haw), with lower leaves glaucous and pale yellow flowers.

East: 01, Sparham; 11, Costessey; 22, Felmingham – all by WGC who noted 'all the Norfolk plants belong to this variety'.

Sempervivum tectorum L.   House-leek

Est. alien. Occasional. Old roofs and walls.

West: 61, W. Winch; 71, Pentney; and undoubtedly elsewhere.

East: 08, 18, 24, 30, 32, 33.

Crassula tillaea L.-Garland   Mossy Tillaea
(*Tillaea muscosa* L.)

Native. Common on heaths on the Greensand and in Breckland; scarcely a track is without it.

West: 60–63, 70–73, 78, 79, 82, 83, 88, 89, 92, 98, 99, 08.

East: 11, 14, 40, 49.

First British record: 1766, Rev. H. Bryant, from Mousehold Heath, Norwich.

SAXIFRAGACEAE

Saxifraga tridactylites L.   Rue-leaved Saxifrage

Native. Common. Dry banks, walls and sandy places.

S. granulata L.   Meadow Saxifrage

Native. Common. Dry basic and neutral grassland.

Chrysosplenium oppositifolium L.   Opposite-leaved Golden Saxifrage

Native. Frequent. Wet woods.

West: 62, 71, 72, 82, 88 (St Helen's Well, Santon, introduced), 90, 91, 93, 01.

East: 00, 03, 04, 11, 13, 14, 23, 29, 32, 39.

C. alternifolium L.   Alternate-leaved Golden Saxifrage

Native. Very rare. In similar places to the last species.

East only: 00, Hardingham, EQB, 1958; 12, Booton Common, FRo, 1960.

PARNASSIACEAE

Parnassia palustris L.   Grass of Parnassus

Native. Frequent in the transition zone between bog and fen.

West: 62, 70–72, 79, 91, 93, 07, 09.

East: 00, 01, 03, 11, 12, 14, 19, 22, 23, 29–32, 40, 42.

GROSSULARIACEAE

Ribes sylvestre (Lam.) Mert. & Koch   Red Currant
(*R. vulgare* Lam.)

Native with us. Frequent. Woodland, scrub, particularly damp woods and fen carr.

The red currant has been cultivated from the early fifteenth century and for a long time remained the wild *R. vulgare* Lam. Later it was subjected to hybridisation and E. A. Bunyard in 'The History and Development of the Red Currant' (*Journ. R.H.S.*, September, 1917) showed that it owed its origin to the inter-hybridisation of three species, *R. vulgare* Lam., *R. rubrum* L. and *R. petraeum* Wulf. Although cultivated forms are seen in hedges as escapes, the wild form, together with the black currant, is a constant feature of carr. As the blackbird is such a voracious feeder on currant bushes, it is strange that the cultivated form is not more frequently seen where these birds live, as in hedgerows, thickets and bushy commons.

R. nigrum L.   Black Currant

Native. Less common than the red currant but in wetter situations.

West: 62, 70–72, 82, 89, 90, 91, 93, 99.

East: 01, 09, 11, 12, 19, 22, 23, 30–32, 41, 42.

R. uva-crispa L.   Gooseberry

Native. Frequent. Woods and hedges.

The small-berried form with glandular hairs (var. *glanduloso-setosum* Koch) appears to be wild but is less common than the glabrous form. Garden-escapes are frequent.

DROSERACEAE

Drosera rotundifolia L.   Sundew

Native. Exclusive to bogs and wet heaths, in which it is constant and locally abundant.

D. anglica Huds.   Great Sundew

Native. Exclusive to bogs, but much rarer than *D. rotundifolia*.

West: 62, 71, 72, 79, 91, 99, 07, 09.

East: 03, 11, 12, 14 (Beeston Bog, frequent and the only species present), 18, 19, 22, 32.

D. intermedia Hayne   Long-leaved Sundew

Native. Exclusive to bogs, intermediate in frequency between the two previous species.

LYTHRACEAE

Lythrum salicaria L.   Purple Loosestrife

Native. Common. Streamsides and fens.

Peplis portula L.   Water Purslane

Native. Common. Margins of ponds especially on bare ground.

THYMELAEACEAE

Daphne mezereum L.   Mezereon, loc. 'Cornelian', Garland Flower
Native. Very rare now. Wet woods on basic soil.
The evidence of old county records, and the occupation of habitats
similar to its European ones, suggest that the species is native here.
West: 91, E. Bilney, DMM, 1962.
East: 03, Edgefield Heath; 10, Hethel, WGC; 22, Bryants Heath near
Felmingham; 30.

Daphne laureola L.   Spurge Laurel
Native. Occasional. Woods on the chalk and boulder clay; absent from
other soils. Plentiful near Sedgeford, W63, where it was recorded by
Moxon in 1842.
West: 60, 63, 74, 80, 82, 84, 94, 98, 99.
East: 00, 01, 03, 04, 10, 12, 14, 18–20, 22, 28–30, 39, 41.

ELAEAGNACEAE

Hippophaë rhamnoides L.   Sea Buckthorn
Native. Locally abundant. Fixed dunes, shingle and cliffs. Rarely inland,
in chalk pits.
W. J. Bean states in *Trees and Shrubs hardy in the British Isles* that 'birds,
no matter how pressed for food, will not touch them'; his observations were
probably confined to the gardens at Kew where more palatable food would
be available. The berries of our plants are greedily eaten by blackbirds;
R. P. Bagnall-Oakley confirms this and adds fieldfares and redwings, but
considers it unusual, adding that waxwings, 'probably the most exclusively berry-eating of the winter migrants do not touch them'. If it were
not for bird-carriage, it would be difficult to account for the spread of this
shrub along the whole of the Norfolk littoral. The east to west movement
of some of the autumn migrants coincides with the dispersal and spread of
the plant.

West: 63, 64, 72, 74, 84, 94.
East: 04, 14, 23, 24, 33, 41, 51.
First record: 1797, Gt. Yarmouth, Dawson Turner.

ONAGRACEAE

Epilobium hirsutum L.   Great Hairy Willow-herb, Codlins and Cream
Native. Common. Streamsides, ponds and fens.
White-flowered variants have been seen at Derby Fen, W72 and in East
Norfolk at Whitlingham, E30, and Halvergate, E40, ETD.

E. hirsutum × parviflorum
West: 09, Hargham Heath, AEE, 1930.

E. parviflorum Schreb.   Small-flowered Willow-herb
Native. Common. Streamsides, marshes and fens.

56 WOOD FORGET-ME-NOT *Myosotis sylvatica Miss D. M. Maxey*

57 VIPER'S BUGLOSS *Echium vulgare A. H. Hems*

58 BINDWEED *Convolvulus arvensis A. H. Hems*

59 SEA BINDWEED *Calystegia soldanella* *A. H. Hems*

60 COMMON DODDER *Cuscuta epithymum* *A. H. Hems*

61 (no English name) *Solanum triflorum* *Miss D. M. Maxey*

62 HOARY MULLEIN *Verbascum pulverulentum* *D. B. Osbourne*

63 BLACK MULLEIN *V. nigrum* *Jarrolds*

64 YELLOW FIGWORT *Scrophularia vernalis* *Miss G. Tuck*

E. montanum L.   Broad-leaved Willow-herb
Native. Common. Woods, wet pastures and a weed in gardens.

E. montanum × *parviflorum*
West: 93, Thursford Common, 1954, conf. G. M. Ash; 60, Runcton Holme.

E. lanceolatum Seb. & Mauri   Spear-leaved Willow-herb
Yet to be recorded in Norfolk although found within a quarter of a mile of the county boundary with Suffolk in 1961, so that, in view of its spread elsewhere, it should soon appear in our county.

E. roseum Schreb.   Small-flowered Smooth Willow-herb
Native. Very rare. Damp places. Our few records are from gardens.
West: 90, Carbrooke, WCFN, 1914; 80, Ashill, FR, 1916.
East: 20, Norwich, TCS, 1959; 39, Ditchingham, MIB, 1954.

E. adenocaulon Hausskn.
Nat. alien. Common, probably now our most abundant species. Woods, gravel pits and cultivated ground.
Unnoticed in the county until 1952 (see 'A New Plant for Norfolk', Swann in *Trans. Nfk. and Norw. Nat. Soc.*, 1952, XVII, **4**, 298).
First record: 1952, Horningtoft, ELS.

E. adenocaulon × *montanum*
The two parents frequently grow together and hybridise.
West: 82, Eastfield Wood near Tittleshall; 92, Horningtoft; 99, Stow Bedon, ALB; oo, Reymerston, ALB.

E. adenocaulon × *obscurum*
West: 61, Tottenhill; 71, E. Walton; 99, Cranberry Rough, Hockham.

E. adenocaulon × *parviflorum*
West: 60, Runcton Holme, conf. G. M. Ash; Crimplesham.
East: 30, Bergh Apton, EAE det. ELS, conf. G. M. Ash.

E. adnatum Griseb.   Square-stemmed Willow-herb
(*E. tetragonum* L.)
Native. Frequent. Damp woods, streamsides and cultivated ground.
West: 60, 61, 63, 72, 73, 78, 79, 82, 90–92, 99, oo.
East: oo, 02–04, 12–14, 18–20, 29, 31, 39.

E. adnatum × *obscurum*
West: 60, Runcton Holme, conf. G. M. Ash.
East: 24, Overstrand, 1928.

E. lamyi F. W. Schultz
Native. Rare or confused with preceding. Margins of woods and dune-slacks.
West: 64, Old Hunstanton, J. P. M. Brenan, 1946; 72, Appleton; 99, Cranberry Fen near Hockham.
East: 21, Frettenham.

E. lamyi × *parviflorum*
West: 64, Old Hunstanton, J. P. M. Brenan, 1946.

E. obscurum Schreb.
Native. Common. Wet places and moist woods.

E. palustre L.   Marsh Willow-herb
Native. Frequent. Characteristic of acid marsh, intermediate between bog and fen, but also in old horse-ponds.

E. palustre × *parviflorum*
East: 42, Hickling, Miss E. S. Todd, 1936.

Chamaenerion angustifolium (L.) Scop. Rosebay Willow-herb, Fireweed.
Native. Common. Clearings in woods and burnt heathland.

Oenothera biennis L.   Evening Primrose
Est. alien. Frequent. Roadsides and waste sandy places in which it spreads freely.
West: 61, 71, 73, 78, 80, 88, 92, 99.
East: 11, 14, 20, 23, 24, 30, 31, 39, 40.

O. erythrosepala Borbás
Est. alien. Occasional. In similar places to *O. biennis*.
West: 61, 64, 78, 80, 88, 89, 92, 99.
East: 04, Wiveton, FMD; 11, Drayton, 'thousands in sandy heath', ETD, 1964; 14, Weybourne, FDSR; 20, Mousehold, ETD; 21, 49.

Circaea lutetiana L.   Common Enchanter's Nightshade
Native. Common. Woods and garden weed on moist soil.

HALORAGACEAE

Myriophyllum verticillatum L.   Whorled Water-milfoil
Native. Far less frequent than the next species. Ditches.
West: 62, Leziate Fen; 69, Wissington; 72, Derby Fen; 93, Thursford, GT.
East: 30; 31, Upton Broad, BSBI Exc., 1938; 32, Barton Turf, PC; Barton Broad, EAE; 39; 42, Heigham Sound, FR, 1914; Calthorpe Broad Nature Reserve, ACJ; 49.

M. spicatum L.   Spiked Water-milfoil
Native. Common. Ditches.

M. alterniflorum DC.   Alternate-flowered Water-milfoil
Native. Apparently rare or overlooked. No mention in 1914 *Flora*.
East: 20, near Norwich, —Talbot, 1919, 'Curious that it should have escaped observation in Norfolk so long'; 42, Calthorpe Broad Nature Reserve, ACJ, 1956; Long Gores Marsh, Hickling, ACJ & ELS, 1966. First record: Norwich (see above).

HIPPURIDACEAE

Hippuris vulgaris L.　　Mare's-tail, loc. 'Fox-tail'
Native. Frequent. Streams and rivers.

CALLITRICHACEAE

The genus *Callitriche* is one of the most difficult as the species are subject to considerable variation, depending on the depth of water and the rate of flow. Land-forms add to the complication.

Dr H. O. Schotsman's work in Holland (abstract in 1955 *Proc., B.S.B.I.*, **1**, 340–1) has helped to clarify the situation and our records are partly based on this work and on the valuable help given by Mr J. P. Savidge. Much study is called for before the distribution is known. The best character is found in mature fruits but, in deep water, these are rarely found.

Callitriche stagnalis Scop.　　Starwort
Native. Common throughout the county, especially on bare mud. The frequency with which this species is credited is somewhat suspect and it has most probably been recorded in mistake for *C. platycarpa*. Very small 'thyme-leaved' forms on bare mud (var. *serpyllifolia* Lönnr.) are the most distinct.

C. platycarpa Kütz.
Native. Frequent. Ponds, streams, rivers and bare mud.
The records for *C. palustris* L.,=*C. vernalis* Koch of the 1914 *Flora* described as common most probably belong here. It will be found to be more abundant.
West: 62, Roydon Common; 71, E. Walton Common; 72, Congham; 84, Burnham Overy Staithe; 99, Breckles Heath, ACJ; 99, Hockham, FR, 1920; Thompson Common, ALB.
First record: 1920, Gt. Hockham, FR det. J. P. Savidge.

C. obtusangula Le Gall
Native. The 1914 *Flora* regarded this as 'apparently rare' but it is the commonest representative of the genus. Ditches, ponds and streams especially from the chalk.

C. hamulata Kütz. (*C. intermedia* Hoffm.)　　Hooked Water Starwort
Status doubtful. Rare. Pools and slow streams.
West: 62, S. Wootton; 71, E. Walton Common; 72, Derby Fen and Gt. Massingham; Hillington; 92, Fakenham; 99, Devil's Punch Bowl, Wretham, FR, 1918.
First record: 1918, Wretham, FR in Druce Herb., Oxford, det. J. P. Savidge.

159

LORANTHACEAE

Viscum album L.   Mistletoe

Native. Rare. On hawthorn, Balsam poplar, but more often on cultivated apples. Kirby Trimmer (*Fl. Norf.*, 1866) recorded it also on whitethorn, oak, ash and maple.

West: 51, Tilney All Saints; 60, Outwell; 00, Reymerston.
East: 01, 18, 20, 29, 30, 39.

SANTALACEAE

Thesium humifusum DC.   Bastard Toadflax

Native. Very rare. A plant of chalk grassland which just reaches the south-west extremity of the county. It has disappeared from most of its recorded habitats.

West: 78, Gooderstone Warren, GT, 1966, several plants; 79, Devil's Dike, Cranwich, in fair quantity.

CORNACEAE

Thelycrania sanguinea (L.) Fourr.   Dogwood

Native. Common. Woods and hedges, usually on chalk.

ARALIACEAE

Hedera helix L.   Ivy

Native. Common. Hedges and woodland, both climbing trees and as a creeping component of the herb-layer where it is often dominant.

HYDROCOTYLACEAE

Hydrocotyle vulgaris L.   Marsh Pennywort, Sheep-rot

Native. Common. Fen and carr on soils of medium acidity.

UMBELLIFERAE

Sanicula europaea L.   Sanicle

Native. Common. Woods, especially on the chalk.

Eryngium maritimum L.   Sea Holly

Native. A constant species of coastal shingle and sand, where it is sometimes abundant.

First record: Sir Thos. Browne *in lit*, c. 1668, 'Ringlestones, a small black and white bird like a wagtayle . . . common about Yarmouth sands. They lay their eggs in the sand and shingle about June . . . *as the eryngo diggers tell me.*'

Chaerophyllum temulentum L.   Rough Chervil
Native. Common. Hedgebanks and woods. Follows *Anthriscus sylvestris* in flowering.

Anthriscus caucalis Bieb.   Bur Chervil, Beaked Parsley
Native. Common. Hedgebanks and waste places.

A. sylvestris (L.) Hoffm.   Cow Parsley, Keck, 'Queen Anne's Lace'
Native. Common. Hedgerows and grassland.

Scandix pecten-veneris L.   Shepherd's Needle
Native. Decreasing considerably as an arable weed of chalky soil.
West: 64, 72, 90–92, 00.
East: 01, 03, 04, 13, 14, 18, 19, 29, 30, 32, 33, 39.

Torilis japonica (Houtt.) DC.   Upright Hedge-parsley
Native. Common. Hedges.

T. arvensis (Huds.) Link   Spreading Hedge-parsley
Colonist. Rare. Fields and waste places. Not seen by authors.
West: 70/80, Swaffham district, ESE.
East: 01, 03, 39 (Maps Scheme).

T. nodosa (L.) Gaertn.   Knotted Hedge-parsley
Native. Less frequent than *T. japonica*, and particularly characteristic of land reclaimed from the sea.

Smyrnium olusatrum L.   Alexanders
Nat. alien. Established widely in hedges and waste places within a few miles of the sea where it is abundant but also occurs inland. Out of forty-four recorded squares, twenty-eight refer to inland localities.
First record: 1780, Norwich, J. E. Smith, where it is still abundant.

Conium maculatum L.   Hemlock
Native. Common. Open woods, hedges, waste places and sea-banks.

Bupleurum tenuissimum L.   Slender Hare's-ear
Native. Characteristic of sea-banks where it is sometimes frequent, extending to banks of tidal rivers.
West: 51, Eau Brink; 62, Wolferton; 63, Snettisham-Heacham; 74, Thornham; 84, Burnham Overy Staithe; 94, Warham, RMB.
East: 04, Cley, RSRF; 50, Gt. Yarmouth, EAE.

Apium graveolens L.   Wild Celery
Native. Frequent in marshes near the sea, and in ditches in reclaimed land.

A. nodiflorum (L.) Lag.   Fool's Watercress
Native. Common. Ditches.

A. inundatum (L.) Rchb. f.   Lesser Water-parsnip
Native. Widely distributed. Ponds and ditches.

Petroselinum segetum (L.) Koch   Corn Caraway
Native. Not common. Grassland on heavy soils, reclaimed land and chalk.
West: 51, Saddlebow and Tilney St Lawrence; 62, N. and S. Wootton;
63, Dersingham, GT; 74, Ringstead; 83, N. Creake, BT.

Sison amomum L.   Stone Parsley
Native. Less common than formerly. Chiefly in hedges in central Norfolk
on the heavier soils.
West: 90, Cranworth, Reymerston and Garveston, ALB; 92, Tittleshall;
00, Hingham; 08, S. Lopham.
East: 00, Deopham, EAE; Hardingham; 01, Mattishall; 02, Themel-
thorpe, GT; 18, 19, 28, 29, 39, 40, 49.

Cicuta virosa L.   Cowbane
Native. Rare in West Norfolk; locally common in Broads district.
West: 90, Scoulton Mere, ERN; 99, Cranberry Rough, Hockham.
East: 30–32, 40–42, 49.

Conopodium majus (Gouan) Loret   Pignut, Earthnut
Native. Common. Grassland and woods on sandy soil.

Pimpinella saxifraga L.   Burnet Saxifrage
Native. Common. Hedges and grassland on chalk.

Aegopodium podagraria L.   Goutweed. Ground Elder, Herb Gerard
Est. alien. Common. Hedges and field-borders usually near dwellings;
persistent garden weed. Perhaps native in wet woods as at N. Wootton.

Sium latifolium L.   Water Parsnip
Native. More frequent in the Broads district; fen ditches.
West: 59–61, 63, 69, 84.
East: 03, 12, 14, 18, 21, 30–32, 39–42, 49.
First record: 1794, 'from Norfolk', Woodward, *English Botany*, t. 204.

Berula erecta (Huds.) Coville   Narrow-leaved Water Parsnip
Native. Common. Ditches, fens and marshes.

Oenanthe fistulosa L.   Water Dropwort.
Native. Common. Ditches and fens.

O. lachenalii C. C. Gmel.   Parsley Water Dropwort
Native. Occasional in ditches near the sea; uncommon in fens.
West: 62, 64, 70, 71, 74, 79, 84, 08.
East: 04, 08, 30–32, 39–42, 49.

O. aquatica (L.) Poir.   Fine-leaved Water Dropwort
Native. Common. Ditches.

O. fluviatilis (Bab.) Coleman   River Water Dropwort
Native. Occasional in the upper- and mid-reaches of the slow-moving
rivers.
West: 78, 88, 91–94, 98, 02.
East: 00, 10, 20, 49.

Aethusa cynapium L.   Fool's Parsley
Colonist. Common. Weed of cultivated land. Most noticeable in autumn stubble fields.

Foeniculum vulgare Mill.   Fennel
Mixed status. Common. Native in its maritime habitats of sea-cliffs etc., but colonist inland.

Silaum silaus (L.) Schinz. & Thell.   Pepper Saxifrage
Native. Frequent in grassland on heavy soil.
West: 61, 70–72, 79, 81, 82, 90–92, 99, 00, 08.
East: 01, 18, 21, 28, 29, 39.

Angelica sylvestris L.   Wild Angelica
Native. Common. Wet places, fen and carr.

Peucedanum palustre (L.) Moench   Milk Parsley, Hog's Fennel
Native. Rare except in the Broads district. Fens and marshes.
West: 60, 61, 90, 98, 99, 07, 09.
First record: 1794, 'Gathered by Dr. Smith in the ditches of a wet, reedy meadow between Norwich and Heigham', *English Botany*, t. 229 as *Selinum palustre*.

Pastinaca sativa L.   Wild Parsnip
Native. Common. Roadsides and grassland, common on chalk, but by no means restricted to it. Particularly abundant in Breckland.

Heracleum sphondylium L.   Hogweed, Cow Parsnip
Native. Common. Hedges and grassland.

H. mantegazzianum Somm. & Lev.   Giant Hogweed
Est. alien. Increasing. Roadsides and waste places as a garden-escape. Not mentioned in the 1914 *Flora*. Probably more than one species is involved.
West: 60, 62, 70, 73, 84, 99, 00.
East: 10, 14, 19, 20, 21, 23, 29, 30, 33.

Daucus carota L., ssp. *carota*   Wild Carrot
Native. Common. Hedges and grassland, especially sea-banks.

CUCURBITACEAE
Bryonia dioica Jacq.   White or Red Bryony, loc, 'Wild Vine'
Native. Common. Hedges, woods and scrub.

ARISTOLOCHIACEAE
Aristolochia clematitis L.   Birthwort
Est. alien. Has persisted in the grounds of Carrow Abbey, Norwich, for many years; most likely descendants from plants used in the nuns' herbal

garden. 'The gardeners regard it as a pestilential weed and barrow-loads are thrown away every year' ETD.

West: 89, Sturston, AEE; ruins of Sturston Hall, abundant, 1967, JEL.

East: 20, Carrow Abbey.

First record: 1793, Carrow Abbey, Rev. Mr Salton, *English Botany*, t. 398.

EUPHORBIACEAE

Mercurialis perennis L.   Dog's Mercury

Native. Common. Woods on sandy soils where it may be dominant; male flowers predominate.

M. annua L.   Annual Dog's Mercury

Colonist. Locally common. Arable land and a frequent garden weed.

Euphorbia lathyrus L.   Caper Spurge

Est. alien. Occasional. Waste places and a spontaneous weed of old gardens. The recurrence of records points to its seeds remaining dormant and viable for a long time.

West: 74, 79, 83, 84, 88, 91–93.

East: 04, 13, 20, 29, 33.

E. platyphyllos L.   Warted Spurge

Native. Very rare.

East: 04, hedgebank at High Kelling, ELS, 1938.

First record: 1913, Cawston, E12, Miss Brenchley.

E. helioscopa L.   Sun Spurge

Colonist. Common in cultivated land.

E. peplus L.   Petty Spurge

Colonist. Common in cultivated land.

E. exigua L.   Dwarf Spurge

Colonist. Less common than the two previous species, in similar situations.

E. paralias L.   Sea Spurge

Perhaps native, though described by V. J. Chapman as a casual (*Scolt Head Island*, p. 96). Locally frequent and exclusive to sea sands and mobile dunes. Absent from apparently suitable dunes elsewhere, but the restriction of another west coast maritime species (*Juncus acutus*) to north Norfolk should be noted. Much of its recent spread followed the 1953 sea-flood.

West: 63, Heacham Harbour, 1959; 74, Scolt Head Island, abundant on the young dunes of the western half of the island and spreading to the mainland at Titchwell, 1954; Thornham, 1959; Brancaster, eighteen plants, ETD, 1963; 84, Holkham, 1955; Burnham Overy Staithe, 1959, one plant.

East: 04, Blakeney, RMB, 1960.

First record: 1913, Brancaster, Miss A. B. Cobbe.

E. amygdaloides L.   Wood Spurge
Native. Infrequent. Subject to severe damage by late spring frosts, especially after coppicing. Damp woods and hedgebanks.
West: 01, E. Dereham, DMM.
East: 01, 04, 10–12, 14, 19, 20, 29–32.

POLYGONACEAE

Polygonum aviculare agg.   Knotgrass
Native. Common. Arable weed and waste places. The aggregate includes the following three species.

> *P. aviculare* L. (*P. heterophyllum* Lindm.)
> Native. The commonest species. Roadsides, tracks and waste places.

> *P. heterophyllum* Lindm., var. *angustissimum* Meissn.
> Native. Frequent weed of stubble fields, not essentially calcicolous. 'This is the plant usually regarded as *P. rurivagum* Jord., with fruits ranging from 2·8 mm. to 2·5 mm. *P. rurivagum* is a very distinct-looking plant from chalky fields in the south of England with fruits of 2·0 mm. or less', C. C. Townsend comm. 1958.

> *P. arenastrum* Bor. (*P. aequale* Lindm.)
> Native. Common but less so than *P. aviculare*, in similar places and especially paths.

P. raii Bab.   Ray's Knotgrass
Native. Very rare. Maritime shingle.
West: 04, only on The Hood, Blakeney Point, 1955, just within the vice-county boundary.
East: 50, Gt. Yarmouth, EFL.

P. bistorta L.   Bistort, Snakeweed
Native. Rare. Moist meadows and wood margins.
West: 72, Congham; 90, Cranworth, ALB; 92, Brisley, DMM; oo, between Hingham and Southburgh, ALB.
East: oo; 02, Wood Dalling, ETD; 03, 04, 14, 20, 30.
First record: 1796, 'from a piece of wet land at Framlingham (*sic* = Framingham), Norwich, where it grows in considerable plenty', Skrimshire, *Catalogue*.

P. amphibium L.   Amphibious Bistort
Native. Common. Ponds and wet places. Non-flowering forms are frequent on dry land.

P. persicaria L.   Red Shank, Willow-weed, Persicaria
Native. Common arable weed of light soil, waste places and pond margins.

P. lapathifolium L.   Pale Persicaria
Colonist. Common as an arable weed.

'*P. nodosum* Pers.'

The doubts concerning the segregation of this species at the time of writing *West Norfolk Plants Today* are now resolved as we agree with Timson that it can no longer be regarded as distinct from *P. lapathifolium* (see Timson, 'The Taxonomy of *P. lapathifolium* L., *P. nodosum* Pers., and *P. tomentosum* Schrank', in *Watsonia*, 1963, v, **6**, 386–95).

P. hydropiper L.   Water Pepper

Native. Common. Wet places, especially on sand, and damp paths over heaths.

P. mite Schrank   Lax-flowered Persicaria

Native. Rare. Ditches.

West: 01, Swanton Morley, PH; 61, E. Winch Common, FR; South Lynn; 62, N. Wootton and Reffley; 72, Congham; 80, Ashill, FR.

East: 13, Bodham; 28, Pulham St Mary, ELS; 33, Ridlington, JHS.

P. minus Huds.   Small Persicaria

Native. Rare, possibly overlooked. Margins of ponds.

West: 61, E. Winch Common; 62, S. Wootton.

East: 33, Ridlington, JHS.

P. convolvulus L.   Black Bindweed

Colonist. A common weed of arable land.

P. cuspidatum Sieb. & Zucc.   Japanese Knotweed

Est. alien. Increasing in frequency and persisting, often in dense colonies, in waste places, as a garden-escape.

West: 61, Islington; 62, King's Lynn; 92, Brisley, DMM; 99, Rockland.

East: 00, 04, 14, 20, 21, 32, 41, 42, 51.

First record: 1927, Diss, waste ground, H. L. Green.

P. baldschuanicum Regel   'Russian Vine'

Est. alien. Persisting and invading hedges and waste places. Possibly under-recorded. A garden escape.

West: 61, Eau Brink; 99, Rockland All Saints.

East: 14, Beeston Regis Common, HAS; W. Runton, TBR; 51, Caister.

First record: 1961, W. Runton, TBR.

Rumex tenuifolius (Wallr.) Löve

Native. Common in the sandy heaths in the Breckland of south-west Norfolk. Not recorded in the 1914 *Flora*.

West: 70, 78–80, 88, 89, 98, 99, 08.

East: 04, 09, 28, 31, 33, 41, 42.

First record: 1950, Cockley Cley, W70, ELS.

R. acetosella L.   Sheep's Sorrel

Native. Common. Heaths and sandy arable land. One of the early colonisers of heaths after burning.

R. acetosa L.   Sorrel

Native. Common. Acid grassland.

R. hydrolapathum Huds.   Great Water Dock
Native. Common. Ditches and streams.

R. hydrolapathum × *obtusifolius*
(*R.* × *weberi* Fisch.-Benz.)
East: 42, between Sea Palling and Ingham, TGT, 1952

R. crispus L.   Curled Dock
Native. Common and variable. Arable land, pastures, waste places and shingle beaches.
The maritime ecotype, a feature of the shingle banks of the north-west coast, with very dense panicles and conspicuous tubercles on all three valves, is locally abundant. It is the var. *littoreus* Hardy and frequently recorded as var. *trigranulatus* Syme.
 A valuable paper, 'Notes on British Rumices' by Lousley appeared in BEC *Reports*, for 1938 and 1939 and we are grateful to the author for his help with some of our herbarium material. Our account is based entirely on this paper.

R. crispus × *obtusifolius*
Is the most frequent hybrid.
West: 62, N. Wootton Heath; 64, Ringstead; 72, Derby Fen; 78, Weeting; 98, Kilverstone.
Dock hybrids are by no means uncommon; indeed, where two species grow together, e.g., *R. crispus* and *R. obtusifolius*, and *R. conglomeratus* with *R. maritimus*, hybrids will almost certainly be present. They are readily distinguished in the field by the partial sterility, the ragged appearance of the panicle following irregular shedding of abortive flowers, and a haphazard mixture of the parents' characters (JEL).

R. obtusifolius L., ssp. *obtusifolius*   Broad-leaved Dock
Native. Common. Cultivated land, roadsides, degenerate grassland and about farm buildings.
Considerable variation occurs, especially in the teeth of the fruiting sepals. Our plants belong to the ssp. *agrestis* (Fries) Danser having prominent teeth and only one well-developed tubercle.

Forma *trigranis* (Danser) Rech. fil., has all three tubercles well developed.
West: 94, Wells-next-the-Sea, Lomas in Lousley, *op. cit.*

R. obtusifolius × *pulcher*
West: 94, West Tofts, J. E. Little, 1929.

R. obtusifolius × *sanguineus*
West: 60, Wallington, J. E. Little, 1923, in WBEC as *R. obt.* × *viridis*.

R. pulcher L., ssp. *eu-pulcher* Rech. fil.   Fiddle Dock
Native. Occasional. Dry places such as roadsides, commons, bases of walls and churchyards.
West: 61–64, 71, 73, 74, 81, 88, 89, 90, 98.
East: 00, 02, 17, 19, 20, 30–32, 39, 41, 51.

R. sanguineus L., var. *viridis* Sibth.   Wood Dock

Native. Frequent in woods and shady places such as under bracken on heathland. The 1914 *Flora* states *R. sanguineus* L. to be rare and var. *viridis* rather common. Presumably *R. sanguineus* refers to the red- or purple-veined dock, var. *purpureus* Stokes. (*R. sanguineus* auct. angl.). We consider this was the species formerly grown as a pot-herb. Unfortunately the character disappears in herbarium material; we have not seen it in a fresh state and can neither deny nor confirm its occurence.

R. conglomeratus Murr.   Clustered Dock

Native. Common. Ditches, fens and marshy pastures.

R. conglomeratus × *maritimus*

West: 62, S. Wootton Common; 99, Thompson Water.

R. conglomeratus × *pulcher*

A rare hybrid as the two parents seldom grow together.

West: 71, E. Walton Common.

R. palustris Sm.   Marsh Dock

Native. Rare. Most records from the old Fenland peat of south-west Norfolk. Continues to be confused with the next species.

West: 60–62, 69, 72, 78, 80, 99.

East: 09, 20, 23, 31, 32, 49.

First record: 1781, 'by Acle Dam', Pitchford in Sm., *English Botany*.

R. maritimus L.   Golden Dock

Native. Less rare than *R. palustris*, in similar situations. Locally frequent in some of the Breckland meres. Variable; ranging from the annual form described by Lousley as forma *humilis* (Petermann) Lousley to the perennial forma *ramosus* Zapal.

West: 60, 62, 78, 79, 82, 84, 88, 90, 94, 98, 99, 08.

East: 04, 14, 31, 32, 40, 42, 49.

URTICACEAE

Parietaria diffusa Mert. & Koch   Pellitory-of-the-wall

Native. Common. Old walls, especially churches.

Urtica urens L.   Small Nettle

Native. Common. Arable weed on light soils; also a characteristic indicator of rabbit-burrows on heaths.

U. dioica L.   Stinging Nettle

Native. Common. Hedges, waste places and woods. Abundant in some woods used by birds as roosting places.

U. pilulifera L.   Roman Nettle

Although Parkinson in his *Theatrum Botanicum* (1640) suggested that the origin of this plant in Kent was due to Caesar's invasion, and that his soldiers used it 'to rubbe and chafe their limbes', we consider the old Norfolk coast records more likely the result of the considerable sea-borne

65 SPIKED SPEEDWELL   *Veronica spicata*   *Miss V. M. Leather*

66 EARLY SPEEDWELL   *Veronica praecox*   *Miss V. M. Leather*

67 SLENDER SPEEDWELL   *Veronica filiformis*   *Miss D. M. Maxey*

68 MARSH LOUSEWORT *Pedicularis palustris* *Miss G. Tuck*

69 YELLOW-RATTLE *Rhinanthus minor* *Miss G. Tuck*

70 CRESTED COW-WHEAT *Melampyrum cristatum* *Dr Campbell*

71 (no English name) *Euphrasia pseudokerneri* *A. H. Hems*

72 PURPLE BROOMRAPE *Orobanche purpurea* *J. E. Lousley*

73 TALL BROOMRAPE *Orobanche elatior* *A. H. Hems*

74 BUTTERWORT   *Pinguicula vulgaris   Mrs D. M. Dean*

75 COMMON BLADDERWORT   *Utricularia vulgaris   Miss G. Tuck*

76 MARSH WOUNDWORT   *Stachys palustris   G. D. Watts*

traffic with ports in Europe and along the Mediterranean coast, where the nettle is both native and abundant.

West: 61, King's Lynn Docks, *c.* 1900, Dr C. B. Plowright, who transplanted it to his garden in N. Wootton where it still persists.

First record: 1745, 'about the walls of Yarmouth', recorded in MS. in a copy of Blackstone's *Specimen Botanicum.*

CANNABIACEAE

Humulus lupulus L.   Hop

Status doubtful. Common. Hedges and scrub. According to common repute, it was introduced in the fourteenth century:

> 'Turkey, heresy, hops and beer
> Came into England all in one year.'

ULMACEAE

The limits of the species in this genus remain in dispute. There is a good collection of Norfolk elms in the herbarium at the Castle Museum, Norwich, made by Mr A. E. Ellis from the Norwich area in 1944. They were named by Dr Melville and students will find there many of his microspecies, hybrids and varieties.

Ulmus glabra Huds., ssp. *glabra*   Wych Elm, Large-leaved Elm

Native. Frequent. Woods and hedges. Reproduces by seed.

West: 62, 70, 71, 78–81, 89, 98, 99, 08.

East: 01, 02, 04, 09, 12, 20, 22, 28, 29, 31, 33, 39, 41, 42, 49, 51.

U. × *hollandica* Mill., var. *hollandica*   Dutch Elm

East: 20, Thorpe; Keswick; 30, Coldham Hall Lane near Surlingham; 31, Ranworth Old Hall; 33, Witton – all by A. E. Ellis.

U. × *hollandica* Mill., var. *vegeta* (Loud.) Rehd.   Huntingdon Elm

East: 13, Mannington Hall, AEE det. RM.

U. procera Salisb.   English Elm

Status in Norfolk doubtful; probably planted. Hedges. Reproduces vegetatively. The south-eastern record cards provided for mapping purposes restricted the genus to two species only, *U. glabra* and this species. It would appear that *U. procera* provided an alternative for any elm differing from the more readily recognised Wych elm, resulting in a frequency we are unable to confirm.

West: 71, Narford, ESE; 78, Weeting; 80, Hilborough; 81, Lit. Dunham, ESE; 99, Snetterton Heath, JHS; between Hockham and Thompson Water, PDS.

East: 20, Martineau Lane, Norwich, AEE det. RM.

U. carpinifolia agg.

Smooth Elm, E. Anglian Elm, E. Anglian Small-leaved Elm

Status doubtful. Frequent and variable; often confused with *U. procera*. Spreads vegetatively and clonal complexes have been accorded specific ranks in the past. A small-leaved form is a feature of the north-Norfolk coastal hedgerows.

U. carpinifolia Gleditsch   Smooth Elm

West: 60, Wallington Park, J. E. Little; 70, Beechamwell; 80, Swaffham, Druce; 99, Cranworth, ALB; 98, Brettenham, ALB; 99, Breckles, ALB; E. Wretham, ALB; 00, Reymerston and Garveston, ALB.

East: 00, Hackford, ALB; 33; 41, between Winterton and Potter Heigham, TGT; 42.

U. angustifolia (Weston) Weston

West: 89, W. Tofts, ELS; 98, Brettenham, ALB.

East: 20, Norwich, AEE; 29, Brooke Mere, EAE, 1955.

U. diversifolia Melville   East Anglian Elm

West: 62, N. Wootton–Castle Rising, det. RM; 72, Pott Row; 80, Swaffham, ESE.

East: 20, Lakenham, JC, plate 1886, *English Botany*, teste RM.

U. coritana Melville   Coritanian Elm

So named by Melville 'from the territory formerly occupied by an ancient British tribe, the Coritanae' (*J. Linn. Soc.*, 1949).

West: 62, S. Wootton; Castle Rising, det. RM; 80, Swaffham, ESE.

East: 21, Spixworth and Hainford; Salhouse; 31, Ranworth, AEE det. RM.

JUGLANDACEAE

Juglans regia L.   Walnut

Est. alien. Occasional. Plantations and single trees in hedges.

MYRICACEAE

Myrica gale L.   Bog Myrtle, Sweet Gale

Native. Locally frequent. Bogs and wet heaths.

First record: 1799, Dersingham Moor, W63, C. Sutton in *English Botany*, t. 562.

BETULACEAE

Betula pendula Roth   Silver Birch

Native. Common. Woods on light soil, in which it is often abundant; invades heath where it is not checked by burning.

B. pubescens Ehrh.   Birch

Native. Common. Wet woods, heaths and a dominant species of carr.

B. pendula × *pubescens*

Intermediates occur in places where the two species grow together.

Alnus glutinosa (L.) Gaertn.   Alder

Native. Common. Streamsides, wet woods and carr.

Var. *macrocarpa* Loudon with large leaves and female catkins, 27 mm. × 18 mm.

West: 60, Wallington, J. E. Little; 62, King's Marsh, Wolferton, J. E. Little; 70, Beechamwell Fen; 78, by the L. Ouse at Hockwold.

CORYLACEAE

Carpinus betulus L.   Hornbeam

Nat. alien. Frequent. Woods and hedges. Its occurrence in Norfolk is largely due to the considerable planting carried out in the nineteenth century. Both hedges and isolated trees occur and the former are frequent in Breckland as wind-breaks. Variable, the var. *quercifolia* Desf., being recorded by Elwes and Henry (*Trees of Gt. Britain and Ireland*) from Cawston, E12.

Corylus avellana L.   Hazel

Native. Common. Woods and hedges.

FAGACEAE

Fagus sylvatica L.   Beech

Perhaps native on chalk, but widely planted.

Castanea sativa Mill.   Sweet Chestnut

Nat. alien. Extensively planted but reproducing freely.

Quercus cerris L.   Turkey Oak

Nat. alien. Occasional. Woods and hedgerows on light soil. Natural regeneration occurs.

West: 61– 63, 71, 74, 84, 91, 92, 98, 99, 01.

East: 03, 09, 13, 14, 20, 23, 33.

Q. ilex L.   Evergreen Oak, Holm Oak

Nat. alien. Occurs only where planted; abundant on the Holkham estate, both as trees and hedges. Regenerating on Ringstead Downs.

West: 64, 80, 83, 90, 93.

Q. robur L.   Common Oak

Native. Common. Woods, hedges and heaths.

Q. petraea (Mattuschka) Liebl.   Durmast Oak

Nat. alien. Occasional in plantations.

West: 62, Whin Hill Covert, Wolferton; 91, Gressenhall, DMM.

East: 03; 13; 21, Sprowston, C. C. Babington; 22, Hevingham Park, E. F. Warburg; 24, Cromer; 31, Burnt Fen Estate, ACJ.

SALICACEAE

Populus alba L.   White Poplar

Nat. alien. Sparingly planted and doubtfully established.

First record: 1790, Lakenham Common near Norwich, J. E. Smith, *English Botany*, t. 1618.

P. canescens (Ait.) Sm.   Grey Poplar

Native. Frequent. Woods and hedges, spreading freely by vegetative means.

First record: 1806, Wells Heath, J. Crowe, *English Botany*, t. 1619.

P. tremula L.   Aspen

Native. Rather common. Woods and hedgerows.

P. nigra L.   Black Poplar

Native. Frequent. Streamsides.

Var. *italica* Duroi, the Lombardy Poplar, occurs occasionally as a planted tree.

P. × *canadensis* Moench var. *serotina* (Hartig) Rehd.
    Black Italian Poplar

Nat. alien. Occasional. More frequent in East Norfolk.

West: 60, Ryston; 70, Beechamwell; 90, Cranworth and Reymerston, ALB; 99, E. Wretham, ALB.

East: 09, 10, 18, 20, 30, 32, 42, 49.

P. gileadensis Rouleau   Balm of Gilead

Nat. alien. Occasional. Streamsides and margins of wet woods.

West: 62, Castle Rising; 72, Grimston; 88, by the Lit. Ouse at Two Mile Bottom, Thetford; 89, Bodney Warren; 93, Thursford.

SALIX   The Willows

The foundation for the classification of this puzzling genus was laid by the famous Norfolk botanist, Sir Jas. E. Smith, as long ago as 1828 in his work, *The English Flora*. Salicologists following him were very unwilling to differ from one who had devoted so much time to their study, 'Full 30 years have I laboured at this task' (*op. cit.*). Unfortunately he did not believe in the existence of hybrids; he 'set aside the gratuitous suppositions of the mixture of species, or the production of new ones, of which . . . I have never met with an instance'. Smith's influence extended over many years and, as late as 1926, in the BEC *Report* of that year, Fraser stated: 'In many cases today, Smith's species remain where he left them at the beginning of the nineteenth century except that the × denotes that they are hybrids and not species.'

The Rev. E. F. Linton, whilst holding the living at Sprowston near Norwich, studied the genus and published his work in the *Supplement* to the *Journal of Botany*, 1913.

It has remained for us to build on this very sound foundation and to bring our records into line with more recent work. Particular attention has been paid to their distribution in West Norfolk (see Swann: 'West Norfolk Willows', in *Proc., B.S.B.I.*, 1957, **2**, 337–45). He acknowledges the help and encouragement given by R. D. Meikle of Kew, R. C. L. Howitt of Farndon and Dr A. Skortsov of the Moscow Botanical Garden. In a few cases the nomenclature differs from the *Flora of the British Isles*.

Salix pentandra L.   Bay willow

Not native. Rare. Wet woods and streamsides.

West: 61, Blackborough End; 70, Cockley Cley, ESE; 98, Shadwell.

S. alba L.   White Willow

Both native and planted. Common. Streamsides, fens and wet woods.

Var. *caerulea* (Sm.) Sm.   Cricket-bat Willow

Frequently planted by the faster-flowing streams. According to Elwes and Henry (*British Trees*) 'known only as a female tree and originated in Norfolk about 1700'.

West: 51, Terrington St John, J. E. Little, 1921, a male tree; 60, Ryston; 71, Westacre and W. Bilney.

Var. *eleyensis* Burtt Davy was first described in 1938 from trees on the Ryston estate near Downham Market where both willows and poplars are cultivated on a large scale. Differs from normal trees by its relatively longer catkins on slightly drooping branches.

Var. *vitellina* (L.) Stokes   Golden Willow

Frequently planted for both use and ornament.

West: 61, Bawsey; 62, N. Wootton; 71, Gayton; 08, Kenninghall, JHS.

S. alba × *fragilis*

West: 69, W. Dereham Fen, 1967, RCLH & ELS, a few old trees.

S. alba var. *vitellina* × *babylonica*   Weeping Willow
(*S.* × *chrysocoma* Doell.)

Occasionally planted for ornament. The true *S. babylonica* L., does not occur in Norfolk.

S. chermesina Hartig.

West: 71, one tree by a pond, Magpie Farm, W. Bilney, 1962, RCLH.

S. fragilis agg.   Crack Willow

The limits of this species have been in dispute for a very long time. It would appear that we have in Norfolk at least four, perhaps five, distinct entities that have been called '*S. fragilis*' by various workers in the past. Our native tree and three, perhaps four, introduced species are involved. The description by Linnaeus in *Species Plantarum*, 1753, as '*SALIX (fragilis) foliis glabris ovato-lanceolatis; petiolis dentato-glandulosis*' is vague and does not separate it with certainty from some allied species.

In 1951, specimens from Sweden were examined at Kew by R. D. Meikle who came to the conclusion that they were identical with *S. fragilis* described by Smith in *English Botany*, ed. 1 and illustrated by Sowerby. They were distinct from both *S. Russelliana* and *S. decipiens*.

These four, with a possible fifth, are:

1 *S. fragilis* L., sec. Smith, *English Botany*
(*S. alba* × *fragilis* forma *monstrosa* sec. Floderus)

A tall, bushy-headed tree; branches set on obliquely, crossing each other, not continued in a straight line, brown to olive in colour. Leaves lanceolate-elliptical with blunt, often unequal, but not coarse serratures. Catkins about 6·5 cm. long, dense. Not as frequent as the next species.

West: 61, E. Winch; Blackborough End; 62, Reffley; Castle Rising; Wolferton.

Var. *latifolia* Moss

Occurs rarely.

West: 60, Wallington; 61, Blackborough End; 62, roadside near Roydon Common; 69, W. Dereham Fen; 88, Kilverstone.

2 *S. russelliana* Sm.   Bedford Willow
(*S. viridis* Fries)

Common and widespread. A tree lighter in colour; branches long, straight and slender (not angular in their insertion like *S. fragilis* Sm.). Leaves lanceolate, never broadly ovate, strongly and coarsely serrate. Fertile catkins more lax and tapering.

West: 62, Reffley Marshes; Wolferton; 69, Dereham Fen, RCLH and ELS.

3 *S. decipiens* Hoffm.   White Welsh or Varnished Willow

Formerly planted in osier holts but now very local. Considered by Floderus and by both German and Russian workers to be *S. fragilis* L.

West: 70, Larch Wood near Beechamwell; 99, frequent at Hockham Mere.

4 *S. basfordiana* Scaling ex Salter, *Gard. Chron.*, **17**, 298.

Basford's willow has been widely planted as an ornamental tree. Has orange bark passing to red in terminal twigs with a shining surface. Long branches coming off at an angle. In spring the golden colour contrasts sharply with the green of *S. Russelliana*.

West: 61, Middleton; 62, Reffley Marshes; 74, Holme-next-the-Sea; 78, Hockwold; 88, near Thetford; 92, Brisley.

5 *S. sanguinea* Scaling   Red Willow

A planted tree. Has the habit of *S. alba* with branches coming off at a closer angle. Close to *S. Basfordiana* but remarkable for its reddish colour.

West: 61, one tree in a gravel pit, Tottenhill, 1961, RCLH.

S. triandra L.   Almond Willow

Both native and planted. Common by streamsides, woods and osier holts. Leaves elliptic-lanceolate, closely serrate, tapering at apex and cuneate at base, glaucous beneath, 7 to 10 cm. long.

Var. *amygdalina* (Sm.) Bab.

Occurs occasionally and is more frequent in osier holts. Leaves broader above, rounded at the base, glaucous beneath. Flowers a second time in the summer.

Var. *hoffmanniana* (Sm.) Bab.

Of rare occurrence as a relict of cultivation, being discarded by growers as too 'spriggy'. 'A shrub of more humble growth', Smith. Leaves narrowly ovate, green beneath, 4 to 6 cm. long.

S. triandra × *viminalis*
(*S.* × *mollissima* Ehrh., *S.* × *hippophaefolia* Thuill.)

Occurs occasionally with the parents.

West: 62, Reffley; 73, Gt. Bircham; 92, Gt. Ryburgh; 99, Knight's Fen, Hockham.

East: 18, Diss, RCLH.

S. purpurea L.   Purple Willow

Native. Frequent. Fens, streamsides and damp woodland. Extensive hedges used as wind-breaks in the peat-land of Methwold Fen.

West: 61, 62, 69–72, 78, 79, 81, 88, 90, 92, 99, 09.

East: 02, 18, 20, 29, 30, 39, 42.

S. purpurea × *viminalis*   Green-leaved Willow
(*S.* × *rubra* Huds.)

Apparently rare but probably overlooked.

West: 62, Bawsey; 69, W. Dereham Fen.

East: 20, R. Yare marshes, Keswick.

S. viminalis L.   Common Osier

Native and planted. Common and widespread, particularly in disused gravel workings. The most frequently grown willow.

Var. *linearifolia* Wimm. & Graeb.

A narrow and small-leaved variety occurring occasionally.

West: 69, West Dereham Fen; 70, Gooderstone; Foulden Common; 71, Westacre; 81, Litcham; 88, Kilverstone.

S. calodendron Wimm.   Grey Willow
(*S. dasyclados* auct., *S. acuminata* Sm.)

Status doubtful. Local. Streamsides and in damp meadows.

West: 63, Boathouse Creek, Wolferton; 64, Holme-next-the-Sea (originating from Holkham); 69, Wallington, J. E. Little; Wretton Fen; W. Dereham Fen; 84, Holkham, frequent, RCLH.

East: 20, marshes at Keswick.

S. caprea L., ssp. *caprea*   Goat Willow, Goat Sallow

Native. Common. Woods, scrub and hedges, particularly on base-rich soils.

S. caprea × *cinerea*
(*S.* × *reichardtii* A. Kerner)

Occurs with the parents rarely. Difficult to distinguish but more often simulates *S. caprea*. Wood faintly striate; lighter in colour and less torulose branches.

West: 71, W. Bilney; 99, Knight's Fen, Hockham.
East: 12, Buxton Heath.

S. caprea × *viminalis*
(*S.* × *sericans* Tausch.)

A frequent hybrid especially in old gravel pits where the two parents are often abundant. Long leaves, wider than *S. viminalis* with a conspicuous complanate or flattened appearance.

West: 61, Cranberry Wood, E. Winch; Bawsey; 62, Roydon Common; 69, W. Dereham Fen, RCLH & ELS; 71, E. Walton; W. Bilney.

S. cinerea L.   Common Sallow
(*S. aquatica* Sm.)

Native. Abundant. Damp woods, marshes, fens and often dominant in carrs. Reproduces freely from seed which is only viable for about a week. Smith described his *S. aquatica* as extremely common in wet habitats, with 'leaves about two inches in length; serrated about the middle and towards the extremity; narrowed at the base . . . of a dull greyish-green . . . glaucous and minutely downy underneath.' These characters are typical of the Norfolk plants.

S. cinerea × *purpurea* × *viminalis*   Fine Basket Willow
(*S.* × *forbyana* Sm.)

Formerly grown in osier holts. Now apparently very rare. Commemorates one of Smith's contemporaries, the Rev. F. Forby, who held a living at Fincham near King's Lynn.

West: 84, Holkham, det. Meikle.

S. cinerea × *repens*
(*S.* × *subsericea* Doell)

A rare and local hybrid. Although its parents are widespread, they have slightly different flowering times.

West: 62, Roydon Common, det. Meikle.
East: 41, Winterton dunes, RCLH det. Meikle.

S. cinerea × *viminalis*
(*S.* × *smithiana* Willd.)

Our commonest hybrid. Its parents are abundant in the pools left in disused gravel workings. It is one of life's ironies that Smith's name should be commemorated by a hybrid!

West: 61, Bawsey; 62, Wolferton; Roydon Common; 69, W. Dereham Fen.
East: 30, Wheatfen Broad, FRo.

S. atrocinerea Brot.

This species does not occur in Norfolk. To credit it with a wide distribution in East Anglia, as has been done in the past, is at complete variance with its known distribution outside the British Isles. It was originally described from Portuguese material and is a member of the Atlantic or Lusitanian Element found throughout western and central France and the Iberian peninsula. It is to be noted that in recent works, for example, the second edition of *The Flora of the British Isles*, it has now been relegated to subspecific rank following the decision that there are few differences separating it from *S. cinerea* and the two are merely geographical races. Dr Skortsov states that it does not occur in Russia. Norfolk records, published in the past, probably refer to the frequent hybrid, *S. aurita* × *cinerea*, with its characteristic ferrugineous indumentum on the lower leaf-surfaces.

S. aurita L.　　Eared Sallow

Native. Frequent on heaths and margins of damp woods.

Forma *pseudohermaphrodita* Gagnep.

Although various species are occasionally seen with aberrations such as odd branches bearing both male and female flowers, it is seldom that the whole bush, as in this instance, bears hermaphrodite inflorescences.

West: 61, E. Winch Common.

S. aurita × *caprea*
(*S.* × *capreola* J. Kerner ex Anderss.)

A rare hybrid, possibly overlooked owing to its resemblance to forms of *S. caprea*.

West: 92, Horningtoft, BSBI Exc., 1954.

S. aurita × *cinerea*
(*S.* × *multinervis* Doell.)

A frequent hybrid.

West: 60, Stow Bardolph, J. E. Little; 61, Wormegay; E. Winch; 62, Roydon Common; Grimston; 70, Foulden Common; 93, Fakenham.
East: 11, Swannington Upgate Common.

S. aurita × *repens* agg.

Appears to be a local hybrid although its parents frequently grow together.

West: 62, Roydon Common.
East: 12, Buxton Heath.

S. repens agg.　　Creeping Willow

Native. Common. Wet heaths, bogs, fens and sand-dunes.

A remarkably polymorphic species. Forms vary according to habitat. Typical prostrate plants are rarely seen. Most of our bushes are decumbent with long, ascending branches, elliptical leaves with recurved points, answering well to Smith's *S. ascendens*. The form in damp hollows of sand-dunes appears to be the var. *argentea* (Sm.). On E. Winch Common strictly erect forms occur.

ERICACEAE

Rhododendron ponticum L.

Nat. alien. Common in woods and on sandy heaths, especially the Green-sand heaths of West Norfolk; extending to dunes at Winterton in East Norfolk.

Calluna vulgaris (L.) Hull   Heather, Ling

Native. Common. Dominant over large areas of heath on Greensand and glacial gravels. Heath fires encourage germination.

Erica tetralix L.   Cross-leaved Heath

Native. Common. Dominant in bogs, spreading to moist heaths, particularly after burning.

E. cinerea L.   Bell-heather

Native. Far less common than *Callina* and never dominant. Dry heaths. Very rare in Breckland.

Vaccinium myrtillus L.   Bilberry

Status doubtful. Recorded by Kirby Trimmer about a hundred years ago "very sparingly . . . Mousehold Heath near Norwich" where it was reported by H. D. Geldart to have been planted.

East: 11, Attlebridge, large colony on a railway bank formerly heathland, ETD, June 1968.

V. oxycoccus L.   Cranberry

Native. Now confined to West Norfolk. Exclusive to bogs, in which it is constant and often abundant. Both red- and brown-speckled fruits (var. *maculatus* Lousley) occur. Adjoining Wolferton Wood (W62) 'is a tract of marsh ground, called Cranberry Fen, in which the *Vaccinium* grows in such profusion as I have never seen equalled . . . villagers sell these for one shilling a pint', Dr John Lowe, 'On the Flora of Lynn & Neighbourhood', *Bot. Soc. Edinb.*, 1866. Cranberry Rough, Hockham and Cranberry Fen, Blackborough, presumably indicate former bog.

West: 61, Blackborough, up to 1949; 62, N. Wootton; Wolferton Fen; Roydon Common; Sugar Fen; 72, Leziate Fen.
East: 31, Horning, until 1913, EAE.

First record: 1780, Dersingham, J. Crowe.

PYROLACEAE

Pyrola minor L.   Common Wintergreen

Native. Very rare. Woods and thickets. Formerly occurred on Roydon Common, material in Herb, Norw. Mus., collected by T. Southwell in 1852 as '*P. rotundifolia*'.

East: 03, Holt Lowes; 20, Thorpe, Mrs Ann Gurney, 1954; 21, Hainford, RPL, 1945; 1952, RMB & ETD.

P. rotundifolia L., ssp. *rotundifolia*   Larger Wintergreen
Native. Rare. Bogs and fens.
West: 62, Roydon Common, Miss Hilbert, 1955, not seen by authors.
East: 20, Racecourse Plantation, Thorpe St Andrew, in Herb. Norw. Mus., 1922; 31, Upton Broad, J. E. Lousley & G. Watts, 1936; Hoveton, Miss Geldart, 1916; 32, Sutton, CG, 1952, one plant; 41, Thurne, EAE, 1953; Shallam Dyke, FRo, 1947; 42, Calthorpe Broad and Stalham, ACJ, 1958; Hickling, A. Bennett, 1900; 51, Winterton, 1967, JHS.

Ssp. *maritima* (Kenyon) E. F. Warburg
Very rare. Persisted until 1953 on dunes at Wells, apparently lost in the sea-floods of that year. Confirmed by A. J. Farmer.
West: 94, Wells-next-the-Sea.
East: 51, Winterton dunes, AJF.

MONOTROPACEAE

Monotropa hypopitys agg.   Yellow Bird's-nest
Native. Rare but possibly increasing in the Breckland pinewoods. Pine plantations and sand-dunes. The distribution of the two segregates is imperfectly known.

M. hypopitys L.
West: 79, Round Plantation, Mundford, 1967, VML; Mundford Covert, ELS.
East: 01, Kelling, ELS; 03, Holt, 1928, CPP; 22, Westwick, C. W. West; 23, N. Walsham, EAE.

M. hypophega Wallr.
West: 84, under pines, Holkham Meols; 94, Wells-next-the-Sea; 98, W. Harling Heath.
First record: 1782, in a pine grove, Stoke near Norwich, Mrs H. Kett in *English Botany*, t. 69.

PLUMBAGINACEAE

Limonium vulgare Mill.   Sea Lavender
Native. Exclusive to salt marsh, in which it is abundant from Weybourne (E14) to Wolferton (W62). The second of the sea lavenders to flower, the extensive sheets of mauve being a conspicuous feature of the marshes in August. Shows considerable variation depending on the age of the marsh. The lower the zone, the nearer it approaches *L. humile* which may probably account for the many reported occurrences of that rare species. The forms of the higher and older zones have been named f. *pyramidale* C. E. Salmon, 'in its extreme form, with enormous trusses of flowers. At the inner edge of the old marshes half-way to Burnham Harbour from the House Hills', Scolt Head Island (Deighton & Clapham, in *Trans. Nfk and Norw. Nat. Soc.*, XII, 1, 102, 1924/25).

L. humile Mill.   Lax-flowered Sea Lavender

Native. Apparently rare. At the time of writing *West Norfolk Plants Today*, 1962, the authors had not seen this species, although recorded for Scolt Head by V. J. Chapman. Specimens from Missel Marsh (W84), 1964, have been confirmed by L. A. Boorman. It appears to be restricted to this locality and hybridises with *L. vulgare* (*L. × neumanii* C. E. Salmon). Plants with a superficial resemblance to this species have been found elsewhere but see under *L. vulgare*.

L. bellidifolium (Gouan) Dum.   Matted Sea Lavender

Native. Frequent. Characteristic of the sandy margins of salt marsh. Distribution westwards from Blakeney (E04) to Wolferton (W63). The first of the sea lavenders to flower, towards the end of July, but the pale lavender-coloured flowers are relatively insignificant.

First record: 1746, 'coast of Norfolk', Henry Scott in Blackstone's *Spec. Bot.*, 47, 1746.

L. binervosum (G. E. Sm.) C. E. Salmon   Rock Sea Lavender

Native. Common. Distribution as in *L. bellidifolium*, but extending to drier situations and to shingle. The last of the species to flower.

Armeria maritima (Mill.) Willd.   Thrift, Sea Pink

Native. Exclusive to salt marshes, where it is common.

PRIMULACEAE

Primula veris L.   Cowslip, Paigle, loc. 'Peggles'

Native. Common. Grassland on chalk or clay.

Reaches its greatest frequency in south and south-west Norfolk but rare in Breckland.

P. veris × vulgaris   Bastard Oxlip
(*P. × variabilis* Goupil)

An occasional hybrid, sometimes confused with the caulescent form of the primrose.

West: 71, 72, 80, 82, 84, 90, 92, 99.
East: 10, 29, 31, 39.

P. elatior (L.) Hill   Oxlip

Native. Very rare. On chalky boulder clay.

Still persists at Dickleburgh (E08), 1963, see S. R. J. Woodell ('*P. elatior* in Norfolk, Immigrant or Relic?' 1965, *Proc., B.S.B.I.*, **6**, 37).

P. vulgaris Huds.   Primrose

Native. Common. Woods, hedgebanks, dykesides and railway embankments.

Forma *caulescens* (Koch) Schinz. & Thell., with a raised umbel on a stalk instead of being sessile and radical, occurs occasionally.

West: 92, Horningtoft, DMM.
East: 01, Hockering Wood.

Hottonia palustris L.   Water Violet
Native. Frequent in ditches in large colonies.

Lysimachia nemorum L.   Yellow Pimpernel.
Native. Occasional. More frequent in East Norfolk woods.
West: 61, 62, 72, 82, 90–93, 03.
East: 01–04, 10, 12–14, 21–23, 29, 30, 39, 40.

L. nummularia L.   Creeping Jenny, loc. 'Herb Twopence'
Native. Frequent. Wet meadows and ditches. Spreads vegetatively.
West: 59–62, 79, 82, 88–92, 98.
East: 03, 08, 18–20, 28–31, 39–41.

L. vulgaris L.   Yellow Loosestrife
Native. Common. Fens, carr and wet woods.

Anagallis tenella (L.) L.   Bog Pimpernel
Native. Occasional in bogs and fens.

A. arvensis L.   Scarlet Pimpernel
Status doubtful. Common as a colonist in cultivated land, also on sand-dunes, its native habitat, but the proximity of arable land to maritime localities in Norfolk makes it suspect. Colour variations are frequent. E. A. Ellis reported blue (distinct from *A. foemina*), salmon and yellow flowers in the Mundesley district and R. Scott writing in *Country Life*, 1953, refers to blue, pale pink and heliotrope plants along with the scarlet type in Norfolk beet-fields.

Var. *pallida* Hook. f., with pale-pink flowers.
West: 93, Binham.
East: 04, Glandford sand-pit, ELS.

A. minima (L.) E. H. L. Krause   Chaffweed
Native. Rare. In damp places on sandy tracks with *Isolepis setacea*.
West: 60, Stoke Ferry, WGC; 62, Whin Hill Covert, Wolferton; Roydon Common.
East: 19, 23, 24, 42.

Glaux maritima L.   Sea Milkwort
Native. Common along the coast in drier parts of salt marshes and damper areas in reclaimed meadows.

Samolus valerandi L.   Brookweed
Native. Frequent. Ditches by the sea and fens inland.

BUDDLEJACEAE

Buddleja davidii Franch.   'Buddleia'
Est. alien. Occasional. Reached its greatest frequency on bombed sites of towns during the Second World War, as at King's Lynn, E. Dereham, Norwich and Thetford; still persists on town walls and rubbish tips.

OLEACEAE

Fraxinus excelsior L. Ash
Native. Common. Woods, hedges, sometimes colonising fens.

Syringa vulgaris L. Lilac
Nat. alien. Common and well-established in hedges, particularly Breckland. Reputed to have been introduced in Norfolk by the Flemish weavers in the reign of Edward III.

Ligustrum vulgare L. Common Privet
Native. Common. Woods, hedges, fen carr and dunes.
Introduced at Scolt Head Island (Privet Hill) and to the dunes from Holkham to Wells.

L. ovalifolium Hassk. Broad-leaved or Garden Privet
Nat. alien. There is little doubt that this species is frequently recorded for the native plant. Occasionally introduced in plantations and bird-sown in places.

APOCYNACEAE

Vinca minor L. Lesser Periwinkle
Denizen. Frequent. Hedges and woodland. Sometimes abundant in damp woods where it spreads vigorously vegetatively, its arching stolons rooting at the tips.

V. major L. Greater Periwinkle
Denizen. Apparently more frequent than *V. minor* and a more certain garden-escape. Hedges and woodland. Spreads as in *V. minor*.
First record: 1798, 'The fruit, which Mr. Curtiss has sent, has not been mentioned and which few botanists have seen, is produced every year in Mr. Kett's garden at Seething, Norfolk' (*English Botany*).

GENTIANACEAE

Cicendia filiformis (L.) Delarbre Slender Cicendia
Scarcely a Norfolk native. Recorded by Benj. Bray of King's Lynn from 'low sandy heath, Roydon, West Norfolk, far from scarce in the locality', 1881 (*Record Club Report*). Material in Herb. Mus. Brit., collected by Colonel Meinertzhagen 'near Brancaster, Norfolk, 1928'. Not seen by authors and probably extinct. Colonel Meinhertzagen admitted he was somewhat doubtful about the station.

Centaurium pulchellum (Sw.) Druce
Native. Frequent on heaths and coastal sand-dunes.
West: 62, S. Wootton Common; 70, Foulden Common; 74, Holme-next-the-Sea (together with the forma *schwartziana* Wittr.); 84, Holkham; Burnham Overy Staithe.
East: 04, Blakeney; 42, Horsey Gap, J. Buckland; 50–51, Gt. Yarmouth.

C. erythraea Rafn   Common Centaury
Native. Common. Heaths. White-flowered forms occasionally in Breck-land.

Blackstonia perfoliata (L.) Huds.   Yellow Wort
Native. Rare now. Woodland on calcareous soil, chalky outcrops, old brick-fields and railway embankments.
East only: 19, Wacton Green, EAE; disused railway, Forncett–Wymond-ham, MBA, 1957; 28, old brick-field, Pulham St Mary, CF, 1962; 29, Brooke Wood, EAE, 1953; 30, marl pit, Surlingham, FRo; Rockland St Mary, EAE, 1950.
First record: 1769, Arminghall, E20, in a 'list supplied by a society of botanists in Norwich for *A Description of England & Wales*', comm. EAE.

Gentiana pneumonanthe L.   Marsh Gentian
Native. Rare. Wet heaths.
West: 61, E. Winch Common; 81, Litcham Common.
East: 12, Buxton Heath; 21, Hainford, DMM, 1960; Horsham St Faiths, HDH, 1923.
First record: 1755, Stratton Strawless Heath, Benj. Stillingfleet in his *Calendar of Flora*, 1762.

Gentianella campestris (L.) Börner   Field Gentian
Native. Very rare now. Chalk grassland.
West: 71, E. Walton Common, thought to be extinct in West Norfolk until refound by Miss B. M. Sturdy, BSBI Exc., 1964; 72, Derby Fen, 1928, CPP; 99, Thompson Common, 1915, FR.
East: 03, Holt Catpits, 1928, CPP; open ground near Lascelles Wood, Holt, 1951, RMB; 14, Greens Common, E. Runton, 1917, WGC.

G. amarella (L.) Börner ssp. *amarella*   Felwort, Autumn Gentian
Native. Typical of chalk grassland, where it is frequent; coastal sand-dunes, Breckland heaths. White-flowered occasionally in Breckland. Small forms from coastal habitats have been recorded as ssp. *axillaris* by the older botanists.
West: 60, 64, 70–72, 74, 78–80, 84, 93, 94, 98, 08.
East: 03, 04, 14, 23.

MENYANTHACEAE

Menyanthes trifoliata L.   Bogbean, Buckbean
Native. Common. Bogs, spreading at times into fen. In dense mats and often spreading vegetatively, as flowering plants form a very small proportion of the total.

Nymphoides peltata (S. G. Gmel.) O. Kuntze   Fringed Waterlily
Native. Rare. Slow-moving rivers of West Norfolk Fenland.
West: 59, R. Delph, Welney; 51, Middle Level Drain, Wiggenhall St Mary the Virgin; 90, Scoulton Mere.
East: 20, R. Yare, Keswick, 1934.

BORAGINACEAE

Cynoglossum officinale L.   Hound's-tongue

Native. Common. Dry banks, wood margins and sand-dunes.

Symphytum officinale L.   Comfrey

Native. Frequent. Streamsides, fens and wet places. Both cream- and purple-flowered forms occur.

West: 59, 60, 62, 69, 70, 72, 73, 78, 88, 91, 99, 00.

East: 01, 03, 04, 08, 09, 12–14, 18–21, 23, 28, 29, 39–42.

S. asperum × officinale   Blue Comfrey

(*S. × uplandicum* Nyman)

Est. alien. Frequent on roadsides.

West: 62–64, 71–73, 88, 98, 99, 08.

East: 08, 10, 18, 20, 23, 24, 30, 32, 33, 39, 42, 49, 51.

S. orientale L.

Est. alien. Frequency increasing. Garden-escape in waste places.

West: 60, 70, 88, 90, 99, 00.

East: 00, 03, 10, 12, 14, 18–20, 51.

S. tuberosum L.   Tuberous Comfrey

Native. Very rare. Damp woods.

West: 88, Kilverstone, HDH.

East: 19, Long Stratton, *c.* 1959, EAE.

Borago officinalis L.   Borage

Est. alien. Occasional. Hedgebanks and waste places as a garden-escape.

West: 73, Sedgeford; 84, Burnham Overy; 00, Reymerston, ALB.

East: 03, Thornage, 1928–60, CPP.

Pentaglottis sempervirens (L.) Tausch   Alkanet

Nat. alien. Frequent. Hedgerows, woods and waste places.

West: 61, 71, 84, 88, 80, 93, 99.

East: 02, 03, 10, 12, 13, 20–24, 29, 30, 33, 39–41.

Anchusa arvensis (L.) Bieb.   Bugloss

Colonist. Common weed of arable land on light soils.

Myosotis scorpioides L.   Water Forget-me-not

Native. Common. Streams and ditches.

We regard the two reported occurrences of *M. secunda* A. Murr. as errors of identification.

M. caespitosa K. F. Schultz

Native. Common. In similar habitats to *M. scorpioides.*

M. sylvatica Hoffm.   Wood Forget-me-not

Native. Rare. Although some records are clearly garden-escapes, the damp woods at Cranworth have a native population.

West: 90, Wood Rising; Potter's Carr, Cranworth, ALB; 99, Cranberry Rough, Hockham, confd. A. E. Wade.

East: 02, Foxley Wood, ALB & ELS, 1968; 03, Holt Lowes, PHS; Edgefield, CPP; 09, near Old Buckenham, Maps Scheme.

First record: 1810, Holt, Rev. R. B. Francis, *English Botany*, t. 2680.

M. arvensis (L.) Hill   Field Forget-me-not

Colonist. Common as a weed of arable land; woodland rides and waste places. The shade form growing in woods, with larger flowers, var. *sylvestris* Schlechtend., is frequently wrongly recorded as *M. sylvatica*.

M. discolor Pers.   Yellow and Blue Forget-me-not

Native. Common. Grass heath and dry banks.

The typical plant with orange-yellow flowers turning to blue appears to be far less frequent than the var. *dubia* (Arrondeau) Wade with white flowers turning to pale blue; the latter is a plant of somewhat damper habitats.

M. ramosissima Rochel   Early Forget-me-not

Native. Common. Characteristic spring ephemeral of grass heath.

Lithospermum officinale L.   Gromwell

Native. Frequent. Hedges and grassland on chalk.

L. arvense L.   Corn Gromwell.

Colonist. Rare weed of arable land.

West: 62, 63, 70, 71, 79, 83, 88, 93, 94, 99, 08.
East: 01, 04, 18, 19, 39.

Echium vulgare L.   Viper's Bugloss

Native. Common. Heaths, pastures, coastal sand-dunes. Abundant in Breckland.

CONVOLVULACEAE

Convolvulus arvensis L.   Small Bindweed, 'Bear-bine'

Colonist. Common as a weed of arable land, roadsides and waste places. Ripe fruits very rarely found. 'I have never seen the capsule or seeds', Sir J. E. Smith, *Eng. Fl.*, 1, 285. 'Ripe fruits are very rarely found in Holland, central Europe . . .', Wilcke, 1949.

In November, 1959, from an old marl pit in Jack's Lane, South Creake, W83, 'mature capsules, subpyriform in shape, up to 6 mm. in diameter, with 2 to 3 seeds in each. Seeds pitted with reticulations, shagreened, and with scattered adpressed hairs' ELS.

Calystegia sepium (L.) R. Br., ssp. *sepium*   Great Bindweed, Bellbine

Native. Common. Hedges, wood margins and climbing reeds in fens.

A colony showing dialysis of the corolla in a hedge at Wiggenhall St Germans, W51 (ELS), and at Wheatfen Broad, Surlingham, E30 (EAE). Pink-flowered forms occur occasionally.

Ssp. *pulchra* (Brummitt & Heywood) Tutin

Est. alien. Garden-escape.

East: 02, Foxley Wood, RMB & EAE; 03, Briston, ETD; 20, Mousehold, ETD; 32, island at Barton Broad, RMB.

Ssp. *silvatica* (Kit.) Maire

Est. alien. Common. Hedges near dwellings and waste places.

C. soldanella (L.) R. Br.   Sea Bindweed
Native. Typical of coastal sands, where it is often abundant.

Cuscuta epithymum (L.) L.   Dodder, 'Strangleweed'
Native. Frequent. Heaths. Hosts: largely on *Calluna*, but also seen on *Ulex europaeus*, *U. gallii*, *Lotus corniculatus* and *Teucrium scorodonia*.

SOLANACEAE

Lycium barbarum L.   Duke of Argyll's Tea-plant
Naturalised alien. Locally frequent in hedges and waste places.

It was with some diffidence that we recorded two species in our *West Norfolk Plants Today*. Further field-work has shown that they cannot be satisfactorily separated and additional confirmation comes from Miss Stella Ross-Craig's *Drawings of British Plants*, part XXI, where there is but the one species. She writes 'Mr. Sealy (Kew Herbarium) who worked on *Lycium* finds that, so far as British material is concerned, it was found impossible to separate as species the material named *L. halimifolium* Mill., from that named *L. chinense* Mill., the differences given by various authorities do not hold good. The British material agrees best with *L. halimifolium* Mill., which, in turn, has been shown by N. Feinbrun and W. T. Stearn to be synonymous with *L. barbarum* L.'.

Atropa bella-donna L.   Deadly Nightshade, Dwale
Status doubtful; perhaps native in woods on the chalk but few such localities. Formerly cultivated. 'It groweth very plentifully . . . at a place called Walsoken neare unto Wisbitch [sic = Wisbech]', Gerard's *Herball*, 1597, its first record.

West: 62, Reffley Wood; Warren Farm, Roydon; 70, Smeeth Wood; Warren Belt, Beechamwell; Caldecote Fen; Gooderstone; 71, Abbey Farm, E. Walton; 72, Hillington; 88, Grimes Graves, Weeting, until 1953.
East: 14, Sheringham, EAE; 20, Carrow, EAE; Postwick, ETD; 21, Horstead, RMB; Coltishall, BCH; 39, Ditchingham, MIB; 42, Horsey Mere, RMB; 50, Southtown, Gt. Yarmouth, EAE.

Hyoscyamus niger L.   Henbane
Status mixed. Occasional in sandy places near the sea, its native habitat, but more frequent in waste places and sand-pits. Sporadic. Varying in numbers from year to year.
West: 60, 62, 63, 70–72, 74, 78, 80, 84, 88, 90, 94, 99.
East: 04, 14, 19, 20, 39, 50.

Solanum dulcamara L.   Bittersweet, Woody Nightshade
Native. Common. Hedges, woods, pond margins, waste places and occasionally in maritime habitats.

S. nigrum L.   Black Nightshade
Colonist. Common and variable. Weed of cultivation. Waste places.

Var. *flavum* Dum., with yellow fruits.

East: 20, Harford, RMB, 1959.

Var. *chlorocarpum* (Spenn.) Boiss., fruits remaining green.

East: 30, Claxton, R. H. Sewell, 1950.

S. triflorum Nutt.

Est. alien. Has persisted on the railway embankment and on sandy ground at Whin Hill Covert, Wolferton, W62, since 1949, CPP.

East: 02, potato field at Bawdeswell, 1964, ETD.

Datura stramonium L.   Thorn-apple

Nat. alien. Frequent. Weed of cultivation and waste places. Sporadic. Seeds remain dormant and viable over long periods. So abundant in some years, such as 1959, as to gain much publicity in the local press. Occasionally plants with lilac flowers = var. *tatula* (L.) Torr., occur.

SCROPHULARIACEAE

Verbascum thapsus L.   Great Mullein, Aaron's Rod

Native. Common. Sandy arable, disturbed heathland and gravel pits.

V. pulverulentum Vill.   Hoary Mullein

Native. Of special interest, being confined to the counties of Suffolk and Norfolk. A local plant of roadsides.

West: 63, Heacham, Dersingham, Snettisham; 72, Harpley, Flitcham; 73, Fring; Sedgeford; Shernborne, GT; 74, Brancaster; 78, north bank, Lit. Ouse, Brandon, D. Dupree; 81, Castleacre, Southacre; 88, railway bridge, Thetford, HDH, 1939, one plant.

East: 10, Bowthorpe, WGC; 11, Easton, WGC; Ringland Hills, WGC; Swannington, RMB; 20, Whitlingham, WGC; Mousehold Heath, RMB; Keswick, WGC; Harford Bridges, ETD; Eaton, F. Long; Norwich, ETD; 21, Salhouse, JHS; Rackheath, RMB; 29, Tasburgh, SA; 31, Acle New Road, S.C. Puddy; 40, Halvergate bridge, PER.

First record: 1745, in MS. in a copy of Blackstone's *Specimen Botanicum* 'about the ditch on the outside of the city walls at Norwich; also, by the river Yare, between Bishopgate Bridge and the ferry-house, both places plentifully'; plate 487 in *English Botany*, 1798, from a Norwich plant. Still plentiful close to the river at Norwich, but much farther downstream than in Blackstone's day.

V. pulverulentum × *thapsus*
(*V.* × *godronii* Bor.)

West: 72, Harpley Dams, J. E. Lousley, RPL & ELS, 1945.

V. nigrum L.   Dark Mullein

Native. Frequent. Dry banks and chalk grassland; abundant in Breckland.

Var. *bracteosum* Pugsl., with 'stems crowded with spreading, long-cuspidate or acuminate leaves or bracts almost to the apex', Pugsley in *J. of Bot.*, 1934, 278.

West: 83, S. Creake.

V. nigrum × *pulverulentum*
(*V.* × *wirtgenii* Franch.)
West: 90, Watton, FR, 1918, in Herb. Norw. Mus.
East: 10, Bowthorpe, WGC, 1915.

V. nigrum × *thapsus*
(*V.* × *semialbum* Chaub.)
West: 78, Weeting Heath, ELS, 1957; 79, junction of the B1106 and Cranwich roads, ELS; 88, Two Mile Bottom, Thetford, RPL, 1953.
First record: 'Found by Mr. Dawson Turner at Swaffham', *English Botany*, 1790–1814.

V. blattaria L.   Moth Mullein
Est. alien. Rare. Waste places.
West: 98, Brettenham, WGC, 1921.
East: 03, Gresham's School rifle range, PHS; 39, gravel pit, Ditchingham, MIB.

V. virgatum Stokes   Twiggy Mullein
Est. alien. Rare. Roadsides and waste places.
West: 72, Massingham-Castleacre, 1948; 89, Lynford-Weeting, 1956; 91, Mileham, JSP; 98, Bridgham-Brettenham, 1955; 98, Garboldisham, 1957.
East: 29, Tasburgh, SA.

Antirrhinum orontium L.   Lesser Snapdragon, Weasel's Snout
Colonist. Frequent. Occasionally as single plants in sandy arable land.
West: 61, 62, 70, 71, 83, 90–92.
East: 01–04, 11, 12, 14, 18, 20–22, 29–31, 39, 40, 49.

A. majus L.   Snapdragon
Est. alien. Established on old walls and waste places as a garden-escape. Under-recorded.
West: 62, Castle Rising castle; 71, Pentney Abbey; 79, Whittington.
East: 20, Harford Bridges; 39, Ditchingham.

Linaria purpurea (L.) Mill.   Purple Toadflax
Est. alien. Walls and waste places.
West: 88, Thetford.
East: 00, Hackford, ALB; 11, Ringland, ETD; 14, Weybourne, CPP; Beeston and Sheringham, PHS; 19, Forncett St Mary, MBA; 20, Harford and various waste places in Norwich; 23, N. Walsham, KHB; 33, Mundesley, KHB.

L. vulgaris Mill.   Toadflax
Native. Common. Hedges, cultivated land and waste places.
Peloric forms are rare but occur at N. Wootton, W62, and Ludham, E31.

Chaenorhinium minus (L.) Lange   Small Toadflax

Colonist. Not infrequent, particularly on railway tracks, waste places and stubble fields.

West: 62, 63, 71–74, 78, 80, 82–84, 90–92, 99, 00.

East: 03, 04, 19, 28, 29.

Kickxia spuria (L.) Dum.   Fluellen

Colonist. Rare arable weed of light soils.

West: 90, Cranworth, ALB; 93, Binham, R. Scott.

East: 29, Tasburgh, SA; 39, Ditchingham, MIB.

K. elatine (L.) Dum.   Fluellen

Colonist. Far more frequent than the last in chalky arable land, often associated with *Legousia hybrida* and *Stachys arvensis*.

West: 60, 62–64, 70–74, 79, 80, 90–92, 98, 99.

East: 01, 03, 04, 12–14, 18, 20, 29, 30, 39.

Cymbalaria muralis Gaertn., Mey. & Scherb.   Ivy-leaved Toadflax

Wall denizen. Common on old walls. Form with white flowers = forma *seguieri* (Beg.) Cuf., on wall at Hillington Hall, W72.

Scrophularia nodosa L.   Knotted Figwort

Native. Common. Woods.

S. aquatica L.   Water Betony

Native. Common. Ditches and streams.

S. umbrosa Dum.   Broad-winged Figwort

Native. Locally frequent, both in West Norfolk, along the rivers Thet, Little Ouse and the upper reaches of the Wissey, and also in the River Yare in East Norfolk. Shows considerable increase since the 1914 *Flora* when it was only known from Scoulton Mere and Watton.

West: 69, Wretton Fen, J. E. Little, 1925; 78, Weeting; Brandon (Norfolk side of river), D. Dupree, 1953; 79, Didlington; 80, Gt. Cressingham; Hilborough; 88, Thetford, 1964; 89, Stanford Water, 1920; Ickburgh; W. Tofts Mere; Bodney; 90, Scoulton Mere; Watton, 1904; 98, Bridgham; 99, Thompson Water, ETD; Rockland All Saints.

East: 00, Seamere near Hingham; 20, Harford Bridges and farther down the R. Yare, EAE, 1964; 30, Surlingham, 1965, EAE.

First record: 1904, Watton and Scoulton Mere, Druce in *BEC*.

S. vernalis L.   Yellow Figwort

Nat. alien. Rare. Plantations as a garden-escape.

West: 84, Holkham pine-woods; 90, Saham Toney, ERN; 94, pine-woods at Wells; Stiffkey, its *locus classicus*.

East: 09, Eccles Hall near Attleborough, abundant, BCH, 1962; gravel pit near Bryant's Bridge, Eccles Heath; 12, Cawston rectory, RMB; 99, Snetterton Heath, 1964, JHS, 'frequent in mixed woodland'.

First record: 1805, Stiffkey, in *Botanists' Guide*.

Mimulus guttatus DC.   Monkey Flower
(*M. langsdorffii* Donn ex Greene)
Nat. alien. Increasing. Streamsides.
West: 62, S. Wootton, 1939; 84, Holkham, BT; Burnham Thorpe to Burnham Overy Staithe; 94, Warham; Wiveton, BT; 98, pond near Ringmere, KD.
East: 04, Cley to Wiveton, FMD; 13, Itteringham; Corpusty, ETD; 21, Horstead Mill, ETD; 22, Buxton Lammas, W. Field, 1939; 32, Barton Turf, PC.

M. guttatus × *luteus*
A sterile hybrid distinguished by large red blotches on the corolla lobes, not small red spots as in *M. guttatus*; pedicels and calyces sparingly pubescent.
East: 20, Keswick, ETD & ELS, 1966, confd. R. H. Roberts (see his 'Mimulus Hybrids in Britain', 1964, *Watsonia*, **6**, 70–81). It is not considerd to be the *M. Langsdorffii* first recorded from this station in 1840, *Fl. Norf.*, 1914.

Digitalis purpurea L.   Foxglove
Native. Frequent in woods on sandy soil. White-flowered forms occur.
West: 60–62, 71, 72, 80–84, 88, 90, 91, 93, 99, 01.
East: 04, 11, 12, 14, 21, 22, 41, 42.
First record: 1568. 'Much in Englande and specially in Norfolke, about ye cony holes in sandy ground and in divers woodes', W. Turner's *Herball*.

Veronica beccabunga L.   Brooklime
Native. Common. Ditches.

V. anagallis-aquatica L.   Water Speedwell
Native. Common. Ditches and streams.

V. anagallis-aquatica × *catenata*
West: 09, Hargham, HJR, det. J. H. Burnett, in Herb. Mus. Brit.

V. catenata Pennell
Native. As common as *V. anagallis-aquatica*, in similar situations.

V. scutellata L.   Marsh Speedwell
Native. Less common than the three previous species.
West: 61, 62, 72, 78, 83, 89, 91, 93, 98, 99, 01.
East: 03, 04, 09, 11–13, 19, 30, 39, 40, 42, 49.

Var. *pilosa* Vahl, a hairy plant growing in dried-up ponds. The indumentum varies and appears to be a response to habitat.
West: 70, Foulden Common; 98, pool near Ringmere; 99, Hockham Mere.
East: 11, Felthorpe.

V. officinalis L.   Common Speedwell

Native. Common. Heaths and open woodland, especially rides in the Breckland pine-woods.

V. montana L.   Wood Speedwell

Native. Frequent. Damp woods.

West: 60, 62, 71, 72, 79, 80, 90–93, 03.
East: 00–04, 10, 12–14, 20, 22, 28, 29, 31, 39, 49.

V. chamaedrys L.   Germander Speedwell

Native. Common. Grassland, woods and hedges.

V. spicata L., ssp. *spicata*   Spiked Speedwell

Native. Very rare and diminishing. Chalk grassland of West Norfolk. One of the Breckland specialities. Was locally abundant in 1915, when W. G. Clarke knew 'a station where there are over 1,400 plants in a very limited space'. It is curious that it was not found until 1910, when it was recorded by Burrell & Clarke at Garboldisham, seeing that it was observed by Ray in his *Catalogus Plantarum* as early as 1660, 'In several closes on New-market Heath' in the adjoining county of Suffolk.

West: 08, Devil's Dyke, Garboldisham, up to 1963; 78, Weeting Heath; 98, E. Harling Heath, still there, JHS, 1965; W. Harling Heath, 1922.

V. serpyllifolia L., ssp. *serpyllifolia*   Thyme-leaved Speedwell

Native. Common. Grassland, heaths and damp arable soil.

V. arvensis L.   Wall Speedwell

Native. Common. Grassland, arable land, walls and dry places. Abundant as a spring ephemeral of grass heaths.

V. verna L.   Spring Speedwell

Native. Very rare now; formerly abundant at Santon Warren. Open sandy places on Breckland heaths, such as rabbit warrens. No recent records. H. Dixon Hewitt last saw it at Santon in 1946. Material in Herb. Mus. Brit., from Croxton, Santon and Thetford; earliest, 1837, J. D. Salmon; latest, 1926, J. E. Little and E. M. Reynolds.

First record: 1828, 'about Thetford', J. E. Smith, *Eng. Fl.*, 1, 26.

V. praecox All.   Early Speedwell

Status doubtful. Very rare. Sandy arable fields of Breckland.

West: 98, Kilverstone, 1935, two plants, Miss M. S. Campbell; 88, Green Lane, Thetford, CPP & ELS, six plants, 1963.

First record: 1935, Kilverstone, above.

V. triphyllos L.   Fingered Speedwell

Native. Very rare. Sandy arable fields of Breckland.

West: 88, Green Lane, Thetford, PAW, 1963.

Specimens in Herb. Mus. Brit., from Thetford, 1836–1916.

First British record: 1670, Rowton [*sic*] ? Roughton, in Norfolk, Thos. Willisell in Ray, *Cat.*, 340.

V. hederifolia L.   Ivy-leaved Speedwell, Bird's Eye

Colonist. Common. Arable weed.

V. persica Poir.   Buxbaum's Speedwell
Colonist. Has now become one of the commonest weeds of cultivated
fields. First recorded for Norfolk about 1860.

V. polita Fr.   Grey Speedwell
Colonist. Frequent. Weed of arable fields.
West: 62, 64, 70–74, 78–80, 83, 88–90, 98, 99.
East: 03, 08, 10, 19, 39.

V. agrestis L.   Green Field Speedwell
Colonist. Formerly described as common, now only occasional. Weed of
arable land and gardens.
West: 70, 71, 73, 79–81, 83, 88, 89, 99, 08.
East: 02–04, 08, 13, 14, 18, 19, 21–23, 31, 39, 41, 42.

V. filiformis Sm.   Slender Speedwell
Est. alien. Often a pest in lawns.
West: 50, Upwell churchyard, 1955; 71, Westacre, 1955; 73, Bagthorpe
Hall, R. Scott; 80, Saham Toney, ERN; 84, Burnham Overy Staithe,
KHB, 1959; 88, river bank, Chisley Vale near Thetford, 1963; Chapel
Farm, Croxton, 1965.
East: 00, Seamere near Hingham; 28, Pulham St Mary, CF, 1963; 31,
Horning, A. G. Spooner, 1951; 42, Calthorpe Broad, TGT, 1956; 39,
Ditchingham, MIB, 1953.
First record: 1951, Horning.

Pedicularis palustris L.   Red-rattle
Native. Common. Fens. White-flowered forms occur occasionally.

P. sylvatica L.   Lousewort
Native. Common. Bogs and wet heaths, in more acid soils than *P. palustris*.
'The Cattell that pasture where plentie of this grass groweth become full of
lice' (Lyte's *Herball,* 1578).

Rhinanthus serotinus (Schon.) Oborny ssp. *apterus* (Fr.) Hyl.
    Greater Yellow-rattle
Reported to be 'abundant in a small area of old established sandhills north
of Holme Lodge', W74, 1958, Maps Scheme. Intensive searching has
failed to confirm its occurrence here and it is considered to be an error of
identification; *R. minor* is abundant.

R. minor L., ssp. *minor*   Yellow-rattle, 'Rattle-baskets'
Native. Common. Grassland and established dunes.

Ssp. *stenophyllus* (Schur) O. Schwartz
Not separated in the 1914 *Flora.* Frequent in damp, neglected pastures
and fens.
West: 64, Old Hunstanton; 70, Oxborough Fen, J. E. Little, 1919, as
R. crista-galli L., var. *stenophyllus* Druce; 71, E. Walton Common; 91,

77 YELLOW ARCHANGEL   *Galeobdolon luteum*   R. Jones

78 BUGLE WITH AN ALBINO   *Ajuga reptans*   *Miss D. M. Maxey*

79 CLUSTERED BELLFLOWER   *Campanula glomerata*   *Miss V. M. Leather*

80 RAYED NODDING BUR-MARIGOLD *Bidens cernua* var. *radiata* *Miss D. M. Maxey*

81 LEOPARD'S BANE *Doronicum pardalianches* *Miss D. M. Maxey*

82 JERSEY CUDWEED   *Gnaphalium luteoalbum*   *Miss G. Tuck*

83 MEADOW THISTLE   *Cirsium dissectum*   *A. H. Hems*

Beetley; 92, Fakenham; 99, Cranberry Rough near Hockham; 08, Low Fen. S. Lopham.
East: 03, Brinton; 12, Booton Common; 32, Barton Turf.
First record: 1905, Filby, Druce, as *R. minor* Ehrh., var. *robustus* Druce.

Melampyrum cristatum L.   Crested Cow-wheat

Native. Very rare. Woods and thickets on clay soils.

West: 91, Rawhall, 1953; 92, Horningtoft.
First record: Gressenhall, Crowe in *Botanists' Guide*, 1805, probably the same locality as Rawhall Farm Wood where it still persists.

M. pratense L.   Common Cow-wheat

Native. Occasional. In similar situations to the last species.

West: 91, Beetley, DMM; Rawhall; 92, Horningtoft; 99, Wayland Wood near Watton, FR, 1919, as f. *ovatum* Spenn.
East: 03; 11, Ringland Hills, WGC, 1917; 20, Eaton Park Wood, ETD, 1963, 'abundant'; 22, Stratton Strawless, EAE, 1950; 29; 31.

Euphrasia   Eyebright

In the early years of the present century, F. Robinson of Watton corresponded with Pugsley and supplied records for the latter's 'Revision of the British Euphrasiae' (*J. Linn. Soc.*, XLVIII, 1930). Our account shows the persistence of some of these. In his *Flora of Swaffham*, unpublished, Edees observes: 'The commonest species is *E. nemorosa*, or at any rate most of the Swaffham plants are best aggregated under that name. I have never seen typical *E. nemorosa* in the district; the common species of the heaths is a very different plant. It varies in size but is usually bushy with small, widely-spaced leaves and showy coloured flowers. A large form found near the old aerodrome on Narborough Field was determined by Pugsley as *E. pseudokerneri* Pugsl., f. *elongata* Pugsl. (see BEC *Report*, 1938); other plants have been named *E. nemorosa* var. *transiens* Pugsl. All these Breckland eyebrights, in spite of individual variation, have a strong family resemblance and further study may show that they deserve a new name.'

E. nemorosa (Pers.) Wallr.

Our commonest species. Heaths, chalk grassland and pastures. Very variable.

Var. calcarea Pugsl.

Chalk grassland and railway embankments.

West: 71, rail. embankt., Swaffham–Narborough, ELS confd. EFW; E. Walton Common, ELS; 73, rail. embankt., Heacham–Sedgeford, ELS.

Var. *transiens* Pugsl.

Appears to be a Breckland type.

West: 70, Gooderstone, FR; 71, Narborough, ELS; between Shingham and Beechamwell, ELS; 89, forest track, Mundford, ELS.

E. confusa Pugsl.

West: 61, E. Winch Common; 71, E. Walton Common, ELS; 79, Cranwich, J. E. Little, 1928, as f. *albida* which is also given for East Norfolk in the *Revision,* locality not specified; 89, Sturston Warren, P. D. Sell.

E. pseudokerneri Pugsl.

Occasional in chalk grassland. Old records for *E. stricta* Host are probably referable here.

West: 64, Ringstead Downs; 70, Marham Fen; Foulden, FR, as f. *elongata*; 71, Narborough, ESE, as f. *elongata*; 72, Leziate Fen; 80, Saham Toney, FR; Swaffham, Druce; 09, Lit. Ellingham, FR.

East: 03, Thornage, G. H. Alston; 14, Beeston Regis, A. R. Horwood.

⎡ *E. rostkoviana* Hayne ⎤

Old records are considered erroneous for this northern and western species and are referable to *E. anglica*.

⎣ ⎦

E. anglica Pugsl.

Occasional. Wet acid grassland overlying peat and associated with *Salix repens* and *Ulex europaeus*.

West: 61, E. Winch Common; 62, Sugar Fen; 71, Westacre, ESE; E. Walton; 80, Necton Common, CPP; 81, Litcham Common; 91, Beetley; 92, N. Elmham; 98, Overa Heath; 99, Stow Bedon; Breckles Heath and Thompson Common.

East: 17, Billingford, DMM.

Odontites verna (Bell.) Dum., ssp. *serotina* (Wettst.) E. F. Warb.
Red Bartsia

Doubtfully native. Common as an arable weed and in waste places.

Ssp. *verna* appears to be rare.

West: 62, West Newton, BSBI Exc., 1949.

OROBANCHACEAE

Orobanche purpurea Jacq.   Purple Broomrape

Native. Rare. Cliff-tops in north-east Norfolk. Parasitic on *Achillea millefolium*.

East: 14, Sheringham, Beeston Regis, W. Runton; 20, Thorpe near Norwich, an inland station, EAE; 23, Trimingham; 24, between Cromer and Overstrand; 33, Mundesley, in railway cutting and on cliffs.

First record: 1779, Northrepps, a single specimen, Mr Scarles in *English Botany*.

O. elatior Sutton   Tall Broomrape

Native. Occasional. Dry calcareous soils. On *Centaurea scabiosa*.

West: 62, W. Newton; 63, Heacham; 64, Hunstanton; 70, Beechamwell and Barton Bendish; 71, E. Walton; 72, Congham; 93, Wighton Halt.

East: 03, Holt; 14, Sheringham.

O. minor Sm.   Lesser Broomrape

Native. Common. On clover, both wild and cultivated, making it difficult to grow this crop in certain fields. In 1915, J. E. Little recorded it on *Erodium cicutarium* and *Echium vulgare*. Also occurs on *Crepis capillaris* and *Hypochoeris radicata* as the var. *compositarum* Pugsl., and has been taken for *O. picridis*.

LENTIBULARIACEAE

Pinguicula vulgaris L.   Common Butterwort
Native. Frequent. Characteristic of bogs, occurring also in the less alkaline parts of fens.

Utricularia spp.   Bladderwort
Considerable work has been done on the genus in Norfolk and all four British species occur. Readers are referred to a very full account by W. G. Clarke and Robert Gurney in 1920–1, *Trans. Nfk. and Norw. Nat. Soc.*, xi, 128–161.

U. vulgaris L.   Common Bladderwort
Native. The commonest of the species occurring in many 'ditches, fen dykes, turf-pits, broads, ponds and swamps', Clarke & Gurney, *op. cit.*
West: 62, 70–72, 79, 92, 99, 07.
East: 08, 11, 13, 19, 20, 22, 30–32, 39, 41, 42, 49.
First record: 1796, Bardolph Fen, W60, Wm. Skrimshire, *Cat.*

U. neglecta Lehm.
Very rare.
West: 70, Foulden Common, FR, 1916; 99, Stow Bedon, FR, 1917.
East: 30, Wheatfen Broad, Surlingham, EAE, 1937, conf. P. M. Hall as *U. major* Schmidel; 41, Martham Dyke, A. Bennett.

U. intermedia Hayne   Intermediate Bladderwort
Native. Rare. 'Anchored in the mud of the sides of pools, among roots of *Juncus*, in *Chara*, *Sphagnum* and *Hypnum*', Clarke & Gurney. Very rarely flowers, 'last two weeks in March critical when the turions are germinating – extra spring warmth rather than summer heat is favourable'. Flowered at Roydon Common in 1910, 'hundreds of square yards . . . treading on it at every step but only one flower' (*op. cit.*). 1921, three plants in flower; 1947, one flower, FRo.
West: 62, Roydon Common and Sugar Fen; 70, Foulden Common; Gooderstone Fen; 71, Marham Fen; 72, Derby Fen, in flower, 1943, ELS.
East: 11, Swannington, WGC; 42, Calthorpe Broad, ACJ, 1956, in flower; 19, Flordon Common, WGC; 32, Barton Turf, WGC; E. Ruston, WGC.

U. minor L.   Lesser Bladderwort
Native. Occasional in bogs and fens.
West: 62, Roydon Common; Sugar Fen; 70, Caldecote; Gooderstone Fen; Boughton, WGC; 71, E. Walton Common; 72, Derby Fen; 92, Guist, WGC; 07, Blo' Norton Fen, WGC; S. Lopham.
East: 03, Holt Lowes, WGC; 11, Swannington; 19, Flordon, WGC; 22, Felmingham, WGC; 23, Southrepps, WGC; 32, E. Rushton, Honing, Barton Turf, WGC; 41, Burgh St Margaret, WGC.

VERBENACEAE
Verbena officinalis L.   Vervain
Native. Frequent. Roadsides and waste places on chalk.

LABIATAE
Mentha   Mint
Considerable attention was given to the Norfolk mints by the older botanists, particularly the Rev. Kirby Trimmer, who published his work in his *Flora of Norfolk* (1866) and *Supplement*, 1885.
In view of the many name changes we confine our records to recent field-work in which we have had the generous assistance of R. A. Graham.

Mentha pulegium L.   Pennyroyal
Native. Very rare. Margins of village ponds, formerly widespread.
West: 61, Tottenhill Row Common, RPL, 1950.
East: 39, Ditchingham, MIB, 1957, both prostrate and erect plants (var. *erecta* Martyn) occurred.

M. arvensis L.   Corn Mint
Colonist. Common. Arable fields, perhaps native in woodland rides and by ponds.

M. aquatica × *arvensis*   Whorled Mint
(*M.* × *verticillata* L.)
Native. Frequent. Damp places; dwarfed forms locally abundant on the margins of some of the Breckland meres.
West: 61, Tottenhill Row Common; 62, N. and S. Wootton; Reffley; 70, Beechamwell Fen, J. E. Little; 71, E. Walton Common; 72, Appleton; Roydon; 79, Foulden; 84, Burnham Thorpe, BT; 88, Fowlmere; Home Mere; 89, West Mere; Bodney Warren; 90, Woodrising, FR; 91, Beetley, EFL; 98, Ringmere; Langmere; 99, Stow Bedon, ALB.
East: 29, Fritton, ETD; 30, Surlingham, EAE.

M. arvensis × *spicata*
(*M.* × *gentilis* L., *M. gracilis* Sole; *M. cardiaca* (S. F. Gray) Baker)
A very variable plant of cottage-garden origin. Formerly grown as the Cardiac mint.
West: 62, King's Lynn tip, det. R. M. Harley; Reffley (det. Graham as either an intermediate between *M.* × *gentilis* and *M.* × *gracilis* or *M.* × *gracilis* var. *cardiaca*); 72, Appleton, 'I agree that this is a form of *M.* × *gentilis*' Graham.

M. aquatica × *arvensis* × *spicata*   Glabrous Red Mint, Black Peppermint
(*M.* × *smithiana* R. A. Graham)
Status doubtful. Apparently a rare plant. First described by Smith as *M. rubra*
West: 88, disused railway, Thetford, 1967, R. Pankhurst; 89, Sturston, 1917, FR.
East: 24, Overstrand, 1959, KHB, 'large patch by disused railway'.

M. aquatica L.   Water Mint
Native. Common and variable. Streams, fens and damp woods.

M. aquatica × *spicata*   Peppermint
(*M.* × *piperita* L.)
Status doubtful. Very variable. Formerly cultivated as a source for menthol.

Var. *piperita* 'Inflorescence spicate; leaves long and narrow, attenuate or rounded at the base', Graham in '*M. piperita* and the British peppermints', 1951, *Watsonia*, **2**, 30–35.
West: 61, Bawsey; 62, Roydon; Wolferton; 63, Dersingham; 72, Grimston; 83, S. Creake; 94, Warham.
East: 14, W. Runton.

Forma *hirsuta* (Fraser) Graham; (lusus *pilosa* Still)
Whole plant hairy; a rare form.
West: 63, Marsh Lane, Dersingham.

Var. *vulgaris* Sole 'Inf. capitate; lvs cuneate or rounded at base'.
West: 62, Roydon.

Var. *subcordata* Fraser, 'Inf. capitate; lvs subcordate at base'.
West: 62, Roydon.

M. spicata L.   Spearmint
Garden plant; occasionally escaping or thrown out.
West: 62, King's Lynn; N. Wootton; 64, Hunstanton, RCLH; 74, Holme-next-the-Sea, RCLH; 90, Carbrooke Fen, FR; 99, Stow Bedon, ALB.
East: 20, Harford Bridges; 31, 49.

M. longifolia (L.) Huds.   Horse Mint
Doubtfully native. Occasional. Cultivated, as are many of the Norfolk mints. Damp, waste places.
West: 70, Marham, FR; 71, Westacre; E. Walton; 79, Whittington; 90, Carbrooke Fen, FR; 99, Shropham, HDH; Stow Bedon, E. M. Reynolds; 00, Reymerston, ALB.
East: 14, Beeston Hill, RMB; 33, Bacton.
Var. *horridula* Briq.
West: 61, Leziate.

M. longifolia × *rotundifolia*   Lamb Mint
(*M.* × *niliaca* Juss. ex Jacq.)
Widespread and abundant in hedges and on roadsides. Often thrown out from gardens by reason of its very vigorous invasive stolons. Our plants are so uniform that their hybrid origin is difficult to believe, although the functionally male flowers support this view. All are the var. *alopecuroides* (Hull) Briq.
West: 61–63, 71, 73, 74, 79, 80, 84, 88, 91, 99, 09.
East: 04, 08, 14, 18, 19, 21, 24, 32, 33.

M. rotundifolia (L.) Huds.   Apple-scented Mint
Native. Very rare. Ditches and waste places.
Although stated that 'In south-west Norfolk they cultivate this exclusively in their gardens for sauce' (*J. Bot.*, LI, 1913, 143) we did not see this
species until 1964 when it was found by the lake at Didlington, W79, by
Mrs B. H. Russell and Miss V. M. Leather and confirmed by R. M.
Harley. This record is of further interest since there is a gathering, 1906,
in Herb. Norw. Mus., made by H. D. Geldart from the neighbouring
village of Northwold.
First record: 1796, Surlingham, Rev. Robt. Forby in *English Botany*, t.
446.

Lycopus europaeus L.   Gipsy-wort
Native. Common. Ditches and fens.

Origanum vulgare L.   Marjoram
Native. Locally frequent. Hedges and roadsides on chalk.
West: 63, Wolferton (on fixed shingle!); 70, Barton Bendish, Beecham-
well, Marham; 71, Narford; Narborough (extensive colonies including
albinos); 72, Flitcham, GT; 73, Sedgeford; 93, Walsingham, ALB; 94,
Wighton.
East: 11, Attlebridge, ETD; 18, Tivetshall; 19, Wreningham, ETD; 29,
Tasburgh, SA.

Thymus pulegioides L.   Wild Thyme
Native. Common in chalk grassland, less so on sand and gravel banks;
frequent on ant-hills.

T. drucei Ronn.
Native. Very rare. Confined to chalk grassland in Breckland. The rarity
of this species in Norfolk is noteworthy considering its relative abundance
in Cambridgeshire. It occurs in that county up to the limits of the more
or less continuous chalk escarpment and disappears when the chalk is
dissected by the Norfolk river valleys. Another plant with similar anomal-
ous distribution is *Viola hirta*. In view of its confusion with *T. pulegioides*
we give only authenticated records.
West: 08, Garboldisham, (C. D. Pigott in *Biol. Flora*); 78, Weeting Heath,
conf. CDP.
East: 03, Holt Lowes, FMD; 11, Alderford Common; railway bank near
Felthorpe woods, ETD.

T. serpyllum L.
Native. 'Confined to very open communities on dry sandy soils in the
Breckland of East Anglia', CDP.
West: 78, Weeting Heath; 88, Santon; and roadside margin near Emily's
Wood, Brandon.

Calamintha ascendens Jord.   Common Calamint
Native. Locally common. Dry hedgebanks.

C. nepeta (L.) Savi   Lesser Calamint

Native. As in *C. ascendens* but far less frequent, being more or less confined to the neighbourhood of Castleacre.

West: 71, E. Walton; Westacre; 81, Castleacre.

East: 29, Tasburgh, SA.

First record: 1804, Saham church, *English Botany*, t. 1414.

Acinos arvensis (Lam.) Dandy   Basil-thyme

Native. Common. Dry places, equally on chalk or sand. Colour variable, white-flowered forms frequent in Breckland.

Clinopodium vulgare L.   Wild Basil

Native. Frequent. Hedges and scrub on chalk.

Melissa officinalis L.   Balm

Est. alien. Rare. Garden-escape, formerly grown for making a herb beer.

West: 63, Dersingham; 89, Stanford Warren; 93, Walsingham, Professor Bullock, 1915.

East: 29, Saxlingham Thorpe, SA.

Salvia pratensis L.   Meadow Clary

Scarcely native; probably best regarded as a denizen. Grassland on chalk, sand-pits.

West: 62, W. Newton, as var. *modesta* Briq.; 71, Pentney, as var. *vulgaris* Reichb.; 88, Kilverstone, ALB.

S. horminoides Pourr.   Wild Clary

Native. Frequent in hedgebanks and dry grassland, particularly on chalk.

Prunella vulgaris L.   Self-heal

Native. Common. Grassland and woods. White- and pink-flowered forms occur.

Stachys arvensis (L.) L.   Field Woundwort

Colonist. Occasional arable weed on chalk and clay, often associated with *Legousia hybrida* and *Kickxia elatine*.

S. palustris L.   Marsh Woundwort

Native. Frequent in ditches, streamsides, and occasional as a weed of root crops.

S. sylvatica L.   Hedge Woundwort

Native. Common. Woods and hedges.

Betonica officinalis L.   Betony

Native. Rare in West Norfolk. Woods and scrub.

West: 71, E. Walton Common, Kew Exc., 1965; 70, Foulden Common, 92, Tittleshall.

East: 03, 04, 13, 14, 19, 20, 29, 30, 33.

Ballota nigra L., ssp. *foetida* Hayek   Black Horehound

Native. Common. Roadsides. White-flowered forms are rare; they have been seen at Watton, W90, FR, 1916; Bawsey Road, King's Lynn, W62, BSBI Exc., 1949.

Galeobdolon luteum Huds.   Yellow Archangel
Native. Frequent in woods on the clay. With a peloric terminal flower, Framingham Pigot, E20, EAE, 1932.
West: 61, 62, 80, 90–92, 99, 01, 03.
East: 00–04, 12–14, 18–21, 28–30, 39.

Lamium amplexicaule L.   Henbit
Colonist. Common. Arable weed on light soils.

L. hybridum Vill.   Cut-leaved Dead-nettle
Colonist. Less common than *L. amplexicaule*. Arable weed.

L. purpureum L.   Red Dead-nettle
Colonist. Common. Weed of cultivated land, hedgebanks and waste places. White-flowered forms occur.

L. album L.   White Dead-nettle
Colonist. Common. Hedgebanks and a weed of cultivated land.

Galeopsis angustifolia Ehrh., ex Hoffm., var *angustifolia*
    Narrow-leaved Hemp-nettle
Colonist. Not common but has been abundant on the shingle of Wolferton beach for at least thirty-five years. This is a critical species and care should be taken to distinguish the 'crowded, coarse papillae of the calyx hairs, its constant character' (see 'Some Notes on *G. ladanum* L., and *G. angustifolia*', C. C. Townsend in *Watsonia*, v, 1962, 143–9).
West: 63, Wolferton, Snettisham, Heacham; 72, Harpley; 73, Ringstead; 78, Weeting; 83, N. Creake, BT; 84, Holkham, Burnham Thorpe, BT.
East: 04, 18, 20, Gt Plumstead, JHS

G. tetrahit L.   Common Hemp-nettle
Native. Common. Arable land, wood margins and rough pastures.

G. bifida Boenn
Native. Rare. Distribution not fully known but appears to favour peaty soils. Has the appearance of a miniature *G. speciosa*.
West: 72, Harpley Dams; 78, Weeting; 98, Overa Heath, RCP; 99, Hockham Rough, Stow Bedon.
East: 04, Blakeney.

G. speciosa Mill.   Large-flowered Hemp-nettle
Colonist. Occasional. Arable weed particularly on the black peat of the Fenland in south-west Norfolk.
West: 59–62, 69, 70, 72, 79, 81, 89, 90, 92.
East: 18, 23, 29, 30.

Nepeta cataria L.   Catmint
Perhaps native. Frequent on roadsides, especially on chalk.
West: 61, 62, 64, 70–73, 83, 88, 94, 08.
East: 04, 14, 19, 20, 23, 28, 30, 41.

Glechoma hederacea L.   Ground Ivy
Native. Common. Woods and grassland.

Marrubium vulgare L.   White Horehound

Denizen. Formerly cultivated as Horehound. Frequent on dry banks, commoner on chalk, but not confined to it.

West: 62, 63; 64, Ringstead Downs, perhaps native here; 70, 74, 79, 84, 88, 99.

East: 04, 12.

Scutellaria galericulata L.   Skull-cap

Native. Common. Ditches, ponds and fens.

S. minor L.   Lesser Skull-cap

Native. Very rare. Wet heaths.

East: 03, Holt Lowes, 1928, CPP; Catpits, Holt, 1928, CPP; Wash Pits, Edgefield, PHS; 12, Blickling; 32, Stalham, PJB.

S. hastifolia L.

Nat. alien. Restricted to rides in semi-natural woodland.

West: 78, Emily's Wood near Brandon, where it was first recorded in 1948 (*Watsonia*, II, 18–21, 1951). Bears very few flowers but spreads vegetatively. Reported from Snake Wood nearby.

Teucrium scorodonia L.   Wood Sage

Native. Common. Woods and heaths, abundant at times on abandoned arable ('breck') land.

Ajuga reptans L.   Bugle

Native. Common. Wet meadows and woods.

PLANTAGINACEAE

Plantago major L.   Great Plantain

Colonist. Common. Waste places, paths, arable land and a weed in lawns.

P. media L.   Hoary Plantain

Native. Common. Grassland, on chalk or clay.

P. lanceolata L.   Ribwort Plantain

Native. Common. Grassland, abundant in Forestry Commission rides and one of the early colonisers following ploughing for fire-breaks.

Var. *anthoviridis* W. Wats. Stamens erect, filaments shorter, anthers greenish yellow and long-elliptical.

West: 72, Hillington, ELS.

East: 04, Cley, ELS.

P. maritima L.   Sea Plantain

Native. Frequent. Upper levels of salt marshes.

P. coronopus L.   Buck's-horn Plantain

Native. Dry grassland; gravelly tracks, both inland and maritime.

Littorella uniflora (L.) Aschers.   Shore-weed

Native. Very rare. Borders of pools.

West: 61, E. Winch Common, until 1943; 88, Home Mere, Wretham, ALB.

East: 11, Swannington Common, EAE, 1939; 22, Perch Lake, Westwick, C. W. Ward; 39, Thwaite Common, WGC.

CAMPANULACEAE

Campanula latifolia L.   Giant Bell-flower

Native. Rare. Woods.

West: 72, Congham; 88, Thetford, HDH; 91, E. Dereham, PDM; 99, E. Wretham, ERN.

East: 00, Seamere near Hingham; Kimberley, ERN; Coston, FRo; 20, Keswick, ETD; 39, Ditchingham, MIB.

C. trachelium L.   Nettle-leaved Bell-flower

Native. Occasional. Woods.

West: 61, 62, 70, 71, 80, 81, 88, 90–92.

East: 03, 14, 39.

C. rapunculoides L.   Creeping Bell-flower

Denizen. Occasional. Wood-margins, hedges, railway embankments and waste places.

West: 63, Ingoldisthorpe; 72, Harpley Dams; 79, Gooderstone; 80, Swaffham, ALB; Snail's Pit Farm, Swaffham, DMM; 84, Burnham Thorpe, GT; 88, Thetford, Miss A. B. Cobbe.

East: 04; 10; 20, factory site, Norwich, ETD; Harford, ETD; 21, Horstead, W. Field; 24, Overstrand, KHB; 42.

C. glomerata L.   Clustered Bell-flower

Native. Locally frequent. Exclusive to chalk grassland, in which it is constant and sometimes abundant as on the old raised banks, 'Devil's Dykes'.

West: 64, Ringstead Downs; 70, Caldecote; Foulden Common; Gooderstone; Beechamwell Fen; Marham; Barton Bendish; 79, Cranworth, ALB; 80, S. Pickenham and Ashill, ESE.

East: 20, Caistor St Edmunds, EAE.

C. rotundifolia L.   Harebell

Native. Common. Dry banks, heaths, chalk grassland and dunes.

Legousia hybrida (L.) Delarb.   Venus's Looking-glass

Colonist. Infrequent but widely distributed arable weed, often associated with *Stachys arvensis* and *Kickxia elatine* on chalk.

Jasione montana L.   Sheep's-bit

Native. Rare in West Norfolk; more frequent in East Norfolk.

West: 62, confined to a strip of heathland bordering Sugar Fen; much more plentiful in the locality fifty years ago, and at a still earlier period apparently widely distributed. 71, Narborough Field, ESE, 1957; ELS, 1966, one plant; Pentney, WGC, 1917.

East: 03, Hunworth, WGC; 20, Harford rail embankment, ETD; 41, Hemsby, ETD; 51, Caister, ETD, 1965, abundant; Winterton dunes, ETD.

RUBIACEAE

Sherardia arvensis L.   Field Madder

Colonist. Common. Arable weed.

Asperula cynanchica L.   Squinancy Wort

Native. Exclusive to chalk grassland, in which it is constant and often abundant.

West: 64, 70–72, 74, 78, 79, 82, 88, 89, 94, 98, 08.

East: 04, Walsey Hills, Cley, FMD.

Galium cruciata (L.) Scop.   Crosswort

Native. Frequent. Woods and roadsides.

West: 61, 62, 71, 72, 78, 80, 82, 90–93, 99, 01.

East: 00–04, 09, 10–14, 18–20, 23, 24, 29–31, 39, 41, 42.

G. odoratum (L.) Scop.   Sweet Woodruff

Native. Locally frequent. Woods on boulder clay.

West: 90, Cranworth and Bradenham, ALB; 91, Rawhall; Honeypot Wood, Wendling; 92, Eastfield Wood, Tittleshall; Horningtoft; 99, Thompson.

East: 01–04, 12, 20, 39, 40, 42.

G. mollugo L., ssp. *mollugo*   Great Hedge Bedstraw

Native. Frequent. Hedges.

Ssp. *erectum* Syme   Upright Hedge Bedstraw

Native. Rare but possibly overlooked owing to the variable forms of the last species.

West: 62, Vincent Hills, W. Newton; 70, Foulden; 71, Bradmoor, Narborough; 81, Sporle gravel pit; 94, Stiffkey; 01, Swanton Morley.

East: 18; 19, Ashwellthorpe, ETD; 21, 39.

G. mollugo × *verum*

(*G.* × *pomeranicum* Retz.)

Apparently a very rare hybrid.

West: 72, near the Waterworks, Gayton, 1945; 79, Whittington church-yard, 1966; 90, Watton, FR, 1916.

G. verum L.   Lady's Bedstraw

Native. Common. Hedgebanks, heaths and dunes.

A prostrate, mat-forming plant with extensive stolons and small, few-flowered cymes is abundant on the older dunes at Burnham Overy Staithe (W84), and appears to be very different from the normal plant of hedgebanks. Does not agree with De Candolle's description of his var. *maritimum*; probably an ecad only.

G. saxatile L.   Heath Bedstraw
Native. Common. Heaths.

G. palustre L., ssp. *palustre*   Marsh Bedstraw
Native. Common. Ditches.

Ssp. *elongatum* (C. Presl) Lange
This plant with longer leaves (up to 3 cm.) and larger flowers (4·5 mm. in diam.) is very striking in the dykes of Welney Wash, W59. Its roots are permanently in water. Associated species are *Schoenoplectus lacustris*, *Rorippa nasturtium-aquaticum* and the pondweeds, *Groenlandia densa* and *P. crispus*. Its distribution is imperfectly known but it will most likely be found to be widespread where standing water is always present.
West: 59, Welney Wash; 62, N. Wootton; 63, Snettisham.
East: 41, Burgh St Margaret.

G. uliginosum L.   Fen Bedstraw
Native. Common. Fens, spreading to more acid soil where bog adjoins fen.

G. aparine L.   Goosegrass, Cleavers, loc. 'Sweet-hearts'
Native. Common. Hedges, arable land, maritime shingle and waste places.

G. parisiense L., ssp. *anglicum* (Huds.) Clapham   Wall Bedstraw
Native. Rare. Old walls. Very rarely in maritime shingle (one station), sand pits (one station) and railway banks (one station).
West: 61, W. Winch church; 63, Snettisham beach; 71, Narborough; 74, Ringstead, GT; 78, Weeting Heath, BFD; Weeting Castle; 79, Foulden, WGC; 81, Castleacre Priory; 89, Sturston Warren; 93, Binham Priory.
East: 03, Briston; 04, Glandford, FMD; 11, Attlebridge, JEL; 41, Stokesby.

CAPRIFOLIACEAE

Sambucus ebulus L.   Danewort, 'Dane-blood', Dwarf Elder
Denizen. Rare. Hedges usually near dwellings.
West: 71, W. Bilney; 72, Flitcham, Grimston; 81, Castleacre, WCFN, 1913; 93, Binham; 94, Wells.
East: 30, Beighton, ETD; 31, Acle, PER; S. Walsham; 32, E. Ruston; 33, Bacton, KHB; 39, Seething, JHS.
First record: 1670, 'in Marshland between Wisbech and Lyn [*sic*] in ye fields there', Plukenet's annotated copy of Ray's *Cat. Plant.*, 1st edit.

S. nigra L.   Elder
Native. Common. Scrub, hedges, woods and dunes.

Var. *laciniata* L., leaflets linear, laciniate.
West: 74, Brancaster Heath, PDM; 78, Weeting; 88, Croxton.
East: 24, Overstrand, KD.

Viburnum opulus L.   Guelder Rose
Native. Common in wet woods, especially characteristic of carr.

Var. *flavum* Horwood, with yellow berries.
East: 30, Wheatfen Broad, A. A. Bullock, 1938.

Symphoricarpos rivularis Suksd.   Snowberry
Est. alien. Originally planted for game coverts. Abundant in many woods and plantations; widespread in Breckland and in hedges in East Norfolk. Spreading vegetatively and often forming large thickets.

Lonicera xylosteum L.   Fly Honeysuckle
Apparently very rare. An introduction.
West: 99, in scrub beside Darkland Plantation, Wretham Park, ALB, 1967, its first Norfolk record.

L. periclymenum L.   Honeysuckle
Native. Common. Woods, scrub, hedges and sand-dunes at Holkham.

L. caprifolium L.   Perfoliate Honeysuckle
Nat. alien. Very rare. Hedges.
East only: 20, Keswick, EAE; Old Lakenham, ETD; 39, Geldeston, EAE, 1934.

ADOXACEAE

Adoxa moschatellina L.   Moschatel, Town Hall Clock
Native. Common. Woods and damp hedgebanks in shade.

VALERIANACEAE

Valerianella locusta (L.) Betcke   Lamb's Lettuce, Corn Salad
Colonist. Frequent. Arable land, hedgebanks, railway embankments, sea-banks and dunes.
West: 62–64, 74, 80, 84, 91, 92, 94, 99.
East: 00, 02, 03, 10, 18–21, 23, 28–31, 33, 39–42, 49.

V. dentata (L.) Poll.   Narrow-fruited Corn Salad
Colonist. Rare arable weed of chalky soils.
West: 64, 70–73, 78, 79, 83, 89, 93.
East: 03, 13, 14, 21, 29, 42.

Var. *eriosperma* (Wallr.) Janch., the hairy-fruited variety.
West: 79, Cold Harbour near Hilborough, VML, 1964, det. Kew. Previously recorded in Breckland as var. *mixta* (Dufn.).

Valeriana officinalis L.   Valerian
Native. Common. Wet woods, marshes and wet places generally, particularly in Broadland.
First record: 1796, Bardolph Fen, W60, W. Skrimshire, *Cat.*

V. dioica L.   Marsh Valerian

Native. Common. Fens, wet meadows and woods.

Centranthus ruber (L.) DC.   Red Valerian

Est. alien. Occasional. Cliffs, walls and escaping from gardens.
West: 62, W. Newton; 64, Hunstanton; 88, Thetford, HDH.
East: 02–04, 09, 18, 31, 41, 49.

DIPSACACEAE

Dipsacus fullonum L., ssp. *fullonum*   Wild Teasel

Native. Common. Riversides, sea-banks, woods and waste places.

D. pilosus L.   Small Teasel

Native. Rare. Damp woods and hedgebanks. The 1914 *Flora* listed many stations.
West: 91, Gorgate near E. Dereham, DMM; oo, Reymerston, ALB.
East: 03, Brinton, PHS; 04, Bayfield Hall near Holt; 11. Attlebridge and Lenwade; 19, bank of R. Tas near Bunwell, ELS; 29, Saxlingham, FR; 39, Ditchingham, MIB.
First record: 1779, Out of St Benedict's Gates, Norwich, JES.

Knautia arvensis (L.) Coult.   Field Scabious

Native. Common. Hedges, arable fields and waste places.

Scabiosa columbaria L.   Small Scabious

Native. Frequent, especially in West Norfolk. Exclusive to chalk grassland, in which it is constant.

Succisa pratensis Moench   Devil's-bit Scabious

Native. Common. Acid grassland and damp heaths.
White-flowered forms very uncommon: E29, Newton Flotman Common, ETD.

COMPOSITAE

Bidens cernua L.   Nodding Bur-marigold

Native. Frequent. Ponds and streamsides.
West: 60-62, 72, 73, 78–80, 82, 83, 89, 92, 98, 99.
East: oo, 02, 09, 11, 13, 18, 20–23, 28, 30–33, 39, 41, 42, 49.

Var. *radiata* DC., with ray florets.
West: 60, Stradsett Lake; 92, Brisley, DMM.
East: 01, Elsing, A. E. Ellis; 20, Thorpe St Andrew, AEE.

B. tripartita L.   Three-cleft Bur-marigold

Native. Less frequent than *B. cernua*. In similar situations.
West: 62, 70–73, 82, 88, 90, 92, 98, oo.
East: 03, 09, 10, 18–20, 30–32, 39.

Galingsoga parviflora Cav.   'Gallant Soldier'

Est. alien. Increasingly frequent. Cultivated ground and waste places.

West: 62, S. Wootton, since 1927; Roydon Common, 1960; 63, Dersingham, 1950; 71, Narborough, 1944; 84, Burnham Overy, 1959; 99, Shropham, 1957, ALB.

East: 03, Holt, PHS, 1948; 20, Dunston, EAE, 1950; 24, Cromer, ALB, 1966; 28, Pulham St Mary, CF, 1963; 50, Gt Yarmouth, EAE, 1943; 51, Caister, GG, 1960.

First record: 1927, S. Wootton, CPP.

G. ciliata (Rafn) Blake

Est. alien. Far less frequent than *G. parviflora*. In similar situations.

West: 62, Whin Hill Covert, Wolferton, 1960; King's Lynn, 1964; 71, Gayton Thorpe, RSC, 1964; 91, E. Dereham, DMM, 1965.

East: 00, Hackford, 1962; 20, Norwich, 'abundant in grounds of W. Norwich Hospital', EAE, 1957; Mousehold Heath, 1957; Harford Bridges, ETD, 1966; 31, Acle new road, ETD, 1964; 50, Gt Yarmouth, EAE.

First record: 1957, Norwich, EAE.

Senecio jacobaea L.   Ragwort, loc. 'Canker-weed'

Native. Common. Grassland, heaths, roadsides and stable dunes.

S. aquaticus Hill   Marsh Ragwort

Native. Common. Marshes, fens, wet pastures and ditches.

S. erucifolius L.   Hoary Ragwort

Native. Locally common on roadsides and field-borders on heavy soil and land reclaimed from the sea.

First record: 1798, at Holm (? Holme), Mr Sutton in *English Botany*, t. 574.

S. squalidus L.   Oxford Ragwort

Est. alien. Increasingly common. Old walls, waste places and railway embankments. Spread rapidly on bombed sites 1941–3.

West: 60–64, 71, 78, 83, 84, 90, 91, 93, 99, 01.

East: 00, 03, 04, 10, 18–24, 30–33, 39–42, 51.

First record: 1850, Eaton, introduced.

S. squalidus × *vulgaris*

East: 20, grounds of Norwich Castle, 1943, H. J. Howard; six plants on bombed sites, 1944, EAE, Norwich.

S. sylvaticus L.   Wood Groundsel

Native. Common. Woods on light, sandy soil and on heaths.

S. viscosus L.   Stinking Groundsel

Colonist. Shingle beaches and waste ground.

In spite of old records, it was unfamiliar in West Norfolk until the Second World War, since when it has been abundant at Snettisham shingle beach. Salisbury in *Trans. Nfk. and Norw. Nat. Soc.*, 1932, map 98, marks it absent from Norfolk; date of introduction to Britain,. 1660.

West: 61–63, 73, 78, 83, 84, 99.

East: 02, 04, 10, 11, 18, 22, 23, 28, 29, 39.

S. vulgaris L.   Groundsel

Colonist. Common. Arable weed and waste places.

Plants with rayed florets, var. *radiatus* Koch, appear to have increased since the 1914 *Flora* when only one record appeared.

West: 61, Middleton, RPL; 62, King's Lynn; 81, Litcham, DMM.

Doronicum pardalianches L.   Leopard's-bane

Nat. alien. Rare. Damp woods.

West: 99, Hockham, ALB.

East: 00, Hardingham, EQB; 03, Gresham's School woods, PHS; 20, Gt Plumstead, JHS; 33, Knapton, EAE; 39, Ditchingham, SA.

First record: 1882, Ranworth, E31, A. W. Preston in Herb. Norw. Mus.

Tussilago farfara L.   Coltsfoot, 'Clote Weed'

Native. Common. Arable weed on heavy soil, railway embankments, river banks and sand-dunes.

Petasites hybridus (L.) Gaertn., Mey. & Scherb.

    Butterbur, 'Wild Rhubarb', also applied to *Arctium* spp.

Native. Frequent. River-banks. Only the male plant is known in Norfolk. Colony with about thirty sterile hermaphrodite florets per head, near the river at Newton-by-Castleacre, W81, EAE.

West: 60, 62, 63, 70–73, 80–82, 88 (bank of Little Ouse, Thetford, on clay introduced from Ely), 91–93, 99, 00, 01.

East: 11, 20, 22, 32, 38.

P. albus (L.) Gaertn.   White Butterbur

Nat. alien. Very rare. Hedgerow.

East: 13, farm at W. Beckham, 1952.

P. fragrans (Vill.) C. Presl   Winter Heliotrope

Nat. alien. Frequent. Roadsides, streamsides and waste places.

West: 62, 63, 73, 74, 80, 81, 84, 88, 90–93, 99, 00.

East: 00, 02–04, 10, 12, 14, 18–20, 22, 20–33, 39, 40, 42, 49.

Inula helenium L.   Elecampane

Nat. alien. Rare. Waste places as a garden-escape.

West: 91, Wendling, 'beside a small stream', A. H. Turner, 1943.

East: 14, W. Runton, KHB; 30, Brundall EAE, 1933.

I. conyza DC.   Ploughman's Spikenard
Native. Practically confined to chalk, where it is not rare in woods, hedges and railway cuttings. Its occurrence on dunes at Holkham and Wells is associated with their content of calcareous shell residues.
West: 61, 63, 64, 71, 72, 74, 78, 81–84, 88, 89, 94.
East: 03, 04, 10–12, 14, 18, 20, 28, 30, 39.

Pulicaria dysenterica (L.) Bernh.   Fleabane
Native. Common. Ditches and wet pastures.

P. vulgaris Gaertn.   Small Fleabane
Native. Very rare. No recent records.
East: 20, Framingham Pigot, 1915, in Herb. Norw. Mus.

Filago germanica (L.) L.   Cudweed
Native. Common. Sandy arable land on acid soils, heaths and waysides.

F. apiculata G. E. Sm.   Red-tipped Cudweed
Native. Rare. Sandy arable land. Flowers a week or two later than *F. germanica* with which it is sometimes associated.
West: 60, S. Runcton; 61, Blackborough End; 63, Snettisham; 91, Hoe, DMM; 02, near Billingford, DMM.
East: 11, Swannington, FRo; 21, Wroxham, GHR.
First record: 1848, Thetford, G. S. Gibson in Herb. Norw. Mus.

F. spathulata C. Presl   Spathulate Cudweed
Native. Very rare. Sandy fields.
West: 70, Gooderstone; 71, Narborough Field; 03, Swanton Novers, ETD.
East: 21, Frettenham.
First record: *c.* 1892, Sheringham, Geldart in Herb. Norw. Mus.

F. minima (Sm.) Pers.   Slender Cudweed
Native. Common. Heaths, 'brecks', and sandy arable land.

Gnaphalium sylvaticum L.   Wood Cudweed
Native. Rare. Woodland rides and heaths.
West: 60, Shouldham Warren; Shouldham Thorpe; 62, S. Wootton; 70, Cockley Cley; 71, Pentney; 78, Snake Wood, Weeting; 80, Swaffham.
East: 04, Kelling Warren; 39.

G. uliginosum L.   Marsh Cudweed
Native. Common. Moist places on sandy arable land and waysides where water has been standing.

G. luteoalbum L.   Jersey Cudweed
Unquestionably a native Norfolk plant. Rare. Damp hollows of sand-dunes and formerly in sandy fields of Breckland, similar to its continental distribution.
    First recorded from fields at 'Larlingford (Larling, W98) by the Rev. G. R. Leathes in Hooker's *Brit. Flora* (1831); in 1882, the Rev. E. F. Linton discovered it in its maritime stations, the Holkham–Burnham Overy Staithe area, W84; in 1913, F. Robinson found it in a sandy field

at Thompson, W99, which is probably the station where W. G. Clarke, in 1915, observed 'an average of 24 plants to the square yard over a considerable area' ('The Breckland Sand-Pall and its Vegetation', *Trans*. Nfk. and Norw. Nat. Soc., x, **2**, 143). There is a specimen in Herb. Norw. Mus., taken by Clarke from Scolt Head in 1923 and, in 1951, A. C. Jermy found one plant on waste ground at Buxton Heath near Hevingham, E12. In 1954, following the sea-flood, Miss Ruth Carey reported it from the small golf course at Wells-next-the-Sea. It still persists in the dunes between Burnham Overy Staithe and Holkham where, in 1967, some forty to fifty plants were counted.

Solidago virgaurea L.   Golden-rod
Native. Rare. Heaths and hedgebanks.
West: 61, E. Winch Common; 72, Derby Fen; 91, Beetley, DMM; Gressenhall, ALB.
East: 03, 09, 11, 13; 14, Beeston Regis Common; 20, 29, 40.

Aster tripolium L.   Sea Aster
Native. Common. In salt marshes at an early stage of development. Extends along river-banks where, in·the softer and less saline mud, the rayed form is very luxuriant.
Var. *discoideus* Reichb., appears to be more frequent now than the rayed form.

Erigeron acer L.   Blue Fleabane
Native. Frequent. Heaths and sand-dunes.

Conyza canadensis (L.) Cronq.   Canadian Fleabane
Nat. alien. Common. Cultivated land on sandy soil, walls and dunes.
'First appeared in the early eighties at Croxton, W88 . . . has become a striking feature in the plant-life of the district' (*In Breckland Wilds*).

Bellis perennis L.   Daisy
Native. Common. Grassland.

Eupatorium cannabinum L.   Hemp Agrimony
Native. Common. Ditches and fens. Small colony of white-flowered plants, Catfield Common, E42, ELS, 1966.

Anthemis arvensis L.   Corn Chamomile
Colonist. Fairly frequent as an arable weed on chalk, less common on sandy soil.

A. cotula L.   Stinking Mayweed
Colonist. Occasional weed of arable land; more frequent on the heavier soils of south Norfolk.
West: 62, N. Wootton; Castle Rising; 70, Beechamwell; Caldecote; Swaffham; 82, Tittleshall; 90, Bradenham, ALB; Ovington, ERN; 91, Wendling; Rawhall.
East: 04, 11, 18, 20, 29, 39, 41, 42.

Chamaemelum nobile (L.) All.   Chamomile
Native. Very rare. Sandy commons.
West: oo, Reymerston, ALB, 1964, the only recent record.

Achillea ptarmica L.   Sneezewort
Native. Occasional. Wet pastures and fens.
West: 61, 62, 70, 72, 79, 81, 91, 99.
East: 02, 11, 19, 22, 29, 30, 39, 41.

A. millefolium L.   Yarrow, Milfoil
Native. Common. Pastures, roadsides and waste places.

Tripleurospermum maritimum (L.) Koch ssp. *inodorum* (L.) Hyl. ex Vaarama
    Scentless Mayweed
Colonist. Common. Arable weed.

Matricaria recutita L.   Wild Chamomile
(*M. chamomilla* auct.)
Colonist. Occasional in the west but more frequent on the heavier soils of
south-east Norfolk.
West: 51, 60–63, 70, 71, 74, 78, 80, 82, 91, 92, 99, 01.
East: 03, 04, 08, 09, 12–14, 18, 19, 21, 22, 29, 31, 32, 39–42.

M. matricarioides (Less.) Porter   Rayless Mayweed, 'Pineapple Weed'
Nat. alien. Widely established and very common as a weed on paths,
tracks and about field gates; weed of cultivated land.
'The popular name of "War Weed" emphasises the fact that it became
noticeable first, during and after the 1914 War, although it existed [*sic*]
prior to 1914', HDH.

Chrysanthemum segetum L.   Corn Marigold
Colonist. Common arable weed showing little of the diminution reported
elsewhere. Sometimes abundant on light acid soils and on former farm
land taken into newly made roads.

C. leucanthemum L.   Oxeye Daisy, Marguerite
Native. Common. Grassland, railway banks and waste places.

C. parthenium (L.) Bernh.   Feverfew
Est. alien. Frequent. Hedges, walls near gardens and rubbish tips.

C. vulgare (L.) Bernh.   Tansy
Colonist. Probably spread by cultivation. Common. Hedges, roadsides
and waste places.

Artemisia vulgaris L.   Mugwort, 'French Tobacco'
Native. Common. Hedges and waste places.

A. verlotorum Lamotte   Verlot's Mugwort
Est. alien. Appeared on Old Hunstanton beach in small quantity in 1950
and has persisted. This was the most northerly extension in the British
Isles at the time of writing the *West Norfolk Plants Today* in 1962.

A. absinthium L.   Wormwood
Status doubtful. Many old records for waste places.
East: 18; 20, Norwich, several hundred plants on a bombed site, 1964,
ETD; rubbish tip at Harford Bridges, 1961; 23, N. Walsham, JHS, 1963;
32, near Dilham, PJB, 1963.

A. maritima L.   Sea Wormwood
Native. Characteristic of the landward side of salt-marshes and sea-banks
where it is common. Forms with short, erect branches (var. *subgallica*
Rouy) are equally common.
First record: 1806, Great Yarmouth, J. E. Smith, *English Botany*, t. 1706
as *A. genuina* Syme.

A. campestris L.   Field Southernwood
Native. Very rare now. Sandy heaths of Breckland. One of the plants
restricted to Norfolk and Suffolk. Two or three dozen plants were intro-
duced to the roadside near Weeting Heath in 1958 but have not persisted.
Evidence points to its disappearance in Norfolk.
West: 79, Devil's Dyke near Cranwich, 1934; Northwold, WGC, 1915.
First record: 1755, Stratton Strawless, Benj. Stillingfleet in his *Calendar
of Flora*, 2nd edit., 1762.

Carlina vulgaris L.   Carline Thistle
Native. Frequent in West Norfolk, less so in East Norfolk. Dry places on
sand and chalk, and on shingle-banks, sometimes in dense colonies.

Arctium lappa L.   Great Burdock, Wild Rhubarb
Native. Occasional. Dykesides on black peat land overlying clay, banks
of the Fenland drains, rivers and roadsides; not in woods.
West: 50, 51, 60, 61, 68–71, 79, 80, 84, 88, 89, 91, 98, 99, 01.
East: 03, 10, 12, 13, 18–22, 30, 32, 33, 39, 41, 42, 49.

A. nemorosum Lej.
(*A. minus* Bernh., ssp. *nemorosum* (Lej.) Syme)
Native. Common. Woods and waste places. The most widespread and
abundant species, intermediate in size between *A. lappa* and *A. minus*.

A. minus Bernh., ssp. *minus*
Native. We consider this a rare species, except in Breckland, where the
typical plant, with small ovoid heads, not exceeding 2 cm. in diameter,
largely replaces *A. nemorosum*. Although reported as occurring in 19
squares in East Norfolk we cannot confirm this. We have no observations
on ssp. *pubens* (Bab.) J. Arènes and consider this segregate may be a source
of confusion.

Carduus nutans L.   Musk Thistle, Nodding Thistle
Native. Common. Grassland, not confined to chalk, but avoiding sands.

C. acanthoides L.   Welted Thistle
(*C. crispus* auct.)
Native. Frequent on roadsides.

Cirsium eriophorum (L.) Scop., ssp. *britannicum* Petrak   Woolly Thistle
Native. Very rare now. Only seen recently on a chalky roadside at Scarning (W91), 1955. The only recent records are for this parish and the nearby one of Longham (Nicholson, *Supplement*, 1923, *Trans. Nfk. and Norw. Nat. Soc.*, XI, 504). It is with some diffidence that we claim the plant as a Norfolk one.
First record: North Pickenham, Rev. Watts, *fl.* 1750–1800.

C. vulgare (Savi) Ten.   Spear Thistle
Colonist. Common weed of arable land and grassland; also in waste places.

C. palustre (L.) Scop.   Marsh Thistle
Native. Common. Wet pastures, fen and carr. White-flowered forms are frequent.

C. arvense (L.) Scop.   Creeping Thistle
Colonist. Common weed of grassland and arable land.

C. acaulon (L.) Scop.   Stemless Thistle
Native. Constant and abundant in chalk grassland, rare on sand, gravel or clay. Caulescent forms are frequent and appear to be much in evidence in a rainy season.

C. acaulon × *dissectum*
(*C.* × *woodwardii* (H. C. Wats.) Nyman), a fertile hybrid.
West: 08, S. Lopham Fen, 1953.

C. dissectum (L.) Hill   Meadow Thistle
Native. Frequent in both bog and fen. Often associated with *Schoenus nigricans*.

Silybum marianum (L.) Gaertn.   Milk Thistle
Est. alien. Rare, persisting at times along roadsides and in waste places.
West: 74, Titchwell; 80, Swaffham, ESE; 84, Burnham Thorpe, BT; 90; 94, Stiffkey.
East: 04; 10, Bawburgh, ETD; 11, Attlebridge; 14, 18, 20, Arminghall, EAE; 28, Brockdish; Harleston, ETD; 29, Tasburgh, SA; 32, 33, Mundesley; 51.

Onopordon acanthium L.   Scotch Thistle
Est. alien. Frequent. Roadsides, field margins, fire-breaks in Breckland and waste places.

Centaurea scabiosa L.   Greater Knapweed
Native. Common. Dry grassland on chalk or gravel. White-flowered forms occasional.

C. cyanus L.   Bluebottle, Cornflower
Distribution in Norfolk suggests alien origin. Once common, now a rare arable weed and garden-escape.
West: 61, E. Winch; 71, Pentney, in a corn field, 1945.
East: 20, 30, 41.

C. nemoralis Jord and *C. nigra* L.

Marsden-Jones and Turrill (*British Knapweeds*, Ray Society, 1954) consider that the situation in the *Centaurea nemoralis-nigra* complex is best understood by regarding *C. nigra* L., and *C. nemoralis* Jord., as two distinct species and intermediates as hybrids between them, the determinations being based mainly on phyllary characters. These concepts have been followed in examining the East Norfolk situation where fairly extensive sampling of populations has been carried out by E. T. Daniels.

Over most of the area investigated the plants were predominantly hybrid swarms of *C. nemoralis-nigra* with which were often mingled true *C. nigra* or true *C. nemoralis*. *C. nigra* was found to be common but almost exclusively on the heavy soils of south Norfolk. It was usually associated with hybrid colonies having strong *nigra* genes, with a very few widely scattered individuals of true *nemoralis*. In the eastern, northern and western parts of the vice-county the influence of *nemoralis* in the hybrid populations increases markedly and frequently full *nemoralis* appears, particularly as the coast is approached, *C. nigra* is extremely rare in these zones.

A very few plants have been found with traces of genes of the alien *C. jacea* L., as evidenced by fimbriation of the margins of some of the phyllary appendages, which are also paler. It is possible the *jacea* ancestor was introduced with chicken food.

No recent examination has been made of the position in West Norfolk, but Marsden-Jones and Turrill, working in this vice-county, mostly in the thirties, found that, in the main, populations were to be referred to *C. nemoralis*, although some showed a mixture of *nemoralis* and *nigra* characters. They considered the situation at Leziate Fen to be the most remarkable ecological occurrence in the vice-county. Here knapweeds were common with such plants as *Cladium mariscus* and *Schoenus nigricans*. Possibly they had been introduced by cattle, but nevertheless were able to grow in fen conditions and to compete successfully.

C. nemoralis Jord.   Knapweed

Native. Common except on very heavy soils where it is extremely rare. Grassland, hedgebanks and fens, in which last habitat it is probably an introduction by way of grazing animals. Forms with enlarged marginal florets are uncommon.

C. nigra L.   Black Knapweed

Native. Locally common in south Norfolk especially on heavy clay soils. Rare elsewhere.

West: 72, Appleton, M-J & T.
East: 18, Tivetshall, Dickleburgh and Burston, Shimpling, Gissing; 19, Long Stratton; 28, Pulham Market; 29, Morningthorpe, Saxlingham Nethergate; 02, Whitwell; 03, Field Dalling; 40, Reedham. All ETD

C.   nemoralis × nigra

A widespread and abundant hybrid occurring with the parents, often with the characters of *C. nigra* dominant, but over most of the area *nemoralis* is the dominant partner.

West: 64, cliffs, Hunstanton, M-J & T; 63, Snettisham; 94, Stiffkey; Wells; 88, Thetford, M-J & T.

East: generally distributed and common.

C. jacea × *nemoralis* × *nigra*

West: 63, Heacham, M-J & T.

East: 04, Wiveton; 20, Norwich; Kirby Bedon. All ETD.

Cichorium intybus L.   Chicory

Est. alien. Widely established on roadsides and field-borders. Much chicory is grown on the peat soil in the valleys of the rivers Wissey and Little Ouse, where escapes are frequent.

Lapsana communis L.   Nipplewort

Colonist. Common arable weed, hedgerows and walls.

Arnoseris minima (L.) Schweigg. & Koerte   Swine's Succory

Colonist. Very rare weed of sandy arable land.

West: 71, Pentney, 1945–53.

East: 04, Walsey Hills in Salthouse parish, RSRF, 1955.

Hypochoeris radicata L.   Cat's Ear

Native. Common. Grassland.

H. glabra L.   Smooth Cat's Ear

Native. Frequent in grass heath, abandoned sandy arable ('brecks') and sand-dunes. Very variable in stature, some small seaside and Breckland forms being minute, barely 2·5 cm. across the rosette. Unbeaked forms (var. *erostris* Coss & Germ.) occur.

Leontodon autumnalis L.   Autumnal Hawkbit

Native. Common. Grassland.

L. hispidus L.   Rough Hawkbit

Native. Common. Moist pastures, fens and railway embankments on calcareous soil.

L. taraxacoides (Vill.) Mérat   Hairy Hawkbit

Native. Frequent. Dry grassland on basic soil and sand-dunes.

Picris echioides L.   Bristly Ox-tongue

Colonist. Frequent arable weed on heavy soil; occasionally in sandy arable land.

West: 51, 61–64, 72–74, 79–81, 84, 90, 91, 00–04.

East: 03, 04, 10, 18, 20, 21, 23, 24, 28–31, 39–42, 49.

P. hieracioides L.   Hawkweed Ox-tongue

Native. Less frequent than the last species. Roadsides, dry banks, railway cuttings, and waste places on chalky boulder clay.

West: 64, 71–74, 80, 83, 90–93, 03.

East: 01, 04, 10, 18, 19, 33, 40.

Tragopogon pratensis L., ssp. *minor* (Mill.) Wahlenb.
    Goat's Beard, 'John-go-to-bed-at-noon'
Native. Common. Roadsides and grassland.

Ssp. *pratensis* appears to be very rare and is probably of adventive origin.
West: 70, near Little Thorns, Swaffham.

T. porrifolius L.   Purple Salsify
Est. alien. Roadsides as a garden-escape.
West: 62, Castle Rising; 72, Harpley; Hillington, RSC.
East: 14, Weybourne; 32, E. Ruston; 50, Gt Yarmouth, EAE.

Lactuca virosa L.   Acrid Lettuce
Status doubtful. Abundant for thirty-five years at least on sand-dunes at
Old Hunstanton, where it appears native, but recently has spread widely
as a weed of waste places. Has been much confused with *L. serriola* which
is a casual and does not persist.
West: 60–64, 70, 73, 80, 84, 88.
East: 03, 04, 14, 24.

L. saligna L.   Least Lettuce
Native. Very rare but in fair quantity on the bank of the Smeeth Lode,
Tilney St Lawrence, W51, in 1953. Perhaps a relict of the old fen flora;
not far away grew *Althaea officinalis*.
Kirby Trimmer (*Fl. Norf.*, 1866) recorded it from Warham, W94, and
Walsingham, W93.

Mycelis muralis (L.) Dum.   Wall Lettuce
Native. Frequent. Woods and old walls.
West: 61, 70–74, 78, 83, 84, 89–91, 93, 94, 99, 02–04.
East: 00, 01, 03, 11, 12, 14, 18, 20–24, 28–30, 39.

Sonchus palustris L.   Marsh Sow-thistle
Native. Has been looked upon as a rare and decreasing species in Britain
but E. A. Ellis concluded in 1944 that it was as frequent in the Broadland
area as it was 220 years ago, when it was first recorded from Lothingland
by Sherard. More recently, Ellis has shown it to be still more abundant and
increasing. 'It is essentially a plant of reed ronds' (strips of marshy land
fringing the lower reaches of rivers), EAE. It does not occur in West
Norfolk and we accept Arthur Bennett's view that the old records for the
vice-county really represent the marsh forms of *S. arvensis*. In East
Norfolk, it occurs 'along the lower reaches of the Waveney and beside the
Yare from Surlingham to Reedham; at Acle and Stokesby on the Bure; at
Horsey Mere and along the Ant at Ludham', EAE. Along the Thurne at
Thurne and Potter Heigham and also spreading along the Chet; especially
abundant along Haddiscoe New Cut, ETD.

84 CHICORY *Cichorium intybus* *A. H. Hems*

85 MARSH SOWTHISTLE *Sonchus palustris* *R. Jones*

86 ARROW-HEAD   *Sagittaria sagittifolia*   *G. D. Watts*

87 WATER SOLDIER   *Stratiotes aloides*   *G. D. Watts*

88 WILD TULIP   *Tulipa sylvestris*   *R. Jones*

89 STAR-OF-BETHLEHEM   *Ornithogalum umbellatum*   *Dr S. Clark*

90 NODDING STAR OF B. *Ornithogalum nutans* Miss D. M. Maxey

91 RAMSONS *Allium ursinum* J. Secker

S. arvensis L.   Corn Sow-thistle

Status doubtful, probably best regarded as a colonist in view of its frequency as an arable weed; also occurs by streamsides and in coastal habitats. Tall plants growing by streams (var. *riparius* Magn.) with the habit of *S. palustris* occur occasionally and have been mistaken for that species.

Var. *angustifolius* Meyer, a distinct-looking plant with long narrow leaves, up to 25 cm. long and not more than 2 cm. wide, with not or scarcely amplexicaul bases, was recorded for the first time as a British plant in 1885 from marshes at Wells-next-the-Sea by F. Long. Material in Herb. Norw. Mus.

Var. *glabrescens* Gunth., Grab., & Wimm., with completely glabrous peduncles and involucres, is very rare.

West: 61, damp, marshy pasture, Setch, ELS.

S. oleraceus L.   Sow-thistle

Colonist. Common as an arable weed; waysides and waste places.

S. asper (L.) Hill   Spiny Sow-thistle

Colonist. Now commoner than the last, in similar habitats.

Var. *glandulosus* (Coss., Germ., & Wedd.)

West: 90, Watton, Druce, 1918.
East: 04, Blakeney, Druce, 1918.

Hieracium   Hawkweed

Norfolk is not rich in hawkweeds and the 1914 *Flora* listed eight so-called 'species', some of which have since been shown to be aggregate species covering more than one taxon. With the exception of the readily identifiable *H. umbellatum*, all other taxa have a breeding system whereby embryos and seeds are produced asexually resulting in apomictic strains of bewildering form. It is such biotypes, differing only in degree, that make this genus very difficult to classify. Our account is based partly on Pugsley's 'Prodromus of the British Hieracia' (*Journ. Linn. Soc.,* LIV, 1948) and the nomenclature follows Sell and West's arrangement in Dandy's *List of British Plants*, 1958.

We are grateful to Mr P. D. Sell for his help with some of our records.

Section Amplexicaulia Fr.

Hieracium speluncarum Arv.-Touv.

Est. alien. Very rare.

West: 72, Abbey Farm, Flitcham, on walls of ha-ha, 1965, RSC & ELS, confd. PDS.

Section Vulgata Fr.

H. scotosticum Hyland.
(*H. praecox* sensu Pugsl.)

West: 62, King's Lynn, has persisted as a garden weed for thirty years, confd. PDS.

H. exotericum Jord. ex Bor.

Native. Occasional. Woodland margins.

East: 11, Felthorpe Woods and Horsford, ETD; Attlebridge, ALB & ELS; 20, Thorpe St Andrew, JHS.

H. grandidens Dahlst.

(*H. exotericum* forma *grandidens* (Dahlst.) Pugsl.)

East: 22, Felmingham, 1916, FR, det. Druce as 'eglandular form of *H. grandidens*'.

H. maculatum Sm.   Spotted Hawkweed

Est. alien. Occasional. Chalk pits and railway cuttings. 'Brought from Westmorland in 1781 by Mr. Crowe from whose garden it has established itself in the neighbourhood of Norwich, spreading extensively by seed', *English Botany*.

East: 01, Hockering Wood, ELS; 10, Hethel airfield, ETD; 14, W. Runton, FR; 19, Silfield, ETD & ELS; 20, Eaton chalk pit, ETD & ELS; Thorpe St Andrews Hospital, JHS; Norwich, ETD; 29, Brooke church-yard, JHS.

H. diaphanum Fr.

East: 03, Edgefield, ELS det. PDS.

H. anglorum (A. Ley) Pugsl.

Native. Occasional. Woodland margins.

West: 62, Gaywood, Pugsley; 89, Scotch Covert, Mundford, ELS.

East: 03, Holt-Edgefield, ELS confd. PDS; Briston, ELS; 12, Oulton near Blickling, ELS; Buxton Heath, ALB; 14, W. Runton, FR; 20, Mousehold Heath, ETD & ELS; Eaton chalk-pit, ETD; 21, Hainford, ETD.

H. strumosum (W. R. Linton) A. Ley

Native. Occasional. Woodland banks.

West: 62, N. Wootton, det. PDS; 70, Foulden Common, det. PDS; 79, Cranwich, GT det. PDS.

H. lachenalii C. C. Gmel.

(*H. sciaphilum* (Uechtr.) F. J. Hanb.)

Native. Rare. Woods and banks.

East: 02, Hindolveston, ETD; 23; 33, Knapton; Swafield to Paston, E. F. Linton as *H. sciaphilum*.

Section Tridentata Fr.

H. tridentatum Fries

Native. Rare. Hedgerows and wood margins.

West: 62, Grimston Warren, ELS; 91, Longham, ALB.

East: 01, Hockering, ALB; 12, Buxton Heath, ELS.

H. eboracense Pugsl.

(*H. tridentatum* sensu Pugsl. pro parte)

Native. Rare. Hedgebank.

West: 92, Whissonsett, ELS confd. PDS.

Section Umbellata Fr.

H. umbellatum L.   Narrow-leaved Hawkweed

Native. Locally abundant. Wood margins, hedgebanks and sand-dunes.

West: 60, Westbriggs Wood, Shouldham; 61, Bawsey; 62, Grimston Warren; 64, Old Hunstanton dunes; 74, Holme-next-the-Sea dunes; 88, 93, 98; 00, Reymerston, ALB.

East: 01; 11, Horsford, ETD; 12, Buxton Heath, ETD; 20, Mousehold Heath, ETD & ELS; 21, Hainford, ETD; 22, Stratton Strawless, ETD; 32, 40, 42; 51, Gt Yarmouth, JHS; Caister-on-Sea, ETD.

First record: 1807, 'gathered on the beautiful wooded hills at the back of Thorpe, Norwich, where it flowers in August', Smith, *English Botany*, t. 1771.

Var. *coronopifolium* Bernh. ex Hornem., occurs frequently with the normal form.

Section Sabauda Fr.

H. perpropinquum (Zahn) Pugsl.

Native. Locally frequent. Woodland margins, hedgebanks and heaths. Many of the older records of *H. sabaudum* are referable here. Was first described as a British plant, under the name of *H. sabaudum*, by Smith in *English Botany*, from specimens collected at Thorpe. The authors follow Sell & West in rejecting Pugsley's *H. bladonii* as a valid species although examples, here treated as *H. perpropinquum*, have been found in the field (ETD & ELS) which agree closely with his description of *H. bladonii* and, in the living state, have a facies which differs subtly from *H. perpropinquum*.

West: 71, Abbey Farm, E. Walton; 73, Docking; 83, Syderstone Common, confd. PDS; 90, Bradenham.

East: 12, Buxton Heath; 20, Mousehold Heath, ETD & ELS; Norwich and Thorpe, ETD; 21, Rackheath, ETD & ELS, det. PDS; 33, near N. Walsham, PJB det. PDS.

Section Pilosellina Fr.

H. pilosella L.   Mouse-ear Hawkweed

Native. Common. Dry banks and grassland on all types of soil from sand-dunes to chalk.

Section Collinina Naeg. & Peter

H. brunneocroceum Pugsl.

Est. alien. Occasional. Garden-escape.

West: 62, Castle Rising, 1939; 91, Beetley gravel pit, 1965.
East: 10, Wymondham, ETD & ELS; 31, Panxworth, 1964, JHS.

Crepis vesicaria L., ssp. *taraxacifolia* (Thuill.) Thell.
    Beaked Hawk's-beard

Colonist. Formerly rare and confined to chalk, now spreading widely, especially in gravel pits, on margins of new roads, and railway banks.

C. biennis L.   Rough Hawk's-beard

Colonist. Has always been a rare plant in Norfolk and is often confused with the last species.

West: 90, Cranworth, ALB, 1961; Watton, FR, 1915.

East: 18, 19, 41.

C. capillaris (L.) Wallr.   Smooth Hawk's-beard

Native. Roadsides (where it is common), arable land, heaths and walls.

Taraxacum officinale Weber   Dandelion

Native. Common weed of arable land; grassland and pastures.

T. palustre (Lyons) DC.   Narrow-leaved Marsh Dandelion

Native. Occasional. Calcareous fens.

West: 91, Scarning Fen; 07, S. Lopham Fen.

East: 03, Holt Lowes; 12, Booton Common; 14, Beeston Bog; 30, near Coldham Hall, ETD.

T. spectabile Dahlst.   Broad-leaved Marsh Dandelion

Native. Rare. In less calcareous places.

West: 89, Bodney, DMM; 91, Beetley, DMM, not seen by authors.

T. laevigatum (Willd.) DC.   Lesser Dandelion

Native. Frequent. Sandy heaths, abandoned sandy arable land and dunes, often associated with *Hypochoeris glabra*.

West: 62, 70–72, 78–80, 84, 88, 89, 98, 99, 04.

East: 02, 09, 10, 12, 20, 21, 23, 24, 31, 33, 42, 51.

MONOCOTYLEDONES

ALISMATACEAE

Baldellia ranunculoides (L.) Parl.   Lesser Water-plantain
Native. Frequent. Fens and wet places.

Luronium natans (L.) Raf.   Floating Water-plantain
Nat. alien. Rare. Broads only.
East: 42, Calthorpe Broad Nature Reserve, ACJ, 1956; 'Is spreading; six colonies in Cottage Dyke, 1964', ACJ.
First record: 1956, the above.

Alisma plantago-aquatica L.   Water-plantain
Native. Common. Ponds and ditches.

A. lanceolatum With.
Native. Very uncommon. In similar places to the last.
East: 20, Thorpe St Andrew, AEE, 1947; 33, Mundesley parish, JHC, 1962.
First record: 1947, see above.

Sagittaria sagittifolia L.
     Arrowhead, loc. 'Serpent's-tongue', Broad Arrow (Broads)
Native. Common. Rivers, streams and ditches.

BUTOMACEAE

Butomus umbellatus L.   Flowering Rush
Native. Frequent. Streams and ditches in deeper water.
West: 59, 61, 62, 70, 74.
East: 04, 18, 20, 28, 30–32, 39, 41, 42, 49.

HYDROCHARITACEAE

Hydrocharis morsus-ranae L.
     Frog-bit, loc. 'Halfpennies and pennies' (Broads)
Native. Common. Ditches and Broads. Spreads successfully by stolons.

Stratiotes aloides L.   Water Soldier
Status doubtful perhaps native. Now restricted to the Broads where it is locally abundant.
West: 61, Setch, until 1930.
East: 30–32, 39, 41, 42.
First record: 1670, 'in a ditch in the way from Lyn to German's bridge in Marshland', Plukenet's annotated copy of Ray's *Cat. Pl. Ang.*, 1st edit.

Elodea canadensis Michx.   Canadian Pondweed
Nat. alien. Ubiquitous in ponds, streams and ditches.

JUNCAGINACEAE

Triglochin palustris L.   Marsh Arrow-grass
Native. Common. Fens and wet places.

T. maritima L.   Sea Arrow-grass
Native. Common. Salt-marshes.

ZOSTERACEAE

Zostera marina L.   Eel-grass, Grass-wrack
Native. Rare. Estuarine mud, towards low-water mark.
West: 63, Wolferton; 74, Holme-next-the-Sea; Brancaster; 84, Scolt Head, RMB; 94, Wells-next-the-Sea; Morston, RSRF.
East: 40, 50, Breydon Water, ETD.

Z. angustifolia (Hornem.) Rchb.
Native. Rare. On mud flats and in shallow water.
West: 74, Titchwell; 94, Morston, RSRF; Stiffkey, ETD.
East: 04, Blakeney, J. F. Peake in Herb. Norw. Mus.

Z. noltii Hornem.
Native. Estuarine mud at higher levels than *Z. marina*, appearing more frequent.
West: 63, Wolferton; 74, Titchwell; 84, Brancaster; 94, Morston, RSRF.
East: 04, Blakeney.

POTAMOGETONACEAE

The wealth of records of the rarer species of pondweed in the 1914 *Flora* testifies to the richness of the Broads at that time. Owing to the havoc wrought by the widespread pollution, many of the rarer species, formerly regarded as locally common, are now very rare.

We are grateful to Mr. J. E. Dandy for his generous assistance with the more critical species.

Potamogeton natans L.   Floating Pondweed
Native. Common. Ponds, ditches and rivers.

P. polygonifolius Pourr.   Bog Pondweed
Native. Frequent. Pools in bog, and streams draining bog.

P. coloratus Hornem.   Fen Pondweed
Native. Occasional. Pools in calcareous fens and old peat diggings.
West: 70, Foulden Common, GHR; Caldecote Fen; Gooderstone Fen; 71, Marham Fen; E. Walton Common; 72, Leziate Fen; 99, Stow Bedon, FR; 08, S. Lopham Fen.

P. lucens L.   Shining Pondweed, loc. 'Bean-weed' (Broads)
Native. Common. Rivers and large drains.

P. gramineus × *lucens*
(*P.* × *zizii* Koch ex Roth)
Native. In the Breckland meres where the parents are frequently abundant, the hybrid is usually present.
West: 88, Fowlmere, RCLH; 89, West Mere; 98, Ringmere; Langmere.
East: 32, Sutton Broad, 1915, Miss M. Pallis; 31, Ludham Bridge, 1923, Druce.

P. lucens × *perfoliatus*
(*P.* × *salicifolius* Wolfg.)
Rare. Recorded by Kirkby Trimmer from Costessey, E11, and from Horsey (?), E42, by A. Bennett, as *P. decipiens* Nolte; the abundance of its parents suggests it may have been overlooked.

P. gramineus L.   Various-leaved Pondweed
Native. Occasional. Ponds, dykes, turf-pits; frequent in the Breckland meres. As its common name suggests, it is a variable species, shallow-water forms having broad coriaceous leaves.
West: 62, Reffley Marshes; 88, Fowlmere, RCLH; Home Mere; 98, Ringmere; Langmere; 99, Thompson Common.
East: 32, Sutton Broads, Miss C. Gurney.

P. alpinus Balb.   Reddish Pondweed
Native. Frequent. Streams.
West: 60, Shouldham; 61, Saddlebow; Wormegay; 62, Reffley; Wolferton; N. Wootton; 09, Swangey Fen.
East: 11, Lenwade gravel pits, ETD; 20, R. Yare near Harford Bridges.

P. praelongus L.   Long-stalked Pondweed
Native. Formerly locally common, now rare.
West: 84, Holkham Lake, RCLH.
East: 22; 32; 39, dykes near R. Waveney at Ditchingham, 1965, GHR; 42, Waxham Cut, R. Gurney, 1948, in Herb. Norw. Mus.

P. perfoliatus L.   Perfoliate Pondweed
Native. Very common. Our most abundant species, occurring in every river and many streams and ditches.

P. friesii Rupr.   Flat-stalked Pondweed
Native. Occasional. Rivers, ponds and ditches.
West: 59, R. Ouse near Hilgay rail-bridge; 61, Puny Drain, Setch; 69, Common Dyke, Methwold Severals.
East: 30, Buckenham Ferry, GHR & ELS; 31, Upton Broad; 32, Sutton Broad, R. Gurney; 42, Hickling Broad, RCLH; Waxham Cut, ELS; 41, Burgh St Margaret, R. J. Burdon.

P. pusillus L.
(*P. panormitanus* Biv.)
Native. Probably common in ditches but easily confused with *P. berch-toldii.*

West: 61, Setch; Wormegay; 62, N. Wootton; 63, Wolferton; 69, Wissington; 74, Thornham; 84, Holkham, J. E. Little; Burnham Overy Staithe; 99, Breckles, ALB; Thompson Water, ELS.
East: 04, Cley; 10; 23; 30; 32, Sutton Broad, R. Gurney; 40, 49.
First record: 1883, Cley, F. J. Hanbury, one of several cited by Dandy & Taylor in their 'Studies of Brit. Potamogetons', XII, in *J. Bot.*, LXXVIII, 1940, 1–11.

P. pusillus × *trichoides*
(*P.* × *grovesii* Dandy & Taylor in 1967, *Watsonia*, VI, 316)
East: 42, between Ingham and Sea Palling, 1897, J. Groves, as *P. trichoides*, its only known station.

P. obtusifolius Mert. & Koch   Grassy Pondweed
Native. Occasional. Ponds and streams.
West: 61, Setch; 72, Leziate Fen; 98, Ringmere, P. M. Hall; 99, Stow Bedon, FR; Thompson Common; 00, Reymerston, ALB.
East: 18, 22; 30; 31; 32, Sutton Broad; 41; 42; 49; 51.

P. berchtoldii Fieb.   Slender Pondweed
Native. Probably commoner than *P. pusillus* but distribution imperfectly known; much confused with *P. pusillus.*
West: 61, Shouldham; 62, N. Wootton; Wolferton Marsh; 68, Sluice Drove, Hockwold; 69, Stoke Ferry Fen; 71, Pentney; 72, Leziate Fen; 88, near Thetford; 89, W. Tofts; 90, W. Bradenham, ALB; 00, Reymerston, ALB.
East: 29, Tasburgh, SA; 31, Upton Broad, BEC Exc.; 42, Horsey, FR.
First record: 1815, Newton St Faiths (see Dandy & Taylor).

P. trichoides Cham. & Schlecht.   Hair-like Pondweed
Native. Rare. Ponds and ditches. East Anglia is the headquarters for this species, which was first discovered in Britain by the Rev. Kirby Trimmer at Framingham Earl, E20, in 1848. The British Museum herbarium contains material from ten stations in E. Norfolk.
West: 84, Holkham Lake, 1946, A. B. Jackson; 98, pond near Ringmere, ELS, 1965, confd. Dandy.
East: 30, Limpenhoe, GHR, 1965; 32, Lessingham, CCT & ELS, 1959; 42, Waxham Cut, R. Gurney, 1948, det. Dandy & Taylor.

P. compressus L.   Grass-wrack Pondweed
Native. Rare. Ponds and ditches.
East: 30, Strumpshaw, TBR; 31, Upton Broad, GHR & ELS; R. Ant above Wayford Bridge, R. Gurney det. Dandy & Taylor, in Herb. Norw. Mus.; Ranworth, J. W. White, 1915; S. Walsham Broad, White & Salmon, 1915.

P. acutifolius Link   Sharp-leaved Pondweed

Native. Very rare now. Ditches.

East: 30, Buckenham Ferry, GHR & ELS, 1959; Limpenhoe, GHR, 1965.

P. acutifolius × *friesii*
(*P.* × *pseudofriesii* Dandy & Taylor)

A hybrid, new to the British Isles, found in a ditch near Buckenham Ferry, E30, along with both parents in 1952 by Miss D. A. Cadbury. Holotype in Herb. Mus. Brit., paratype in Herb. Kew. See Dandy in 'Plant Notes', 1958, *Proc., B.S.B.I.*, **3**, 49.

P. crispus L.   Curled Pondweed

Native. Common. Streams and ditches.

P. pectinatus L.
    Fennel-leaved Pondweed, loc. 'Grass-weed', 'Lace-weed' (Broads)

Native. Abundant in brackish streams and ditches.

Groenlandia densa (L.) Fourr.   Opposite-leaved Pondweed

Native. Common. Rivers, ditches and ponds.

RUPPIACEAE

Ruppia spiralis L., ex Dum.   Greater Ruppia

Native. Rare. In ditches near the sea.

West: 63, Snettisham; Heacham; 64, Hunstanton; 74, Thornham. East: 04, Cley, FMD; Salthouse, PHS.

R. maritima L.   Tassel Pondweed

Native. Dominant aquatic species of brackish ditches and of drainage channels in reclaimed land as long as saline influence exists, i.e., up to fifty years after exclusion of the sea.

ZANNICHELLIACEAE

Zannichellia palustris L.   Horned Pondweed

Native. Common. Streams and ditches.

A variable species. In the brackish ditches at Burnham Deepdale, W84, the var. *pedicellata* (Fr.) Wahlenb. & Rosen occurs with normal plants.

NAJADACEAE

Najas marina L.   Holly-leaved Naiad

Native. Rare. In soft mud, in more or less clear and open water, and in reedswamps. The only known British stations are in a few broads of north-east Norfolk.

East: 31, Upton Broad, 1951; Alderfen Broad; Hoveton Little Inner Broad, 1952; 32, Barton Broad; 42, Hickling and other Thurne Broads, all ACJ.

First record: 1883, Hickling Broad, Arthur Bennett, who discovered it.

225

LILIACEAE

Narthecium ossifragum (L.) Huds.   Bog Asphodel

Native. Exclusive to bogs, in which it is constant and often abundant. West Norfolk only.

West: 61, Cranberry Fen, E. Winch; 62, Roydon Common; Grimston Warren; Sugar Fen; Dersingham Fen; Wolferton Fen; N. Wootton Heath.

Convallaria majalis L.   Lily of the Valley

Native. Locally frequent. Dry woods.

West: 62, Wolferton Wood (since 1870); 71, Mill Covert, Westacre; 72, Congham; 92, Horningtoft; 93, Kettlestone, FRo; 03, Swanton Novers. East: 01, Hockering Wood; 02, Foxley; 03, The Dale, Edgefield and Triangle on Holt-Kelling road; Twenty Acre Hill near Holt; Gresham's School Woods, PHS; 12, Buxton Heath (under bracken), ETD; 28, Denton, CF; 29, Tasburgh, SA.

Polygonatum multiflorum (L.) All.   Solomon's Seal

Native. Very rare. Woods.

East: 02, Wood near Reepham, ETD; 03, Selbrigg Wood near Kelling; 13, Hempstead, 'another familiar garden plant grows wild at Hempstead', B. A. F. Pigott in *Flowers and Ferns of Cromer and Neighbourhood*, 1885; 14, 20, 40.

Maianthemum bifolium (L.) Schmidt   May Lily

Doubtfully native. Very rare. Only at Swanton Novers Great Wood, W03, in native woodland and remote from dwellings. Not recorded in the 1914 *Flora*.

First record: the above, 1955, RFB-O.

Asparagus officinalis L., ssp. *officinalis*   Asparagus

Est. alien. Widespread as an escape from cultivation. Considerable acreage of the light, sandy soils in Breckland is used for growing asparagus. Pigeons feed on the berries in autumn and are chiefly responsible for its dispersal. Reputed to have been brought to Norfolk, together with lilac, by the Flemish weavers in the reign of Edward III.

West: 70, Cockley Cley; 74, Holme-next-the-Sea Bird Observatory; 78, Weeting; 84, Holkham; 89, Mundford; 90, Cranworth; 98, Kilverstone; Ringmere; 99, Thompson.

East: 49, Haddiscoe New Cut, ETD.

Ruscus aculeatus L.   Butcher's Broom

Nat. alien. Occasional in woods and hedgerows. Very rarely fruiting.

West: 84, Burnham Thorpe, GT; 91, E. Dereham, JBE.

East: 03, 10, 13, 20, 23, 24, 30, 39, 42, 49.

Lilium martagon L.   Martagon Lily

Nat. alien. Rare. Woods. Has been known in the Harleston area in south-east Norfolk since 1886.

East: 28, Denton near Harleston, 'many hundreds in woodland and some in a ditch by the roadside', CF, 1963.

Tulipa sylvestris L.   Wild Tulip
Native. Very rare. Grassland.

East Norfolk only. 'First detected growing wild towards the end of the eighteenth century in old chalk workings at Whitlingham near Norwich. In the following century, it was noticed at Carrow, Lakenham, Trowse, Kirby Bedon, Seething, and in the woods at Cromer. Today its distribution seems to be very much the same with regard to districts, and it grows on the Norfolk side of the Waveney at Harleston and Bressingham', E. A. Ellis.

First record: 1790, first discovered in Britain by Hugh Rose, the Norwich apothecary, in chalk pits near Norwich.

Gagea lutea (L.) Ker-Gawl.   Yellow Star of Bethlehem
Native. Very rare. Wood on boulder clay. Few flowers at any time.
West: 99, Wayland Wood near Watton.
First record: Castle Rising, Dr John Lowe, *fl.* 1830–1902.

Ornithogalum umbellatum L.   Star of Bethlehem
Undoubtedly native in Breckland where it is frequently abundant in field-margins but escapes notice as it has few flowers, a character of the truly wild plant. Elsewhere, many of the records are most likely garden-escapes and probably refer to *O. augustifolium* Bor., which flowers a month earlier.
West: 79, 88–90, 93, 98, 01
East: 01, 03, 04, 08, 10–12, 18–20, 28–31, 39–41, 49.

O. nutans L.   Drooping Star of Bethlehem
Est. alien. Rare. Relict of cultivation.
West: 89, Bodney, on the grass bank of the churchyard whence F. Robinson recorded it fifty years ago; Stanford Battle Area, ETD; 94, Warham, where it has persisted for many years on the site of the house used by the Coke family before Holkham Hall was built; also, at Binham nearby.
East: 20, weed in Messrs Daniels' Nursery, Norwich, ETD.

Endymion non-scriptus (L.) Garcke   Bluebell
Native. Common. Woods, and hedges in formerly wooded areas. Very rare in the Thetford area.

Allium vineale L.   Crow Garlic
Native. Rare. Fields, roadsides and sea-banks. All with bulbils only = var. *compactum* (Thuill.) Bor.
West: 74, Thornham sea-bank; 83, Stanhoe railway cutting; 88, Thetford; Croxton, ETD; 93, Wighton, BT; 94, Wells, BT.
East: 10, 20, 30, 32, 39, 40.

A. paradoxum (Bieb.) G. Don
Est. alien. Naturalised and persisting on roadsides and banks of ditches where it is locally abundant in the Broads district.
West: Walsingham, BT, 1960.
East: 20, Bramerton; 21, Wroxham; 31, Salhouse, EAE; 39, Loddon, EAE.

A. ursinum L.   Ramsons, Garlic
Native. Common in woods on clay.

TRILLIACEAE

Paris quadrifolia L.   Herb Paris
Native. Rare. Oak-hazel woods on the clay of central Norfolk.
West: 90, Cranworth; Bradenham, ALB; 91, Honeypot Wood, Wendling; Rawhall Wood; Horse Wood, Mileham; 92, Horningtoft Wood; 99, Merton Park, HDH.
East: 00, Hardingham, EQB; Caston, FRo; 01, Hockering Wood; 02, Hindolveston, T. W. Irvine; 11, Honingham, FRo; 14, Weybourne Springs, PHS; 21, Newton St Faiths, R. Burn; 29.

JUNCACEAE

Juncus squarrosus L.   Heath Rush
Native. Common. Damp heaths.

J. tenuis Willd.
Est. alien. Grassland. Noted for the first time in West Norfolk in 1953, apparently introduced in making the cricket pitch on E. Winch Common.
West: 61, E. Winch, RPL.
East: 29, 42, 51.
First record: 1920, sandhills at Caister, Miss A. B. Cobbe.

J. compressus Jacq.   Round-fruited Rush
Native. Probably frequent, often confused with *J. gerardii*. Pastures near the sea and reclaimed land.
West: 51, 61–64, 74, 84.
East: 03, 04, 08, 18, 39, 41, 42, 49.

J. gerardii Lois.   Mud Rush
Native. In similar places to *J. compressus*, but commoner in salt marshes.
West: 62–64, 74, 84, 94.
East: 04, 40–42, 49.

J. bufonius L.   Toad Rush
Native. Common. Pond margins, damp sand-pits, paths and wet places in arable land.

J. inflexus L.   Hard Rush
Native. Common. Wet pastures, especially on the heavier soils.

J. effusus L.   Soft Rush
Native. Common. Wet pastures, bog and fen.
The var. *compactus* Hoppe is frequent, and often confused with *J. conglomeratus*.

J. effusus × inflexus
(*J. × diffusus* Hoppe)
West: 62, Reffley Marshes; Sugar Fen, BSBI Exc., 1949; 71, E. Walton, GHR; Narborough; 72, Derby Fen, BSBI Exc., 1949; 94, Wells, J. E. Little.
East: 08, Roydon Fen near Diss, AC det. TGT; 41, Martham, GHR.

J. conglomeratus L.   Clustered Rush
Native. Far less common than *J. effusus* and restricted to acid soils.
Var. *subuliflorus* (Drej.)
East: 02, Foxley Wood, GHR det. Dony.

J. maritimus Lam.   Sea Rush
Native. Abundant in the drier parts of sandy salt marshes but not extending to estuarine mud.

J. acutus L.   Sharp Rush
Native. Very rare and apparently diminishing, since the older records indicate a wide distribution along the north Norfolk coast. Now confined to a few colonies on coastal sand at Brancaster, W74, Thornham, W74, and Burnham Deepdale, W84.
First record: 1828, Brancaster, Crowe in *Engl. Fl.*, II.

J. subnodulosus Schrank   Blunt-flowered Rush
Native. Common. Dominant in fen where peat is not deep enough for *Cladium mariscus*; frequent in wet pastures.

J. acutiflorus Ehrh., ex Hoffm.   Sharp-flowered Rush
Native. Common. Wet pastures, woods and carr on acid soils, often dominant in a zone between bog and fen.

J. acutiflorus × articulatus
(*J. × surreajanus* Druce)
Is a fairly frequent hybrid. It bears fascicles of leaves with swollen bases, which assist in vegetative propagation in the absence of seeds.

J. articulatus L.   Jointed Rush
Native. Common. Ditches and ponds. Occasionally luxuriant plants are found, stiffly erect with more numerous flowers, corresponding to the form with eighty chromosomes described by Timm and Clapham (*New Phytologist*, **39**, 1–16).

J. bulbosus L.   Bulbous Rush
Native. Common. Bogs and streams draining from them.

J. kochii F. W. Schultz
Native. In similar places to the last but more frequent in woods on acid soils.

Luzula pilosa (L.) Willd.   Hairy Woodrush

Native. Occasional. Woods on the clay of central Norfolk.

West: 62, Roydon Common, one plant; 90, Wood Rising; Cranworth; 91, Mileham; Wendling; Rawhall; 92, Horningtoft; 03, Swanton Novers Great Wood.

East: 01, Hockering; 02, 03, 14, 29, 30, 33, 41.

L. sylvatica (Huds.) Gaud.   Greater Woodrush

Native. Rare. In similar situations to *L. pilosa*.

West: 92, Tittleshall; Horningtoft; 03, Swanton Novers Great Wood.
East: 13, near Edgefield; 20, Thorpe, ETD; 22, Felmingham.

L. campestris (L.) DC   Field Woodrush

Native. Common. Grassland, particularly on sandy soil.

L. multiflora (Retz.) Lej.   Many-headed Woodrush

Native. Frequent. Heaths on damp acid and peaty soils; extending to bogs. Plants with subsessile flowers in compacted heads = var. *congesta* (DC.) Lej., are frequent.

AMARYLLIDACEAE

Leucojum vernum L.   Spring Snowflake

Est. alien. Rare. Hedgebanks and woods.

West: 91, E. Dereham, since 1927.
East: 10, Cringleford, since 1911; 29, Shotesham, GG, 1963.

Galanthus nivalis L.   Snowdrop

Status doubtful. Locally abundant. Damp woods and hedgebanks. Several of the modern garden cultivars have been developed from plants of Norfolk origin.

West: 71, 72, 79, 81, 83, 90, 99.
East: 10, 11, 20, 28–30, 39.

Narcissus pseudonarcissus L.   Wild Daffodil

The truly wild daffodil is rare in Norfolk; the only recent records are from the Broads district, 30 and 41, and from meadows at Hethel, E10, 'where it has been known for many years', ETD. The 1914 *Flora* records also *N. biflorus* Curt., and *N. poeticus* L., both very rare garden-escapes. Double-flowered forms occur on the Ryston estate, W60.

East: 10, Braconash; Hethel and Swardeston, ETD; 19, Tacolneston, ETD; 29, Saxlingham Nethergate, ETD.

IRIDACEAE

Iris foetidissima L.   Gladdon, Stinking Iris

Native. Rare. Woods, hedgerows and frequently in churchyards.

West: 71, Southacre; Westacre; 92, N. Elmham, DMM.
East: 00, Kimberley; 04, Kelling Warren; 10, Bawburgh, JHS; Swardeston; 14, Weybourne, EAE; 20, Arminghall, JHS; 23, Sidestrand, KHB;

28, Harleston, ETD; 29, Tasburgh, SA; Bedingham; 39, Ditchingham, JHS; abundant on Bath Hills, ETD.

I. pseudacorus L.   Yellow Flag

Native. Common. Marshes, swampy woods, streams and river-banks.

Crocus purpureus L.   Purple Crocus

Nat. alien. Very rare. Was formerly abundant at Harleston where it had been known for well over 200 years. A few flowers were seen and photographed by J. E. Lousley in 1937. None seen in 1965. Considered to be extinct in the county until E. T. Daniels found a flourishing colony in a wood at Lakenham, E20, in 1967, an important re-discovery, confirmed by J. E. Lousley.

DIOSCOREACEAE

Tamus communis L.   Black Bryony

Native. Common. Hedges and scrub.

ORCHIDACEAE

Epipactis palustris (L.) Crantz   Marsh Helleborine

Native. Frequent in fens where the soil is decidedly alkaline and in damp hollows of sand-dunes.

West: 61, 64, 70–72, 74, 79, 80, 83, 84, 91–94, 99, 07, 09.
East: 00–03, 08, 09, 11, 12, 14, 18, 19, 22, 23, 29–32, 40–42.

Var. *ochroleuca* Barla, with yellowish-white flowers.

West: 84, Burnham Overy Staithe dunes.
East: 14, Beeston Bog.

E. helleborine (L.) Crantz   Broad-leaved Helleborine

Native. Rare. In old-established woods in central Norfolk.

West: 60, Runcton Holme, CPP; 79, Mundford, ELS; 88, Weeting; 91, Baker's Wood, Gressenhall, DMM; 92, Horningtoft; 03, Swanton Novers Great Wood.
East: 03, Edgefield; 11, Attlebridge, ETD; Felthorpe, ALB; 30, Buckenham; Surlingham Wood, EAE.

Spiranthes spiralis (L.) Chevall.   Autumn Lady's Tresses

Native. Very rare now. Chalk grassland.

West: 64, Ringstead Downs; 70, Caldecote Fen, 1950; Foulden, FR, 1915.
East: 03, Holt, D. C. Lang; 1957; Gresham's School and Old Racecourse, PHS; 14, Beeston Regis, A. R. Horwood, 1915; 31, 32.

Listera ovata (L.) R. Br.   Twayblade

Native. Common. Woods, scrub and fens.

Neottia nidus-avis (L.) Rich.   Bird's-nest Orchid

Native. Very rare. Dense shade in humus-rich woods on calcareous soil.

West: 92, Horningtoft, 1953.

East: 03, wood on Old Race Course, Holt, PHS, 1954; 10, Hethel, WGC, 1915; 18, Furze Covert, Rushall, R. E. Emms, 1954.

Goodyera repens (L.) R. Br.   Creeping Lady's Tresses

Nat. alien. Rare. In litter of pine needles and moss; rarely in open heathland. As this orchid is never seen in natural woodland we do not claim it as a native plant. It has slowly extended its range but not its numbers; formerly locally abundant, particularly in the Holt area. Reputed to have been introduced with trees from Scotland. We understand it has been introduced to localities near Norwich and near Brandon but with what success we do not know.

First and subsequent records are:

1885   Decoy Wood, Westwick, Miss Southwell.
1890   Holt, Miss A. M. Barnard.
1891   Hempstead, Miss A. M. Barnard.
1900   Bodham; Beeston Common, F. J. Spurrell.
1906   Holt, Hemsley in *Kew Bulletin*.
1910   Cawston, W. A. Nicholson.
1928   Holt, abundant, C. P. Petch.
1954   Heath House Wood and Old Race Course, Holt, P. H. Simon.
1957   Holt, 'many thousands', D. C. Lang.
1963   Near Burrow Gap, Holkham Meols, Miss G. Tuck and ELS.
1964   Near High Cape, Holkham Meols, A. L. Bull.
1965   Cranwich, several colonies, Miss S. Corbet, comm. Miss Leather.

Hammarbya paludosa (L.) O. Kuntze   Bog Orchid

Native. Very rare. Bogs.

West: 62, Roydon Common, since 1909; varying in numbers each year.

East: 18, Roydon Fen near Diss, 1937; 22, Bryants Heath near Felmingham, BEC Exc., 1938; two plants in 1945.

First record: 1769, Felthorpe, Rev. H. Bryant.

Liparis loeselii (L.) Rich.   Fen Orchid

Native. Rare. Wet fen peat.

West: 71, E. Walton Common, GHR, 1958 to 1964; 07, Blo' Norton Fen, WGC, 1917; S. Lopham Fen, 1957, three plants.

East: 09, Old Buckenham Fen, GHR, 1953, one plant; 31, Upton Broad, BEC Exc., 1938, one plant; 32, Dilham, 106 plants, GHR and others; Catfield, Col. Meinhertzhagen in Herb. Mus. Brit.; Sutton Broad; 41, Thurne, FRo.

First record: 1767, Newton St Faiths, John Pitchford in *English Botany*.

Coeloglossum viride (L.) Hartm.   Frog Orchid

Native. Very rare; sporadic in appearance. Pastures.

West: 62, Reffley, 1944; 91, Brisley, 1954, JSP.

East: 03, Holt, CPP, 1926; 10, Hethel, WGC, 1915.

First record: 1779, Newton St Faiths, J. E. Smith.

Gymnadenia conopsea (L.) R. Br.   Fragrant Orchid

Native. Locally frequent. Fens, marshes and damp heaths, rarely in drier habitats. Until recently, the distinction between this species and its sub-species *densiflora* has not been made and the following records are probably referable to the latter taxon. Dr Francis Rose tells us that he has not seen this species, *sensu stricto*, in Norfolk.

West: 62, 70–72, 79, 80, 91, 97, 08.

East: 00, 01, 03, 12, 14, 18–20, 23, 29–31, 39, 40.

Ssp. *densiflora* (Wahlenb.) Lindl.   Marsh Fragrant Orchid

Unless stated otherwise, the records here are those of Dr Rose.

West: 62, Roydon Common; 70, Foulden Common; 71, E. Walton Common, ELS; 72, Derby Fen; 91, Potters Fen, E. Dereham, ELS; 07, Blo' Norton; S. Lopham Fen; 09, Swangey Fen.

East: 12, Buxton Heath; 14, Beeston Bog, ELS; 18, Roydon Fen near Diss; 19, Flordon Common, SA & ELS; 30, Surlingham marl pit and Wheatfen; 40, Acle Decoy Carr.

Platanthera chlorantha (Cust.) Rchb.   Greater Butterfly Orchid

Native. Rare. Woods on calcareous soil.

West: 84, Holkham, BT; 91, Wendling; Scarning, ALB; Rawhall; 92, Horningtoft; 94, Wells, PHS; 99, Griston, FR, 1914.

East: 23, Felbrigg, PHS; 14, Sheringham, PHS; 28, Hardwick airfield, SA; 29, Brooke Wood, RMB.

P. bifolia (L.) Rich.   Lesser Butterfly Orchid

Native. Fairly frequent on bog and wet heath, always on acid soils.

West: 62, S. Wootton Common; Roydon Common; Sugar Fen; 72, Derby Fen; Leziate Fen; 91, Scarning Fen.

East: 03, Holt Lowes; High Kelling, PHS; 11, Swannington Common, PER; 12, Booton Common; 22, Buxton Heath; 31, Upton Broad, BEC Exc., 1938.

Ophrys apifera L.   Bee Orchid

Native. Widespread but sporadic in appearance. On chalky boulder clay and sand-dunes.

West: 62, 64, 70–72, 74, 80, 81, 84, 89, 91, 92, 98, 00.

East: 01, 03, 11, 14, 18–20, 28, 30, 39.

O. insectifera L.   Fly Orchid

Native. Formerly locally common, now very rare.

East: 14, Weybourne Springs, 1949; 24, Overstrand, 1933, KD.

Himantoglossum hircinum (L.) Spreng.   Lizard Orchid

It first appeared in Norfolk in 1923 when it was found at Haddiscoe, E49, by H. K. Airy Shaw of Kew. Further occurrences are:

1932   Harleston, E28, Dr F. N. Maidment.
1936   Cringleford, E10, Miss B. Foster.
1937   Holme-next-the-Sea, W64, A. Hitchcock.
1952   Saxlingham, E03, in Herb. Norw. Mus.
1956   Long Stratton, E19, E. A. Ellis

233

Orchis morio L.   Green-winged Orchid
Native. Common. Wet meadows and fens.

O. mascula (L.) L.   Early Purple Orchid
Native. Frequent in woods on heavy soils, hence commonest in central Norfolk.
West: 62, 70, 72, 74, 82, 83, 90–92, 99, 01, 03.
East: 01–04, 13, 14, 18–20, 22, 23, 28–30, 39, 40.

Aceras anthropophorum (L.) Ait. f.   Man Orchid
Native. Very rare. Chalky places.
East: Ingmote Hill near Holt, 1928 until 1950, CPP.
Twelve stations were listed in the 1914 *Flora*, most of which are older records. There is a specimen in Herb. Univ. Leeds, taken by F. C. Newton from Thetford in 1912.
First record: 1785, Ashwellthorpe, Jas. Crowe.

Dactylorhiza spp.   Spotted and Marsh Orchids
The many fens and wet heaths still to be found in Norfolk provide characteristic habitats for this genus. In field populations, where two or more species grow together, it is always possible to find a number of plants showing characters either distinctly intermediate or favouring one of the parents. Occasionally one parent may be absent as in *Orchis pardalina* Pugsl., reputed to be the hybrid *D. fuchsii* × *praetermissa*. So common are these hybrids that we do not list them separately.
Nomenclature follows Hunt and Summerhayes' account in 1965 *Watsonia*, VI, 128–133. We are indebted to Mr Summerhayes for his great help extending over many years.

Dactylorhiza incarnata (L.) Soó ssp. *incarnata*   Early Marsh Orchid
Native. Frequent and widespread in fens of base-rich soils and waterlogged pastures. True albinos, distinct from ssp. *ochroleuca* occur occasionally.
West: 62, 70–72, 79, 80, 90, 91, 98, 99, 08.
East: 00, 02, 09, 11, 12, 14, 18, 20, 24, 30–32.

Ssp. *ochroleuca* (Boll.) P. F. Hunt & Summerh.
Rare. In fens where the soil is markedly calcareous. Flowers pale-straw coloured.
West: 79, Foulden Common, EQB conf. J. Heslop-Harrison; 07, Blo' Norton Fen, J. Hes.-Har.; 08, S. Lopham Fen, FRo.
East: 00, Seamere near Hingham; 18, Roydon Fen near Diss, J. E. Lousley, the first British record.

Ssp. *pulchella* (H.-Harrison f.) Soó
Plants with bright reddish-purple flowers, similar to the northern *D. purpurella* in colour, occur on damp, acid heaths.
West: 62, Sugar Fen; 70, Foulden Common; 71, E. Walton Common; 72, Derby Fen; 99, Thompson Common; 07, Blo' Norton Fen, FRo; 08, S. Lopham Fen, FRo.
East: 03, Holt Lowes, DMM; 18, Roydon Fen near Diss, Pugsley and others, 1928; 32, E. Ruston.

D. praetermissa (Druce) Soó   Common Marsh Orchid
Native. Common. Wet pastures, bogs and fens, tolerant of a wide range of soils. Albinos occur. Follows *D. incarnata* in flowering.
West: 60–62, 64, 70–72, 80, 82, 83, 90–93, 99.
East: 00, 03, 11, 12, 14, 18, 19, 22–24, 30–32, 40.

D. traunsteineri (Sauter) Soó   Pugsley's Marsh Orchid
Native. Calcareous fens. A recent segregate. See 'Orchis traunsteineri in the British Isles' (J. Heslop-Harrison in *Watsonia*, **2**, 371, 1953). The author's account includes two West Norfolk stations, Foulden Common and Snetterton. He suggested it would occur in other Norfolk localities and some of the following records have been confirmed by him.
West: 70, Foulden Common; 71, Marham Fen; E. Walton Common; 91, Worthing; near Hoe, ALB; 01, Swanton Morley, ALB; 08, S. Lopham Fen, FRo; S. of Garboldisham Heath, FRo.
East: 03, Holt Lowes; 11, Costessey, BFD; 12, Buxton Heath; 14, Beeston Bog, FRo; 32, E. Ruston, FRo.
First record: 1922, Foulden Common, J. E. Little in Herb. Univ. Camb., det. J. Heslop-Harrison.

D. maculata (L.) Soó ssp. *ericetorum* (Linton) P. F. Hunt & Summerh.
    Heath Spotted Orchid
Native. Locally frequent. Acid soils on dry and wet heaths. Not formerly separated from *D. fuchsii*.

D. fuchsii (Druce) Soó   Common Spotted Orchid
Native. Common. Wet pastures, fens and damp woods on basic soil.

D. fuchsii × *praetermissa*   Leopard Orchid
(*Orchis pardalina* Pugsl.)
Occasional; sometimes in absence of one parent.
West: 60, Stoke Ferry, J. E. Little; 62, Sugar Fen; 70, Caldecote Fen, J. E. Little; 71, Lambs Common, E. Walton; 72, Derby Fen; 99, Watton, Druce; 91, Rush Meadow, E. Dereham; E. Bilney, DMM; Hoe, ALB; 92, Fakenham; 99, Thompson Common; Stow Bedon.
East: 00, Seamere near Hingham; 11, Swannington Common, ALB & ELS; 12, Booton Common.

Anacamptis pyramidalis (L.) Rich.   Pyramidal Orchid
Native. Occasional. Chalk grassland, railway cuttings, chalk pits and sand-dunes.
West: 70, Beechamwell; 71, Gayton Thorpe; Narborough Field; 72, Congham; Hillington; Gayton chalk pit; 79, Cranwich; 84, Scolt Head Island; Burnham Overy Staithe dunes; 88, Thetford, ALB; 94, Danish Camp, Warham.
East: 03, Thornage chalk pit; Old Race Course, Holt, PHS; 04, Cley; Weybourne; 29, Tasburgh, SA; 39, Ditchingham, MIB.

235

ARACEAE

Acorus calamus L.   Sweet Flag

Denizen. Rare in the West but locally frequent in East Norfolk. Margins of ponds and riversides.

This plant has had a long and interesting history in Norfolk. In his paper on '*Acorus Calamus* in England' (*J. Bot.*, XLI, 1903, 23), Arthur Bennett quotes from a letter written by Sir Thomas Browne to Merrett in 1668, 'Some 25 yeares ago I gave an account of this plante unto Mr. Goodyere. . . . This elegant plante groweth very plentifully and beareth its Jules yearly by the bankes of Norwich river, chiefly about Claxton & Surlingham, & also between Norwich and Hellsden bridge, so that I have known Heigham Church in the suburbs of Norwich strewed all over with it. It hath been transplanted and set on the sides of Marish ponds in severall places of the county, where it thrives & beareth ye Jules yearly.'

First Norfolk record: 1643 (*lit. cit.*).

West: 59, 69, along the R. Ouse at Hilgay and Southery; 72, Houghton lake; 84, Holkham lake; 98, pool near Ringmere.

East: 12, 20, 23, 30, 32, 39, 49.

Arum maculatum L.   Lords and Ladies, Cuckoo Pint

Native. Common. Woods and hedges in shade.

LEMNACEAE

Lemna polyrhiza L.   Great Duckweed

Native. Frequent. Ponds and ditches.

West: 61, 62, 78, 83, 84, 92, 93.

East: 03, 20, 30–32, 40–42, 49, 51.

L. trisulca L.   Ivy-leaved Duckweed

Native. Common. Ponds and ditches.

L. minor L.   Duckweed

Native. Common. Ponds and ditches.

L. gibba L.   Gibbous Duckweed

Native. Frequent. Ponds and ditches.

West: 41, 60–62, 63, 71, 74, 84, 94.

East: 04, 30–32, 42, 51.

SPARGANIACEAE

Sparganium erectum L., var. *erectum*   Bur-reed

Native. Common. Ponds, ditches and rivers.

Little work has been done on the distribution of the four varieties, a full account of which appears in *Watsonia*, V, 1–10 ('*Sparganium* in Britain', C. D. K. Cook 1961).

The var. *microcarpum* (Neuman) Hylander appears to be more frequent; var. *neglectum* (Beeby) Schinz & Thell., has been recorded from Bryants Heath, Felmingham, E22, BEC Exc., 1938, Hockham Mere, W99, and from the River Thet at Bridgham, W98.

S. emersum Rehm.   Unbranched Bur-reed
Native. Common. Ponds and ditches.

S. minimum Wallr.   Small Bur-reed
Native. Formerly widespread, now rare. Ponds and ditches.
West: 70, Foulden Common; 99, Thompson Common; 07, S. Lopham Fen.
East: 32, E. Ruston; 33, Witton; 42, Potter Heigham; Long Gores Marsh, Hickling, ELS & ACJ; 51, near Caister.

TYPHACEAE

Typha latifolia L.   Great Reedmace, 'Pokers'
Native. Common. Ponds and reedswamps.

T. angustifolia L.   Lesser Reedmace
Native. Less common than *T. latifolia*, in similar situations.

CYPERACEAE

Eriophorum angustifolium Honck.   Common Cotton-grass, Bog Cotton
Native. Common. Wet heaths, bogs in which it may be dominant, and acid parts of fens.

Var. *brevisetum* Druce in BEC, 1925, 789. A form with very short bristles (10–15 mm.).
West: 71, E. Walton Common, ELS; 72, Leziate Fen, ELS; 91, Badley Moor near E. Dereham, JBE.

E. gracile Roth
Native. Very rare. Found in a wet, acid bog near Acle, E40, GHR, 1955, the only record.

E. latifolium Hoppe   Broad-leaved Cotton-grass
Native. Occasional. In fen and bog, in less acid places then *E. angustifolium*.
West: 62, Roydon Common; 70, Oxborough Fen; Gooderstone Fen; 71, E. Walton Common; 72, Leziate Fen.
East: 12, Booton Common; 40, Acle.

E. vaginatum L.   Hare's-tail
Native. Rare. Now restricted to two stations in West Norfolk.
West: 62, Sugar Fen; Roydon Common (locally abundant).

Trichophorum cespitosum (L.) Hartm., ssp. *cespitosum*   Deer-grass
Native. Characteristic of bogs and wet heaths, where it is common. The question of the subspecies needs working out.

Eleocharis quinqueflora (F. X. Hartmann) Schwarz
    Few-flowered Spike-rush

Native. Occasional. Bogs.

West: 62, Roydon Common; Grimston Warren; 71, E. Walton Common; 72, Derby Fen; 91, Scarning Fen; 93, Thursford Common.
East: 03, Holt Lowes; 12, Booton Common; Buxton Heath; 14, Beeston Bog; 22, Bryants Heath, Felmingham.

E. multicaulis (Sm.) Sm.   Many-stemmed Spike-rush
Native. Frequent. Bogs.

West: 62, N. Wootton; Reffley Marshes; Dersingham Fen; Roydon Common; 71, E. Winch Common; 72, Derby Fen; 73, Gt. Bircham; 91, Scarning Fen.
East: 03, Holt Lowes; 12, Buxton Heath; 14, Beeston Bog; 20, R. Yare marshes, Keswick; 22, Bryants Heath, Felmingham.

E. palustris (L.) Roem. & Schult.   Common Spike-rush
Native. Common. Marshes, ponds and ditches.
The distribution of the two subspecies is imperfectly known but ssp. *vulgaris* has been seen in both vice-counties; ssp. *palustris* recorded by S. M. Walters from Derby Fen, W72, 1949, *J. Ecol.*, **37**, 192.

E. uniglumis (Link) Schult.
Native. Rare. Fens in the west and coastal marshes in the east.
West: 62, Wolferton, RPL; Roydon Common; 72, Derby Fen.
East: 42, near Horsey; 51, Winterton, RMB.

Scirpus maritimus L.   Sea Club-rush.
Native. Frequent. Margins of brackish ditches and ponds.

Blysmus compressus (L.) Panz. ex Link   Broad Blysmus
Native. Occasional. In wet flushes such as spring-heads in open communities of damp pastures, associated with *Carex nigra*; rare in dune slacks.
West: 64, 71, 72, 80, 83, 91–93, 99, 02.
East: 00–02, 10–12, 14, 20, 23, 39.

Schoenoplectus lacustris (L.) Palla   Bulrush, 'Bolder' (Broads)
Native. Common. Ponds, streams and broads.

S. tabernaemontani (C. C. Gmel.) Palla   Glaucous Bulrush
Native. Common. Ponds and ditches, replacing *S. lacustris* almost entirely in the north-western part of Norfolk.

Isolepis setacea (L.) R. Br.   Bristle Scirpus
Native. Frequent. Wet heaths, wet sand and wet, acid pastures.
West: 62, 70–72, 74, 80, 81, 92, 93, 98–00.
East: 01, 03, 12, 14, 22, 31, 41, 42.

I. cernua (Vahl) Roem. & Schult.   Nodding Scirpus
Native. Rare. Wet sandy places in fens and bogs.
West: 91, Scarning Fen, FR, 1915; Potter's Fen, E. Dereham, JSP.
East: 03, Holt; 12, Booton Common; Buxton Heath; 14, Beeston Bog,

abundant; 22, Bryants Heath, Felmingham, BEC Exc., 1938; Hevingham, WGC.

Eleogiton fluitans (L.) Link   Floating Scirpus
Native. Frequent. Fen and carr.

West: 61, Wormegay; 62, Roydon Common; Sugar Fen; Dersingham Fen; 70, Foulden Common; 72, Derby Fen; 07, S. Lopham Fen, EAE. East: 22, Bryants Heath; 41, Burgh St Margaret, R. J. Burdon; 42, Hickling.

Cyperus longus L.   Galingale
Nat. alien. Rare. Lakesides and by streams.

East: 10, Cringleford, FTJP; 12, Blickling.

Schoenus nigricans L.   Bog-rush
Native. Common in both fen and bog, apparently tolerant of a wide range of soil reaction.

Rhynchospora alba (L.) Vahl   White Beak-sedge
Native. Exclusive to bogs, in which it may be abundant.

West: 62, N. Wootton; Wolferton Fen; Dersingham Fen; Roydon Common; 71, East Winch Common.
East: 03, Holt Lowes; 12, Booton Common; Buxton Heath; 22, Bryants Heath, Felmingham.

Cladium mariscus (L.) Pohl   Sedge
Native. Common. The dominant species of true fen; indicates all relict fen in West Norfolk. Abundant over many acres in the Broads and dominant in 'thack' marshes where it is cut regularly for thatching.
First record: 1796, as *Schoenus mariscus*, Bardolph Fen, Skrimshire, *Cat.*

Carex distans L.   Distant Sedge
Native. Frequent in wet grassland near the sea and edges of shingle pits; rare inland.

West: 61–63; 71, 72, 74, 84, 94.
East: 03, 04, 18, 32.

C. hostiana DC.   Tawny Sedge
Native. Frequent. Bogs and fens.

West: 61, 62, 70–72, 79, 80, 90, 91, 99, 00.
East: 00, 03, 04, 11, 12, 14, 30.

C. hostiana × *lepidocarpa*
(*C.* × *fulva* Gooden.)

West: 70, Gooderstone Fen; 71, E. Winch Common.

C. binervis Sm.   Ribbed Sedge
Native. Occasional. Heaths.

West: 61, 62, 70, 71, 79–82, 93, 99, 03.
East: 03, 33, 40.

C. lepidocarpa Tausch
Native. Common in the very wet parts of fens indicating more alkaline soils.

239

C. demissa Hornem.

Native. Occasional. Wet places on heaths and fens on more acid soils than the previous species.

West: 61, 62, 71, 72, 81, 82, 89, 92, 08.

East: 03, 11, 12, 14, 19, 22, 31, 41, 49.

C. demissa × *hostiana*

West: 71, E. Winch Common.

C. demissa × *serotina*

West: 62, Grimston Warren, 1967, ELS.

C. serotina Mérat

Native. Rare. Open places in fens on base-rich soils.

West: 61, E. Winch Common (west side); 62, Grimston Warren; Sugar Fen; 69, Wretton Fen, J. E. Little; 72, Derby Fen; 82, Helhoughton Common; 89, W. Tofts, ELS; 98, Overa Heath, RCP.

East: 14, Beeston Bog; 18, Roydon Fen near Diss, AC; 22, Bryants Heath, Felmingham, EAE.

C. extensa Good.   Long-bracted Sedge

Native. Occasional. Edges of salt marsh and brackish ditches.

West: 63, Snettisham; Heacham; 64, Old Hunstanton; 74, Holme-next-the-Sea; Thornham; Titchwell; 84, Brancaster; Scolt Head Island; 94, Morston, ETD.

East: 04, Cley.

C. sylvatica Huds.   Wood Sedge

Native. Frequent. Woods on the heavier soils of central Norfolk.

West: 60–62, 72, 80–82, 90–92, 00, 03.

East: 01, 02, 10, 14, 20, 22, 29, 39, 40.

First record: 1801, Earsham; Smith in *English Botany*, t. 995.

C. pseudocyperus L.   Cyperus Sedge

Native. Frequent. River-sides, ditches, ponds and wet shady places.

West: 60–63, 73, 79, 89, 91, 98, 99, 00.

East: 08, 10, 18–20, 29–33, 39–42, 49.

C. pseudocyperus × *rostrata*

West: 99, Cranberry Rough, Hockham, its only British locality; see 'A Hybrid Sedge from West Norfolk', Petch & Swann, 1956 *Proc., B.S.B.I.*, **2**, 1–3.

C. rostrata Stokes   Beaked Sedge, Bottle Sedge

Native. Common. Bogs and streams draining from them.

C. vesicaria L.   Bladder Sedge

Native. Frequent in West Norfolk but rare in East Norfolk.

West: 61, 62, 69, 71, 72, 74, 99.

East: 03, 12, 22, 30.

C. riparia Curt.   Great Pond-sedge

Native. Common. Banks of rivers, streams, ditches and ponds.

92 HERB PARIS *Paris quadrifolia* *Miss D. M. Maxey*

93 MARSH HELLEBORINE *Epipactis palustris* *A. H. Hems*

94 CREEPING LADY'S TRESSES   *Goodyera repens   A. H. Hems*

95 BOG ORCHID   *Hammarbya paludosa   Dr M. George*

96 FEN ORCHID   *Liparis loeselii   Dr M. George*

97 MARSH FRAGRANT ORCHID   *Gymnadenia conopsea* var. *densiflora*   *A. H. Hems*

99 LESSER BUTTERFLY ORCHID   *Platanthera bifolia*   *A. H. Hems*

98 GREAT BUTTERFLY ORCHID
Platanthera chlorantha   Miss D. M. Maxey

100 GREEN-WINGED ORCHID
Orchis morio   Miss G. Tuck

C. acutiformis Ehrh.   Lesser Pond-sedge
Native. Common. Ditches and ponds.
The forma *spadicea* (Roth) Aschers. & Graebn., occurs occasionally and appears to be a state of drier habitats.
C. acutiformis × *riparia*
West: 62, Roydon Common and Grimston Warren, CPP.
East: 23, Southrepps Common, JBE det. BM.
C. pendula Huds.   Pendulous Sedge
Native. Rare in the west, though frequent in East Norfolk. Woods on clay.
West: 90, Bradenham, ALB; Carbrooke, RCLH; 91, E. Bilney; Gressenhall, DMM; Horse Wood, Mileham; Dillington Carr.
East: 01–03, 10–13, 32, 33, 41.
C. strigosa Huds.   Loose-spiked Wood-sedge
Native. Very rare now. Wood margin.
East: 02, Foxley Wood, RPL, 1954; locally frequent, 1968, ALB & ELS.

C. pallescens L.   Pale Sedge
Native. Rare. In woods on the heavier Norfolk soils.
West: 91, Scarning, WGC; Honeypot Wood, Wendling; Rawhall Wood; 92, Horningtoft Wood.
East: 02, Foxley Wood, ALB & ELS; 18, Gissing; 29, Brooke Wood, ETD; Tasburgh, SA; Ditchingham, MIB.

C. panicea L.   Carnation-grass
Native. Common. Bogs and fens.

C. limosa L.   Mud Sedge
Native. Rare. Very wet bogs.
West: 62, Roydon Common, last seen in 1924 by W. H. Mills and A. H. Evans.
East: 32, Smallburgh, WGC, 1917; Sutton Broad, GHR; 41, Shallam Dyke near Thurne, GHR.

C. flacca Schreb.   Glaucous Sedge
Native. Common. Grassland on chalk or clay and in fens.

C. hirta L.   Hairy Sedge
Native. Common. Wet or dry grassland.
Forms growing amongst marginal vegetation of ponds frequently have both leaves and sheaths more or less glabrous and are forma *subhirtaeformis* Kneucker, habitat forms only; the drier the ground the greater the pubescence.

C. lasiocarpa Ehrh.   Slender-leaved Sedge
Native. Occasional in fens and reedswamps, sometimes escaping notice as it bears few flowering spikes.
West: 61, 70–72, 93, 98, 99, 02, 07, 09.
East: 00, Seamere near Hingham; 02, Guist, FRo; 30, Wheatfen, EAE; 32, Sutton; Barton Turf; Dilham, FRo; 42, Hickling, EAE.

C. pilulifera L. Round-headed Sedge
Native. Common. Heaths.

C. ericetorum Poll. Silvery Sedge
Native. Although stated in the 1914 *Flora* to be very rare and but one station listed, recent field-work has shown this species to be widespread in the chalk grassland-heath of the south-west Norfolk Breckland. It has most likely been confused with *C. caryophyllea* with which it is often associated.
West: 70, Foulden Common; 78, Weeting Heath and Devil's Dike; 79, Cranwich; 80, Cockley Cley Heath; 89, Grimes Graves, FRo; Bodney Warren; 98, E. Harling, HDH; 08, Garboldisham.
First record: 1880, Santon, A. Bennett.

C. caryophyllea Latour. Spring Sedge
Native. Common. Dry grassland on chalk or sand.

C. elata All. Tufted Sedge
Native. Abundant. Broads and fens, dominant in the absence of *Cladium*. A variable species. Plants with female glumes pitch-black throughout = f. *nigrans* (*C. stricta* Good., var. *nigrans* Beck) occur occasionally (see Lousley's note on variation, *BEC*, 1936, 230–231).

C. elata × *nigra*
(*C.* × *turfosa* Fr.)
West: 70, Foulden Common, conf. ACJ; 99, Breckles Heath, ACJ.
East: 30, Wheatfen Broad, ELS, 1943, det. ACJ.

C. acuta L. Tufted Sedge
Native. Not seen in West Norfolk; frequent in East Norfolk. Dykesides.
East: 00, Hackford, ALB; 30, Surlingham, EAE; Brundall, GHR & ELS; 31, Thurne, GT; 32, 39, 42.

C. acuta × *elata*
West: 71, E. Walton Common, det. ACJ.

C. nigra (L.) Reichard Common Sedge
Native. Common. Damp places, chiefly on acid soils.

C. paniculata L. Great Panicled Sedge
Native. Common. Wet shady places. Characteristic of carr developing on peat, in which it is often dominant.

C. appropinquata Schumach.
Native. Occasional, in similar situations to *C. paniculata*.
West: 61, Mow Fen, Shouldham; Shouldham Warren, up to 1943; 80, Gt. Cressingham; 90, Scoulton Mere, HDH; 98, Overa Heath; 99, Cranberry Rough, Hockham; Thompson Common; 09, Swangey Fen, GHR; 02, Guist Fen.
East: 00, 09, 30, 31, 32, 40, 42, 49.

C. appropinquata × *paniculata*
(*C.* × *solstitialis* Figert)
East: 31, near Upton Broad, J. E. Lousley, 1936.

C. diandra Schrank   Lesser Panicled Sedge

Native. Exclusive to fens, where it may be frequent.

West: 61, 62, 71, 72, 91, 99, 07, 08.

East: 00, 02, 12, 18, 19, 30, 32.

First record: 1792, '*prope* Norwich', Crowe; Goodenough in *Trans. Linn. Soc.*, **11**, 163.

C. otrubae Podp.   False Fox Sedge

Native. Common. Pond margins and dykes in heavy soil. The true Fox Sedge (*C. vulpina* L.) has not been found in Norfolk; old records for this refer to *C. otrubae*.

C. otrubae × *remota*

(*C.* × *pseudoaxillaris* K. Richt.)

Occurs rarely when both parents are present.

West: 62, N. and S. Wootton; Reffley Marshes.

East: 29, Hempnall, ELS.

First record: 1801, Earsham, T. J. Woodward in *English Botany*, t. 993 as *C. axillaris* Good.

C. disticha Huds.   Brown Sedge

Native. Frequent. Fens, wet heaths and acid grassland.

C. arenaria L.   Sand Sedge

Native. Common. Coastal dunes, Breckland and Greensand heaths. Forms large pure societies on heaths where rabbits were formerly numerous.

C. divisa Huds.   Divided Sedge

Native. Frequent. Fresh-water marshes near the sea, bordering ditches and depressions in pastures, never in brackish water but never far from the coast. One of a group of 'para-maritime' species which includes *Ranunculus sardous* and *Oenanthe lachenalii*. On Wolfterton marsh in 1940, among many normal plants, one was found showing a monstrosity in the form of proliferous utricles. Such aberrations are not uncommon and are the normal features of *C. microglochin* Wahl.

West: 61, S. Lynn; 62, N. and S. Wootton; 63, Wolferton; 84, Burnham Overy Staithe; 94, Wells, FR.

East: 04, Blakeney; 40, 41.

C. muricata agg.

This group of species has long been a source of confusion. We follow Nelmes in his interpretation and have used his paper ('Two Critical Groups of British Sedges', *BEC*, 1945, 95–105) in the study of the Norfolk representatives. We were much indebted to him for help in field-work.

The following key is based on Norfolk material.

Inflorescence very long, up to 17 cm., increasingly interrupted, lower branched. Ligule as broad as long. Female scales whitish. Utricles to 4 mm.   ....*divulsa*

Inf. to 5 cm., more or less contiguous, never branched. Ligule broader than long. Female scales brownish. Utricle to 5 mm.

............*polyphylla*

Inf. to 4 cm., contiguous. Ligule much longer than broad. Utricles to 6 mm., opening star-shaped when mature. Lower sheaths always reddish. ....*spicata*

Inf. to 3 cm., closely contiguous. Ligule as long as broad. Utricles to 4 mm. ....*pairaei*

C. divulsa Stokes   Grey Sedge

Native. Rare. Hedgebanks on calcareous soil and wood-margins. Although there are some two dozen records for this species, we suspect many may be referable to *C. polyphylla*, which is the more frequent.

West: 60, Downham Market, J. Gilmour & CPP, 1931; Wimbotsham, J. E. Little, 1929; Stradsett Lake, ELS; 90, Cranworth, ALB; 91, Rawhall Wood; 92, Eastfield Wood.

East: 11, Alderford Common, ETD; 12, Haveringland, ELS; 18, Diss and Roydon, AC; 39, Chedgrave, ELS.

C. polyphylla Kar. & Kir.

Native. Frequent. In isolated tufts on dry roadside banks of alkaline soil but not markedly calcicole. Somewhat variable and rarely as robust as plants from Cambridgeshire and Gloucestershire.

West: 60–63, 70–74, 78, 79, 82–84, 88, 89, 94, 98, 99, 08.

East: 02, Whitwell, ETD; 11, Alderford Common, ETD; 12, Gt. Witchingham, ETD; 18, Roydon near Diss, AC det. TGT; 32, Ingham, TGT.

First record: 1946, Snettisham, ELS confd. Nelmes.

C. spicata Huds.   Spiked Sedge

Native. Occasional. Damp grassland and wood margins, tolerant of a wide range of soils; very rarely on maritime shingle.

West: 62, 63, 71, 72, 79, 89, 90, 99.

East: 03, 10, 12, 18, 29, 42.

C. pairaei Schultz   Prickly Sedge
(*C. muricata* auct.)

Native. Frequent. Dry hedgebanks on acid and gravelly soils.

C. echinata Murr.   Star Sedge

Native. Frequent in bogs.

C. remota L.   Distant-spiked Sedge

Native. Common. Damp shady places.

C. curta Good.   White Sedge

Native. Locally common. Marshes, bogs and carr developing from bog. More frequent on the acid soil of the Lower Greensand.

West: 61, Wormegay; Ashwicken, WCFN; Setch; Leziate, Miss I. M. Roper, 1930; 62, N. Wootton; Reffley Marsh; Babingley; Roydon Common.

East: 12, Booton Common; 31, Woodbastwick, GHR & ELS; Ludham, EAE; 32, E. Ruston Common, RMH; Sutton Broad; 40.

Var. *tenuis* Lang. N. Wootton, W62, CPP det. A. Bennett.

C. ovalis Good.   Oval Sedge
Native. Frequent. Bogs and wet heaths.

C. pulicaris L.   Flea Sedge
Native. Frequent. Bogs and wet heaths.

C. dioica L.   Dioecious Sedge
Native. Rare. Bogs and fens.
West: 62, Roydon Common; 72, Derby Fen; 80, Houghton Springs; 91, Scarning Fen.
East: 00, Runhall, EAE; 03, Holt Lowes; 12, Buxton Heath; 14, Beeston Bog.

GRAMINEAE

For his generous help, both in the field and revising the typescript, adding records and annotations, we are considerably indebted to Dr C. E. Hubbard, C.B.E., who is himself a Norfolk man. The taxonomy and nomenclature are based on his book, *Grasses*, (Pelican Books, A295, 1954). This account includes both natives and casuals.

Bromus sterilis L.   Barren Brome
Native. Common. Both waste- and cultivated land and hedgebanks.

B. tectorum L.   Drooping Brome
Nat. alien. Locally abundant on sandy tracks and fire-breaks throughout the West Norfolk Breckland. The hairy-spiculate variant (var. *longipilus* Borbas) has, so far, not been detected.
West: 78, Weeting; 88, Thetford; Croxton; Santon; 89, Fowlmere Farm; 98–99, Kilverstone.
East: 20, Norwich, WGC, 1915.
First record: 1889, Thetford, H. D. Geldart; same year, King's Lynn, Dr C. B. Plowright.

B. diandrus Roth   Great Brome
Nat. alien. Frequent. Has shown a remarkable increase in recent years in West Norfolk. Hedgerows, sand-pits and rubbish tips. Recorded in the 1914 *Flora* as *B. maximus*; one record only.
West: 70, 71, 73, 74, 78–80; 83, 84, 88–90: 94, 98, 99.
East: 11, Hellesdon, 1967, ETD; 18, Diss, 1961, AC; 20, Norwich, ETD; 50, Gt Yarmouth, 1914, FR.

B. ramosus Huds.   Hairy or Wood Brome
Native. Frequent. Shady places.
The increasing number of records of *B. benekenii* (Lange) Trimen elsewhere suggest it should also occur on the chalk in beechwoods in Norfolk. It has shorter or no hairs on the sheaths and a smaller, contracted, nodding inflorescence.

245

B. unioloides H. B. K.   Rescue Grass

Casual. Rare. Waste places and near grain mills.

West: 62, King's Lynn Harbour; ash-tip, King's Lynn, 1944; 88, Thetford, FR, 1916.

East: 11, Drayton, WGC, 1922.

B. erectus Huds.   Upright Brome

Native. Not common in Norfolk. Dry banks on calcareous soil. Var. *villosus* Leight., occurs with the normal plant in places.

West: 62, Vincent Hills, W. Newton; 72, Paston's Clump, Flitcham; 73, Ringstead; Shernborne; 81, Westacre; 83, Wighton; 84, Holkham, BT; 89, Buckenham Tofts Park; Bodney Warren; 94, Wells-next-the-Sea; 98, Bridgham; Rushford; Kilverstone.

East: 19, Forncett St Peter, ELS.

B. inermis Leyss.   Hungarian or Awnless Brome

Nat. alien. Rare. Sandy soils. With the exception of the Sandringham colony all our plants belong to the awned variety, var. *aristatus* Opiz.

West: 63, Snettisham beach, 1951; 70, Beechamwell, 1956; 72, Sandring-ham; 98, E. Harling, 1950.

First record: 1950, E. Harling, CPP.

B. arvensis L.   Field Brome

Casual. Rare. Weed of cultivation. No recent records.

West: 80, Holme Hale, FR, 1914; 98, Roudham, Miss A. B. Cobbe, 1920.

First record: 1889, King's Lynn Docks, C. B. Plowright in Herb. Norw. Mus.

B. mollis L.   Soft Brome or Lop Grass

Native. Very common. Roadsides, grass- and arable land.

B. thominii Hard.

Native. Widespread in West Norfolk. Hayfields, roadsides and waste places.

West: 61, Bawsey; 62, S. Wootton; Wolferton sea-bank; 63, Snettisham beach; 71, Pentney; 74, Titchwell; 90, Cranworth, ALB.

East: 04, Blakeney; 10; 18, Roydon near Diss, AC; 22, 30; 32, E. Ruston.

Var. *hirsutus* (Holmb.) with hairy spikelets.

West: 62, Wolferton sea-bank, CEH & ELS, 1962.

B. lepidus Holmb.   Slender Brome

Native. Locally frequent in or near cultivated land on light soils.

West: 60–63, 70, 72, 74, 80, 88, 90, 94, 98.

East: 02, 09, 11, 12, 19, 20, 29, 31, 41, 49.

Var. *micromollis* (Krosche) C. E. Hubbard with hairy spikelets.

West: 62, W. Newton, CEH & ELS; 63, Wolferton beach, CEH & ELS.

B. interruptus Druce   Interrupted Brome

First recorded for the county in 1884 from Thetford by E. F. Linton.

Appeared in crops of sainfoin in 1946 but, like many of the alien bromes, has not persisted.

West: 62, W. Newton; 72, Anmer; 99, Rocklands, FR, 1916.

B. racemosus L.   Smooth Brome

Casual. Rare. Arable and waste land.

West: 80, Swaffham, ESE det. Howarth, 1954; Saham Toney, WCFN, 1917.

East: 42, Hickling.

B. commutatus Schrad.   Meadow Brome

Nat. alien. Weed of arable land and hedgerows. We add the years to the records as this species is not likely to persist.

West: 62, W. Newton, 1946; 71, E. Walton, 1940; Narborough, 1949; 72, Anmer, 1946, CEH; Grimston, 1954; 80, Swaffham, ESE, 1954.

East: 04, Blakeney, 1948; Salthouse, CEH, 1956; 08, Bressingham, 1960, AC; 10, 39.

Var. *pubens* Wats. Has softly hairy spikelets.

West: 63, Snettisham, 1954, ELS.

B. commutatus × *mollis*

East: 04, Blakeney, 1948, ELS det. CEH.

B. secalinus L.   Rye Brome

Nat. alien. Arable weed, hay-fields, field-margins and around stack-bottoms. All our plants have hairy spikelets=var. *hirtus* (F. Schultz) Aschers. & Graebn. A form with hairy basal sheaths and small florets= var. *hirsutus* (Kindb.) occurred at Appleton, W72, 1936, CEH.

West: 61, E. Winch, 1947; 62, W. Newton; Wolferton, 1946; 72, Anmer, CEH, 1946.

East: 11, Drayton, 1922, WGC; the glabrous form, in Herb. Norw. Mus.

First record: 1889, King's Lynn Docks, C. B. Plowright.

Brachypodium sylvaticum (Huds.) Beauv.
      Slender or Wood False Broom

Native. Common. Woods.

B. pinnatum (L.) Beauv.   Chalk False Brome

Native. Rare. Chalk grassland. Although regarded as questionably native in the 1914 *Flora* and only one old record given from the Earsham area (E38), we see no reason to doubt its status.

West: 99, Broom Covert, Tottington, 1967, ALB; here it is locally dominant.

Var. *pubescens* S. F. Gray

West: 98, along the Thetford-Rushford road, 1954, RCP.

Agropyron caninum (L.) Beauv.   Bearded Couch

Native. Rare. Shady places in hedgerows and woodland margins.

West: 60, Fincham; 80, Necton, ALB; 90, Cranworth, ALB; 91, Scarning; Bradenham, CPP; Rawhall Wood; 92, N. Elmham.

East: 00, 20, 30; 31, Upton Broad, BEC Exc., 1938; 39, 40.

A. repens (L.) Beauv.   Couch or Twitch
Native. Very common and a pest in cultivated land; roadsides and waste places. Exceedingly variable; glabrous or hairy; green or glaucous; awned or awnless. We recognise plants with long-awned lemmas as var. *aristatum* Baumg., which are frequent and often mistaken for *A. caninum*.

A. pungens (Pers.) Roem. & Schult.   Sea Couch
Native. Common. All along the coast on dunes, shingle, salt marsh, sea-banks and banks of tidal rivers. Variable; awned and mucronate forms being frequent.

Var. *pycnanthum* (Gren. & Godr.)

West: 63, Wolferton beach; Snettisham beach.

A. pungens × *repens*
(*A.* × *oliveri* Druce)
West: 63, Wolferton, CEH & ELS, 1962; 84, Burnham Overy Staithe, CEH & ELS.

A. junceiforme (A. & D. Löve) A. & D. Löve   Sand Couch
Native. Far less common than *A. pungens* and confined to coastal sand, in which it is a pioneer dune-builder. Plants with larger spikelets = forma *megastachyum* (Fries) occur occasionally in richer soil.
West: 62, King's Lynn; 63, Wolferton-Snettisham; 64, Holme-next-the Sea; 74, Titchwell; Thornham; 84, Holkham.
East: 04, Blakeney; Cley; 14, 24, 33, 41, 42, 51.

A. junceiforme × *pungens*
A. × *obtusiusculum* Lange; *A. acutum* auct.)   Hybrid Sea Couch
Native. Occurs very commonly where the parents are found and in an intermediate zone, but is often overlooked as the more frequent erect forms simulate the *A. pungens* parent.

Elymus arenarius L.   Lyme Grass
Native. Common on coastal sand and sometimes dominating large areas.

Hordeum murinum L.   Wall Barley
Native. Common. Roadsides and waste places.

H. marinum Huds.   Sea Barley
Native. Common. Sea-banks and waste places along the coast.

H. secalinum Schreb.   Meadow Barley
Native. Common. Grassland; noticeably abundant near the sea.

H. jubatum L.   Fox-tail Barley
Casual.
East: 14, carrot-field, Weybourne, 1967, ALB; 20, tip at Harford, CEH, 1967.

Glyceria declinata Bréb.   Glaucous Sweet-grass
Native. Less frequent than *G. plicata* or *G. fluitans*, but moderately common around pond margins and in shallow ditches.
West: 60, 62, 63, 71, 72, 80, 82–84, 89–92, 99, 00.
East: 00–03, 10, 14, 18, 20, 22, 28, 30, 33, 39, 41, 42.
First record: 1915, Flegg Burgh Common, and Salmon White

G. fluitans (L.) R. Br.   Floating Sweet-grass
Native. A common aquatic species in ditches and sluggish streams where the water is shallow.

G. fluitans × *plicata*   Hybrid Sweet-grass
(*G.* × *pedicellata* Towns.)
Native. This male-sterile hybrid grows occasionally with its parents.
West: 61, Ashwicken; 62, N. Wootton; 63, Heacham; 72, Appleton, CEH; 93, Wighton, AC.
East: 08, 21; 30, Strumpshaw, AC; 39, 42.

G. plicata Fries   Plicate Sweet-grass
Native. Common, but less frequent than *G. fluitans*, in ponds, ditches, streams and swampy places.
West: 60–64, 70–73, 80, 83, 90, 94, 98, 99.
East: 00, 01, 04, 18, 20, 28, 30, 33, 39, 41, 49.

G. maxima (Hartm.) Holmb.   Reed Sweet-grass
Native. Common and growing in deeper water than the other species, often forming pure stands in the fens and Broads.

Festuca tenuifolia Sibth.   Fine-leaved Sheep's Fescue
Native. Not a rare species but has been overlooked. Margins of woodland on sandy, acid soils and heaths. Recorded in the 1914 *Flora* as *F. ovina* var. *capillata* (Lam.).
West: 62, 70, 72, 81, 89, 91, 98, 99, 08.
East: 01, 11, 32, 40.

Var. *hirtula* (Hack.) Howarth, with minutely hairy lemmas.
West: 72, White Hills, Hillington, CEH.

F. ovina L.   Sheep's Fescue
Native. Common. Grassland on chalk or sand, at times dominant. Plants with shortly hairy lemmas = var. *hispidula* (Hack.) Hack., often grow with normal plants.

F. longifolia Thuill.   Hard Fescue
Nat. alien. Rare. Road margins, banks and waste places.
West: 61, Leziate; 62, W. Newton, CEH; 72, White Hills, Hillington, together with var. *villosa* (Schrad.) Howarth, CEH.
First record: 1934, W. Newton, CEH.

F. rubra L., subsp. *commutata* Gaud.   Chewings' Fescue
Nat. alien. Rare. A component of some grass-seed mixtures escaping to roadsides and waste ground.
West: 62, Roydon Common, det. CEH; 70, Beechamwell, det. CEH.

F. rubra L., ssp. *rubra*   Red or Creeping Fescue
Native. Common. Grassland, heaths, open woodland, sand-dunes and raised parts of salt marshes.

Var. *barbata* (Schrank) Richt., with hairy spikelets, is widespread and occurs in every habitat.

249

Var. *glaucescens* (Hegets. & Heer) Richt., with bluish-green leaves and panicles and hairy spikelets, is locally common in salt marshes.

Var. *pruinosa* (Hack.) Howarth, similar to the last but with glabrous spikelets, is also common in salt marshes.

Var. *arenaria* Fries   Sand Fescue

Native. Common on coastal sand-dunes and has been confused with *F. juncifolia* which is far less common.

F. juncifolia St.-Amans   Rush-leaved Fescue

Native. Occasional. On the older sand-dunes.

West: 63, Snettisham, CPP conf. CEH; Heacham, CEH; 64, Hunstanton, D. C. Clouston in Herb. Kew; 74, Thornham, ELS confd. CEH; 84, Burnham Overy Staithe, CEH & ELS.

East: 42, near Horsey Corner, PJB confd. ELS; also from 41 and 51 but not seen by authors.

F. pratensis Huds.   Meadow Fescue

Native. Common. Grassland on heavy soils and roadsides.

F. pratensis × *Lolium multiflorum*
(× *Festulolium braunii* (Richt.) A. Camus)

West: 72, Appleton, on heavy soil in damp situations, CEH, 1933.

F. pratensis × *Lolium perenne*
(× *Festulolium loliaceum* (Huds.) P. Fourn.)

Native. This sterile intergeneric hybrid is fairly frequent in damp meadows on heavy soil.

West: 71, 72, 78, 79, 83, 84, 93, 99.
East: 20, 28.

F. arundinacea Schreb.   Tall Fescue

Native. Frequent. Meadows and streamsides. Variable; tall, luxurious forms occur on the heavier soils in damp places.

F. gigantea (L.) Vill.   Giant Fescue

Native. Frequent. Damp open woodland associated with *Bromus ramosus* and *Brachypodium sylvaticum*.

F. gigantea × *pratensis*
(*F.* × *schlikumii* Grantzow)

West: 72, Appleton; Flitcham, CEH, 1952.

Lolium perenne L.   Perennial Rye-grass

Native. Common. Grassland, roadsides and waste places. Has a number of curious forms showing modification of all parts of the inflorescence. Many of the newly-made roadside paths have been sown with this species but it is not likely to persist by reason of competition from native vegetation.

Var. *longiglume* Grantz.
West: 63, Wolferton, on fixed shingle, 1946; 84, Burnham Overy Staithe, sea-bank, CEH & ELS.

L. multiflorum Lam.   Italian Rye-grass

Nat. alien. Not known to the early botanists; was introduced by the Edinburgh firm of Peter Lawson & Sons in 1831–2. Established as an arable weed and in waste places.

L. multiflorum × perenne
(*L. × hybridum* Hausskn.)

Should be far more common than its single record suggests as local farmers are making increasing use of this fertile hybrid as 'HI' or 'Short Rotation Rye-grass'.

West: 72, Appleton, CEH, 1947.

L. temulentum L.   Darnel

Casual. Formerly known as a weed of arable land and waste places from the imports of foreign barley during the First World War; now rare and restricted to rubbish dumps. Var. *arvense* Lilj., without awns, is the usual form.

West: 62, King's Lynn tips; 88, Thetford, FR, 1916.
East: 20, Norwich, WGC, 1915; Harford tip, CEH, 1967; 39, Ditchingham, MIB.

Vulpia bromoides (L.) Gray   Squirrel-tail Fescue

Native. Common. Heaths, wall-tops and sandy arable land.

V. myuros (L.) C. C. Gmel.   Rat's-tail Fescue

Native. Occasional. Waste places on gravel and sand.

West: 60–62, 70, 80–82, 84, 90, 92, 98–00.
East: 00, 08, 18, 20, 29, 39.

V. megalura (Nutt.) Rydb.   Foxtail Fescue

Casual. Very rare. Railway sidings and a maritime habitat.

West: 62, Grimston Road station, 1945, det. CEH; 63, mud and shingle bank, Heacham, CEH.
East: 21, railway track near Wroxham, Miss C. M. Goodman, 1958, det. CEH, in Herb. Norw. Mus.

V. ambigua (Le Gall) A. G. More   Bearded Fescue

Native. Frequent on tracks on sandy heaths, warrens and waste places throughout the south-west Norfolk Brecklands and extending north-east to Little Ryburgh and Wroxham in East Norfolk.

West: 61, 62, 70, 72, 78–81, 88, 89, 91, 92, 94, 98, 99.
East: 21, Wroxham.

V. membranacea (L.) Dumort.   Dune Fescue

Native. Very rare. Sand-dunes.

West: 64, Holme-next-the-Sea, 1950, still present 1968, CEH.

Nardurus maritimus (L.) Murb.

Status doubtful. New to Norfolk. Two chalk pits in Breckland. Occurs elsewhere in the British Isles in open places overlying chalk or limestone.
West: 89, Ling Heath, Tottington, June 1967, J. M. Schofield and T. C. E. Wells, confd. CEH.

Poa annua L.   Annual Meadow-grass
Native. Ubiquitous. Flowers throughout the year.

P. bulbosa L.   Bulbous Meadow-grass
Native. Very rare except at Great Yarmouth where, at one time, it formed 'the principal part of the herbage of our denes', Dawson Turner in *Botanists' Guide*, 1805.
West: 94, Wells-next-the-Sea, 1963, an interesting extension of its range.
East: 04, Blakeney, site of old golf course, 1948, ELS; 49, Haddiscoe, 1936, EAE; 50–51, Gt. Yarmouth, still frequent there.

P. nemoralis L.   Wood Meadow-grass
Native. Locally common in shady woodland.
West: 61–63, 71–73, 78–80, 84, 89, 91, 92, 99, 00.
East: 01, 08–10, 12, 14, 18, 20, 21, 29, 30, 33, 39.

P. palustris L.   Swamp Meadow-grass
Status doubtful, perhaps a denizen. Damp habitats.
West: 81, ditch at Castleacre Priory, 1946–54.
East: 18, Roydon Fen near Diss, AC, 1962; 20, tip at Harford near Norwich, CEH, 1967.

P. chaixii Vill.   Broad-leaved Meadow-grass
Nat. alien. Rare.
East: 12, Blickling, AC, 1960, abundant in woodland rides.

P. trivialis L.   Rough Meadow-grass
Native. Common. Woods, grass- and arable land on moist soil. Smooth-sheathed forms (var. *glabra* Doell.) are not infrequent.

P. angustifolia L.   Narrow-leaved Meadow-grass
Native. Rare or overlooked. Dry hedgebanks.
West: 60, Shouldham, WGH; 61, near Leziate Golf Course, WGH; 62, N. and S. Wootton; 63, Wolferton, CEH & ELS; 72, Congham; 78, Weeting Heath, FRo; 89, W. Tofts.
East: 18, Roydon near Diss, AC.

P. pratensis L.   Meadow-grass
Native. Common and very variable. In a variety of habitats.

P. subcaerulea Sm.   Spreading Meadow-grass
Native. Frequent. Widespread in a variety of habitats including fens, coastal sands and Breckland conifer plantations.
West: 62, 63, 70–72, 79, 84, 89, 90–92, 99.
East: 04, 11, 12, 51.

P. compressa L.   Flattened Meadow-grass
Native. Rare. Usually on walls.
West: 61, Setch, WGH; Leziate; 62, King's Lynn, BSBI Exc., 1948; 79, Whittington; 80, Swaffham, ESE; 81, Castleacre; 93, Gt. Walsingham, BT; 94, Wells-next-the-Sea, BT.
East: 04; 20, Cathedral Close, Norwich, AC; 23, 39, 49.

Puccinellia fasciculata (Torr.) Bickn.   Borrer's Salt-marsh Grass
Native. Occasional. Brackish marshes, often colonising bare mud.
West: 62, N. Wootton; 74, Holme-next-the-Sea; Thornham; Brancaster; Titchwell; 84, Burnham Overy Staithe; Holkham; 94, Wells-next-the-Sea, J. E. Little.
East: 04, Cley, FMD; Salthouse, CEH.

P. distans (L.) Parl.   Reflexed Salt-marsh Grass
Native. Widespread in the higher parts of salt marsh, reclaimed land and pastures near the sea.

P. distans × fasciculata
East: 04, Salthouse, A. L. Still in Herb. Lousley, 1932.

P. distans × maritima
West: 84, Burnham Overy Staithe, 1967, CEH & ELS.

P. pseudo-distans (Crép.) Jans. & Wacht.
Native. Rare or overlooked.
West: 84, Burnham Overy Staithe, 1967, CEH & ELS.
East: 04, Salthouse, CEH.

P. maritima (Huds.) Oarl.   Common Salt-marsh Grass
Native. Widespread in salt marsh and forming a pure sward in the mature grazed marsh. Remains dominant for up to five years after reclamation from the sea.

P. rupestris (With.) Fern. & Weath.   Stiff Salt-marsh Grass
Native. Very rare. Margins of salt marsh.
East: 50, behind wall of R. Bure, Gt. Yarmouth, 1954, EAE.

Catapodium rigidum (L.) C. E. Hubbard   Fern Grass
Native. Common. Walls, paths and sandy soil (both coastal and inland).

C. marinum (L.) C. E. Hubbard   Stiff Sand-grass
Native. Occasional. In maritime sand and shingle.
West: 63, Wolferton: Snettisham; Heacham; 64, Hunstanton; 74, Holme-next-the-Sea; 84, Holkham, BT.
East: 04, Cley; 14, Sheringham; 24, Cromer; 33, Mundesley; 50, Gt. Yarmouth.
First record: 1794, north coast of Norfolk, Rev. H. Bryant in *English Botany*, t. 221, as *Triticum loliaceum*.

Briza maxima L.   Large Quaking Grass
Casual. Rare. Garden-escape; rubbish tips and waste places.
West: 61, Blackboro' End sand-pit, E. A. Vine, 1964; 79, Whittington, RCLH, 1961.
East: 10, Hethersett, ETD; 11, Drayton, WGC, 1922.

B. media L.   Common Quaking Grass or Totter Grass
Native. Frequent. Grassland on calcareous soil, old meadows and fens.

Dactylis glomerata L.   Cocksfoot

Both native and cultivated on a large scale. Roadsides, grassland and open woodland.

Cynosurus cristatus L.   Crested Dog's-tail

Native. Common. Old grassland.

C. echinatus L.   Rough Dog's-tail

Casual. Rare. Rubbish tips and waste places.

West: 64, Hunstanton, Gambier Parry, 1918; 88, Thetford, FR, 1916; 94, Wells-next-the-Sea, FR, 1917.

East: 11, Drayton, WGC, 1922.

Catabrosa aquatica (L.) Beauv.   Water Whorl-grass

Native. Frequent as a marginal plant beside streams and ditches.

Melica uniflora Retz.   Wood Melick

Native. Locally plentiful in woods and on shady banks.

West: 62, 70, 80, 81, 90–92.

East: 00, 01, 03, 10, 11, 14, 19, 20, 28.

First record: 1797, Wissonsett [*sic*] = Whissonsett, Skrimshire, *Cat.*

Helictotrichon pubescens (Huds.) Pilger   Hairy Oat-grass

Native. Common in dry grassland overlying chalk.

H. pratense (L.) Pilger   Meadow Oat-grass

Native. Less common than the last species, in similar places.

Arrhenatherum elatius (L.) J. & C. Presl   False Oat-grass

Native. Abundant and widespread along roadsides, in hedgebanks and waste places; forming more or less pure stands in old shingle at Wolferton. Forms with both florets awned, hence 2-awned spikelets = forma *biaristatum* (Peterm.) Bartram, are found occasionally.

Var. *subhirsutum* (Aschers. & Graebn.)

West: 63, brackish creek, Heacham, CEH; 72, old pastures and roadsides, Appleton, CEH; Hillington, CEH & ELS; 91, Gressenhall, CEH.

Avena fatua L.   Spring or Common Wild Oat

Nat. alien. Established for a long time as a pestilent weed of cultivated land. Very variable. Var. *fatua* (var. *pilosissima* Gray) with densely hairy lemmas becoming reddish brown when mature and var. *pilosa* Syme with slightly hairy lemmas becoming grey, are both common; var. *glabrata* Peterm., with hairless lemmas remaining yellow appears to be rare.

West: 74, Holme-next-the-Sea, FMD; 80, Saham Toney, ELS.

East: 04, Walsey Hills, Cley, FMD.

So far, the Winter Wild Oat (*A. ludoviciana* Durieu), a pest in the Midlands, has not been seen. It would appear that the old Norfolk system of husbandry militates against its establishment but the increasing use made of modern methods, including combine harvesting, will most likely result in its appearance.

Koeleria gracilis Pers.   Crested Hair-grass
Native. Frequent on dry calcareous grassland, sand-dunes and consolidated shingle.

Trisetum flavescens (L.) Beauv.   Yellow Oat-grass
Native. Widespread. Roadsides and grassland particularly on chalky soils.

Deschampsia setacea (Huds.) Hack.   Bog Hair-grass
Native. Very rare. Peat bog.
This species was originally described from Norfolk material by Hudson in his *Flora Anglica*, 1762, under *Aira setacea*. It was sent to him by Benj. Stillingfleet who found it at Stratton Strawless, E22, in 1755. In 1776 it was found at Cawston Decoy, E12, some six miles away, by Rev. H. Bryant. In 1957 we found several plants in a bog pool on East Winch Common, W71.

D. flexuosa (L.) Trin.   Wavy Hair-grass
Native. Frequent on heaths overlying gravel, much less so on the Greensand, sometimes occupying large areas as a local dominant.

D. caespitosa (L.) Beauv.   Tufted Hair-grass
Native. Very common. Rough grazing pastures and fens.

Corynephorus canescens (L.) Beauv.   Grey Hair-grass
Native. Rare. Coastal sands.
West: 64, Holme-next-the-Sea, not until 1955 following the sea-floods of 1953; 74, Brancaster, EAE; 94, Wells-next-the-Sea, HDH; 04, Blakeney Point.
East: 20, Bramerton, EAE; 1954; 42, Horsey; 50, Gt. Yarmouth, ETD, abundant, 1964; 51, Caister, E. Nelmes; near Winterton.

Aira caryophyllea L.   Silvery Hair-grass
Native. Common. In dry gravelly and sandy soils, especially on the Greensand.

A. praecox L.   Early Hair-grass
Native. More common than *A. caryophyllea* and in similar habitats.

Holcus lanatus L.   Yorkshire Fog
Native. Common. Grassland, arable land and waste places, both wet and dry. Albinos occur.

H. mollis L.   Creeping Soft-grass
Native. Common in open woodland.

Anthoxanthum puelii Lec. & Lam.   Annual Vernal-grass
Casual. Very rare.
East: 20, Eaton, EAE, 1939, the first and only record.

A. odoratum L.   Sweet Vernal-grass
Native. Common in a considerable range of habitats from coastal sands to woodland.

Phalaris canariensis L.　　Canary Grass
Casual. Frequent in waste places and on rubbish tips.
West: 61, Setch; 62, King's Lynn; S. Lynn; 71, E. Winch; 79, North-wold; 84, Burnham Overy; 88, Thetford; 98, E. Harling, DMM.
East: 00, Hackford; 12, Aylsham; 14; 20, Earlham; Harford, Norwich; 24, 49.

P. minor Retz.
Casual. Rare weed of American Lend-Lease carrot seed.
West: 61, Bawsey, 1945, det. CEH; 72, Houghton, 1950.

P. paradoxa L., var. *praemorsa* Coss. & Dur.
Casual. Very rare. Rubbish tip.
West: 62, King's Lynn, 1945, det. CEH.

P. arundinacea L.　　Reed Canary-grass
Native. Common. Ponds, streams and ditches. Forming pure com-munities around some of the Breckland meres.
First record: 1819, from the Podyke between Outwell and Downham, Skrimshire, *Cat.*

Milium effusum L.　　Wood Millet
This native species has been sown in woodland in the past and was known to Hudson, Ray and Stillingfleet. Established in woodlands on the heavier soils.
West: 62, N. Wootton; 81, Necton; 90, Wood Rising; Cranworth, ALB; 91, Wendling; 92, Eastfield, Wood, Tittleshall.
East: 00, 01, 02, 09, 18, 19, 29, 32, 39.
First record: 1796, woods in Norfolk, Skrimshire, *Cat.*

Calamagrostis stricta (Timm.) Koel.　　Narrow Small-reed
Native. This rare grass was first discovered by F. Robinson on the site of the former Hockham Mere in 1911. (The mere, formerly 280 acres, was drained in 1795.) It is still abundant in the area in carr, open fen and pond margins. 'For about a day it is spreading and not unlike robust *Agrostis alba*; after flowering the branches are closely adpressed to the stem, giving it a spike-like appearance', FR.
West: 99, Hockham; Shropham; Stow Bedon; Thompson Common, ERN, 1955; marsh near Breckles Heath, ALB, 1966.

C. canescens (Weber) Roth　　Purple Small-reed
Native. Frequent in fens and open woodland, and characteristic of carr developed over peat.
Naming them var. *pallida* Lange, Salmon and White found plants 'of a delicate pale yellow tint' near Horning Ferry, E31, in 1915. Recently (1966), similar plants, the pale colour replacing the usual purple-tinted inflorescences, were found at Stoke Ferry Fen, W79, ELS. These are albinos, a form only, and Dr Hubbard suggests they be placed under f. *canescens* Prahl (Junge, *Die Gramineen Schleswig-Holsteins* 178, 1913).
West: 61, 62, 70–72, 78, 79, 82, 89–92, 99, 08.
East: 02, 08, 21, 22, 30–33, 40–42, 49.

C. epigejos (L.) Roth   Wood Small-reed or Bush-grass

Native. Frequent. Ditches and wet woods on heavy soils; sometimes frequent on dry Breckland heaths.

West: 60–63; 70–72; 78–80; 84, 88, 89, 91, 92, 94, 98, 99.
East: 18, 20, 30–32, 42.

Ammophila arenaria (L.) Link.   Marram Grass
Native. Abundant along the coast and dominant in established dunes.

Ammophila arenaria × *Calamagrostis epigejos*   Hybrid Marram
(× *Ammocalamagrostis baltica* (Schrad.) P. Fourn.)

Native. This intergeneric hybrid has long been known to occur naturally along the east coast from Caister to Waxham where it is abundant on the dunes, often where *C. epigejos* is absent. It was introduced to the dunes at Holme-next-the-Sea and Brancaster following the sea-floods of 1953.

Apera spica-venti (L.) Beauv.   Loose Silky-bent
Status doubtful. Frequent as a weed in cultivated land on light sandy soil, sand-pits and waste places. Sporadic in appearance.

West: 61–63, 70, 74, 78–80, 88, 98.
East: 20–21.

A. interrupta (L.) Beauv.   Dense Silky-bent
Native in some localities but range much increased by human activity. Confined to West Norfolk. More frequent than *A. spica-venti* and sometimes abundant in fire-breaks in the open in Breckland. Large plants have been confused with *A. spica-venti*.

West: 61, 62, 70–72, 74, 78, 80–82, 88, 89, 98–00.
First record: 1849, Thetford, Gibson and Newbould.

Agrostis canina L., ssp. *canina*   Velvet Bent
Native. Widespread in damp places.

Ssp. *montana* Hartm.   Brown Bent
Native. Abundant on heaths on sand or peat; associated with *A. tenuis*.

A. tenuis Sibth.   Common Bent or Brown Top
Native. Common. Grasslands, heaths, wood-margins and roadsides. At times dominant or co-dominant on grass heath or abandoned arable land ('brecks').

A. gigantea Roth   Black Bent or Red Top
Native. Common. Troublesome weed on arable land; rough grassland and waste places.

A. stolonifera L., var. *stolonifera*   Creeping Bent
Native. Common in a variety of habitats including grassland, fens, salt marshes and coastal sands. The salt-marsh variant, *ecas salina*, is of frequent occurrence on the shores of The Wash.

Var. *palustris* (Huds.) Farw., with extensively creeping stolons is frequent in more or less permanent water.

Agrostis stolonifera × *Polypogon monspeliensis*   Perennial Beard-grass
(× *Agropogon littoralis* (Sm.) C. E. Hubbard)
Native. A very rare intergeneric hybrid. Sand-dunes.
East: 42, Horsey Corner, TGT, 1958.
First record: 1777, on northern coast of Norfolk, Rev. H. Bryant.

Polypogon monspeliensis (L.) Desf.   Annual Beard-grass
Native. Very rare. Salt marshes.
East: 04, Cley; 42, Horsey, 1947.
First record: 1805, Cley, Humphrey in *Botanists' Guide*.

Phleum arenarium L.   Sand Cat's-tail
Native. Occasional on coastal sands, rarely inland in Breckland.
West: 63, 64, 73, 74, 78, 80, 84, 88.
East: 04, 14, 24, 31, 42, 50.

P. phleoides (L.) Karst.   Purple-stem Cat's-tail
Native. A characteristic Breckland species, often abundant on chalk grass-
land in south-west Norfolk. Two forms occur: the keels of the glumes being
ciliate in the commoner (forma *blepharodes* (Aschers. & Graebn.) and
scabrous in the less common plant.
West: 72, Gayton chalk pit, known there since 1844 (Herb. Druce); 78,
Weeting Heath; 79, Cranwich (Devil's Dike); 80, Swaffham, FR; 88,
Santon; 98, Kilverstone.

P. bertolinii DC.   Smaller Cat's-tail
(*P. nodosum* auct. non L.)
Native. Common in dry grassland.

P. pratense L.   Timothy Grass
Both native and cultivated forms occur. Common. Field-margins, roadsides and meadows.

Alopecurus myosuroides Huds.
      Slender Foxtail, Black Grass, Black Twitch
Native. Infrequent weed of arable land and waste places.
West: 60–62, 71, 72, 79, 83, 84, 88, 90, 91, 98.
East: 00, 08, 10, 18, 19, 23, 28, 33, 39, 42.

A. aequalis Sobol.   Orange or Short-awn Foxtail
Native. Rare but occasionally abundant in some of the Breckland meres;
pools and ditches in shallow water.
West: 61, E. Winch Common; 71, E. Walton Common; 72, Appleton,
CEH; Congham; Grimston; 82, Gt. Massingham; 88, Langmere; 98,
Ringmere; 99, Home Mere; Hockham; Fowlmere; Stow Bedon, ALB.
East: 11, Horsford, J. Bowra; 12; 28, Wortwell; 39, Ditchingham.
First record: 1805, Swainsthorpe, Mr Stone in *English Botany*, t. 1467.

A. aequalis × *geniculatus*
(*A.* × *haussknechtianus* Aschers. & Graebn.)
West: 72, Appleton, CEH.

A. bulbosus Gouan   Tuberous or Bulbous Foxtail
Native. Apparently very rare. Tidal salt marshes and brackish meadows.
West: 84, Holkham, A. E. Ellis, 1938.

A. geniculatus L.   Marsh or Floating Foxtail
Native. Common in wet places.

A. geniculatus × *pratensis*
(*A.* × *hybridus* Wimmer)
West: 72, Appleton, CEH, 1948; 99, Breckles Heath, FR, 1919, in Herb.
Norw. Mus.
East: 31, Acle, Miss E. S. Todd, 1925.

A. pratensis L.   Meadow or Common Foxtail
Native. Common in grassland on moist soils.

Parapholis strigosa (Dum.) C. E. Hubbard   Sea Hard-grass
Native. Frequent along the coast on the margins of salt marshes and banks
of tidal rivers. Abundant in *Puccinellia maritima* sward on landward side of
sea-bank at N. Wootton associated with *Juncus gerardii, Glaux maritima*
and *Triglochin maritima*. Flowers after the next species.
West: 51, 52, 62–64, 73, 74, 84, 94.
East: 04, 41, 42.

P. incurva (L.) C. E. Hubbard   Curved Sea Hard-grass
Native. Rare. Gravelly mud banks of salt marshes in relatively drier places
than *P. strigosa*.
West: 63, Wolferton; Snettisham; Heacham; 74, Holme-next-the-Sea.
East: 04, Salthouse, CEH; 41, Acle Bridge, PER in Herb. Norw. Mus.
First record: 1937, Salthouse, CEH.

Nardus stricta L.   Mat-grass
Native. Common. Heaths.

Phragmites communis Trin.   Common Reed
Native. Abundant. Broads, ditches, ponds, fens and carrs. Considerable
quantities cut for thatching, etc., in the Broads. Occasionally forms with
very long stolons, due to mechanical injury, are found in drying-up ponds.

Molinia caerulea (L.) Moench   Purple Moor-grass
Native. Widespread and very variable in bogs, wet heaths and carr;
frequently dominant in large colonies. Inflorescences range from more or
less spicate to very loose panicles.

Sieglingia decumbens (L.) Bernh.   Heath Grass
Native. Frequent on heaths where the flowers are usually cleistogamous.
We have not observed the normal flowering occurring mostly in wet
places.

Spartina maritima (Curt.) Fernald   Cord-grass
Native. Rare and diminishing. Fringing pools in the higher parts of salt marshes.
West: 63, Wolferton; 74, Norton Creek, Brancaster; Holme-next-the-Sea, FMD; 84, Burnham Overy Staithe; Missel Marsh, Scolt Head.
East: 04, Blakeney; Salthouse, RMB.

S. alterniflora × *maritima*
('*S.* × *townsendii* H. & J. Groves'; *S. anglica* ined.)
Established and abundant in mobile estuarine mud; plays an important role in the early stages of colonisation of large areas, making eventual reclamation possible.
   This fertile species arose by hybridisation between the introduced American *S. alterniflora*, and the native *S. maritima*. Dr Hubbard has recently shown that, although later collections were fertile, the early material, including the type-specimen, consisted of sterile plants. Its present name, therefore, covers two plants but strictly speaking the fertile plant is without a name.
   The fertile plant was introduced into the Lynn Cut in 1909; a few plants were noted in 1925 at Wolferton; by 1938 this grass formed a continuous zone up to half a mile wide along the seaward edge of the salt marsh between these places and had taken over the primary role of colonising bare mud. Outlying colonies were seen at Brancaster in 1938 and Holme-next-the-Sea in 1948; it is now abundant in both marshes and is recorded from every Norfolk salt marsh.

S. × *townsendii* H. & J. Groves *sensu stricto*   Townsend's Cord-grass
Compared with the fertile plant, it has smaller, indehiscent anthers of about 7 mm., smaller ligules not exceeding 1·5 mm., narrower, more erect leaves which are less glaucous, and shorter and narrower spikelets.
West: 62, east bank of R. Ouse, King's Lynn; 63, Wolferton; 84, Burnham Overy Staithe.
First record: 1962, Wolferton, CEH.

Cynodon dactylon (L.) Pers.   Bermuda Grass
Casual. Very rare. Persisted for twelve years following its discovery at Castle Rising, W62, in 1945. It probably owed its origin to dispersal of seed from a bombed tobacco warehouse on the docks at King's Lynn.

Echinochloa crus-galli (L.) Beauv.   Cockspur Grass
A casual, frequent as a weed in carrot-fields during the Second World War but not persistent, owing its more recent records to fresh intro-ductions. Occasionally on rubbish tips.
Recent records:
West: 69, onion field, Southery, 1961, G. R. Chadwick; 71, Narford, 1967, ALB; Narborough, 1958; 80, Swaffham, 1961; 91, Dereham, garden weed, KD; near N. Elmham, 1967, ALB.
East: 14, Weybourne, 1967, JHS; 20, Harford tip; 30, Rockland St Mary, RMB.

E. crus-pavonis (HBK) Schult.　　Peacock-spur Grass
Casual. Very rare. Carrot-field weed.
West: 73, Gt. Bircham, 1949, det. CEH.

E. pungens (Poir.) Rydb.
Casual. Very rare; probably confused with *E. crus-galli*.
West: 73, Gt. Bircham, 1949, det. CEH.

Setaria viridis (L.) Beauv.　　Green Bristle-grass
Casual. A frequent weed in carrot-field and waste places.
West: 60, 61, 70–72, 80–82, 88, 89.
East: 14, Weybourne, ELS; 20, Harford; 24, Cromer, RCLH.

S. italica (L.) Beauv.　　Foxtail Millet
Casual. From cage-bird seed on rubbish tips and waste places.
West: 62, King's Lynn, 1964.
East: 18, Diss, 1959; 20, Harford, 1959, RMB; Sprowston, EAE; 23, Trimingham, 1959; 28, Pulham St Mary, 1967, CF; 32, Stalham, 1962, RMB.

S. glauca (L.) Beauv.　　Yellow Bristle-grass
Casual. A rare carrot-field weed.
West: 61, Bawsey; 70, Cockley Cley; 73, Gt. Bircham, 1949, det. CEH; 89, Ickburgh, 1944, RPL; 91, near N. Elmham, 1967, ALB.
East: 14, Sheringham, RMB.

Digitaria ischaemum (Schreb.) Muhl.　　Smooth Finger-grass
Nat. alien. Very rare. Sandy fields.
West: 80, Gt. Cressingham, 1954, in a field of annual blue lupins, continental seed source.
East: 39, Bath Hills, Ditchingham, 1952, in some quantity; has been known there since 1840.
First record: 1805, near Brandon, J. Lightfoot in *Botanists' Guide*.

D. sanguinalis (L.) Scop.　　Hairy Finger-grass
Casual. Rare. Garden weed and cage-bird seed alien.
East: 10, Hethersett, H. W. Back, 1947; 11, Attlebridge, 1959, M. E. Smith; 39, Ellingham Hall near Bungay, 1962, RMB.

Panicum miliaceum L.　　Common Millet
A frequent bird-seed escape.
West: 62, King's Lynn tip, 1964; 88, Thetford; 93, Binham.
East: 20, Harford tip, 1958, RMB; Sprowston; 32, Stalham, RMB; 33, Paston.

P. capillare L., var. *occidentalis* Rydb.　　Witch Grass
A very rare carrot-field weed.
West: 73, Gt. Bircham, 1948, det. CEH.

EXTINCT SPECIES

In common with other counties, Norfolk has suffered the loss of several species, both true natives and some the status of which is questionable. The increasing effects of drainage, intensive cultivation, use of cleaner seeds, afforestation, building, herbicidal spraying, change of climate and human pressure have all added their quota to the diminution and eventual loss.

Modern drainage has undoubtedly accounted for the loss of *Senecio paludosus* and *S. palustris*. Now that cleaner seeds are used, *Melampyrum arvense* and *Roemeria hybrida* are no longer to be found as weeds of cultivation. The conspicuous orchids, all of which were very rare, most probably never gained a foothold and were chance introductions.

The list of extinctions, taken from the earliest times since plant-recording was begun, represents not more than 5 per cent of the present-day flora. The losses have been more than compensated for by the additions of new discoveries of native species, modern segregates and alien plants that have become established.

The comments quoted are those of the 1914 *Flora*.

List and Abbreviations of Authorities quoted.

AB	Arthur Bennett, 1843–1929.
BEC	Botanical Exchange Club, 1869–1946.
Bell	Miss A. Bell, recorded for H. C. Watson and Kirby Trimmer, *fl.* 1836.
BG	*Botanists' Guide*, Turner and Dillwyn, 1805.
Br. Fl.	*British Flora*, W. J. Hooker, 1830–1.
CS	Rev. Chas. Sutton, 1756–1846.
DS	Daniel Stock, 1828–66.
DT	Dawson Turner, 1775–1857.
EB	*English Botany*, Smith and Sowerby, 1790–1814.
Eng. Fl.	*English Flora*, J. E. Smith, 1824–8.
FL	Dr Frederick Long, 1840–1927.
Fl. Br.	*Flora Britannica*, J. E. Smith, 2nd. edit., 1800–4
Fl. Pl. H.	*Flowering Plants of Harleston*, F. W. Galpin, 1888.
FR	Fred. Robinson, *fl.* 1914.
GM	Rev. Geo. Munford, 1794–1871.
GRL	Rev. G. R. Leathes, 1778–1836.
Hb. Salm.	Herbarium of J. D. Salmon in Norwich Castle Museum.
HB	Rev. Henry Bryant, 1721–99.
HDG	H. D. Geldart, 1831–1902.
HDH	H. Dixon Hewitt.
HGG	H. G. Glasspoole, 1825–87.
HJM	H. J. Mennell.
HR	Hugh Rose, 1717–92.
HT	H. Trimen.
J. Bot.	*Journal of Botany*, 1863–1942.
JC	James Crowe, 1751–1807.
JEM	J. E. Moxon.

JES	Sir J. E. Smith, 1759–1828.
JL	Dr John Lowe, 1830–1902.
JP	John Pitchford, 1737–1803.
KT	Rev. Kirby Trimmer, *Flora of Norf.*, 1866 and *Suppt.* 1885.
LW	Lilly Wigg, 1749–1828.
MCHB	Rev. M. C. H. Bird.
NBG	*New Botanists' Guide*, H. C. Watson, 1835.
RBF	Rev. R. B. Francis, 1768–1850.
RF	Rev. Robert Forby, 1759–1855.
RW	Robert Wigham, 1785–1855.
SPW	S. P. Woodward, 1821–65.
TJW	T. J. Woodward, 1745–1820.
WA	*Withering's Arrangement of British Plants*, 1776–1820.
WBEC	Watson Botanical Exchange Club, 1884–1934.
WET	Rev. W. E. Thompson, *fl.* 1900.
WHB	W. H. Burrell, 1865–1945.

Lycopodium selago L.   Fir Club-moss
No recent records. Last seen in the Holt area, WHB, 1903.

L. clavatum L.   Common Club-moss
No recent records. Last seen in the Sheringham area, WHB, *c.* 1903.

Thelypteris robertiana (Hoffm.) Slosson   Limestone Fern
Very rare. Only once recorded.
West: 88, Santon, WHB, 1909, on the railway bridge; disappeared about 1929 and the bridge was rebuilt 1939–40.

Equisetum hyemale L.   Dutch Rush
'Very rare, perhaps extinct. Marshy places.' Last recorded from Ditchingham, E39, *Fl. Pl. H.*

Anemone pulsatilla L.   Pasque Flower
Has not been seen since the early years of the nineteenth century and now considered extinct. Formerly recorded for Sporle, W81, and on the Tulip Hills, W81, near Lexham, the latter name most likely derived from this plant rather than the wild tulip which is exclusive to East Norfolk.

Roemeria hybrida (L.) DC.   Violet Horned-poppy
'Very rare. No recent records'. Formerly a cornfield weed.
East: 12, 'four miles from Aylsham towards Cromer', B. Stillingfleet in *EB*.

Raphanus maritimus Sm.   Sea Radish
'Native. Very rare.'
West: 94, Wells-next-the-Sea, FL in Herb. Norw. Mus., 1888.
East: 24, Cromer, KT; 33, Mundesley, KT.

Iberis amara L.   Wild Candytuft
'Colonist. Very rare.'
West: 61, E. Winch, NBG; 88, Two Mile Bottom near Thetford, HDH, 1943; 90, Ovington, FR in *J. Bot.*, 1913.

Draba muralis L.   Wall Whitlow Grass
'Very rare.'
West: 62, Gaywood, JL. Usually regarded as an error.

Erophila praecox (Stev.) DC.   Early Whitlow Grass
'Apparently rare, but may have been overlooked.'
West: 64, Hunstanton, WBEC, 1904–5.

Viola stagnina Kit.   Fen Violet
Native. One of our losses through drainage. Seen by J. E. Lousley in
West Dereham Fen in 1936.

Holosteum umbellatum L.   Jagged Chickweed
'Extinct.'
First British record: On the city walls of Norwich, JP, 1765. Last record:
Wall near the New Mills, Norwich, HDG, 1887.
'It used to occur on thatch in the Breckland district, and as it is a well-
marked plant of the North German heaths, which apparently show the
nearest affinity with the Breckland heaths, it seems probable that it was
essentially a Breckland plant so far as this country is concerned, and that
it got from the heaths into the rye-fields, perhaps, and so on to the roofs
with the straw.' A. G. Tansley to W. G. Clarke, *in lit.* (*Trans. Nfk. and
Norw. Nat. Soc.*, 1915–16, 142).

Sagina subulata (Sw.) C. Presl   Awl-shaped Pearlwort
'Locally common.'
West: 63, Dersingham, KT; 73, Gt. Bircham, KT; 83, Stanhoe, KT; 92,
Hempton, KT; 94, Wells-next-the-Sea, FL, 1884; 99, Thompson, FR,
1914. (The gatherings from Thompson and Wells in Herb. Norw. Mus.,
do not appear to be this species.)
East: 11, Drayton, KT; 24, Cromer, HT.

Chenopodium botryodes Sm.   Many-spiked Goosefoot
'Native. Rare.'
First British record: near Gt. Yarmouth, LW in *EB*, 1811.
East: 33, Bacton; 40, Halvergate and Reedham; 41, Winterton and
Runham; 49, Haddiscoe; 50, Gt. Yarmouth – all by KT.

Halimione pedunculata (L.) Aell.   Pedunculated Sea-orache
'Native, perhaps.'
West: 62, King's Lynn, *EB*, 1778; 74, Thornham, KT; Holme-next-the-
Sea, CS, 1800.
East: 41, Runham, HB and in Herb. Mus. Brit.; 50, Gt. Yarmouth, LW.

Geranium sylvaticum L.   Wood Cranesbill
'Not native. Rare. Woods and thickets.'
West: 61, Leziate, JC; 80, Swaffham, RF in *BG*, 1805.
East: 20, Framingham Earl, A. Mayfield; 21, Spixworth, W. Humphrey,
1771.

101 EARLY MARSH ORCHID   *Dactylorhiza incarnata*   *H. J. Jarrold*

102 EARLY MARSH ORCHID, INDIAN RED FORM   *Dactylorhiza incarnata forma*   *A. H. Hems*

103 COMMON SPOTTED ORCHID *Dactylorhiza fuchsii* *A. H. Hems*

104 COMMON MARSH ORCHID *Dactylorhiza praetermissa* *A. H. Hems*

105 PUGSLEY'S MARSH ORCHID *Dactylorhiza traunsteineri* *Miss G. Tuck*

Erodium maritimum (L.) L'Hérit.   Sea Storksbill
'Native (?). Rare.'
West: 74, Brancaster, KT.
East: 31, Acle, JC in *WA*, 1808; 33, Mundesley, KT: 42, Palling, KT.

Trifolium squamosum L.   Sea Clover
'Native. Rare.'
West: 63, Snettisham, JES; Heacham, KT; 74, Brancaster, WET, 1902;
84, Burnham Deepdale and Burnham Norton, KT; 94, Wells-next-the
Sea. JES.
East: 04, Blakeney, KT; 20, Thorpe, in Herb. Mus. Brit.; 24, Cromer,
JES; 50, Gt. Yarmouth, LW in *EB*.

Lathyrus montanus Bernh.   Tuberous Pea
'Apparently very rare.'
West: 63, Dersingham, JEM.

Lythrum hyssopifolia L.   Grass Poly
'Colonist (?) in Norfolk. Very rare. Damp stagnant places.'
East: 42, Heigham, Miss A. M. Barnard; 32, Brumstead, MCHB.

Carum carvi L.   Caraway
'Native (?). Rather rare.'
West: 60, Stow, NBG; 62, north of Lynn, GM; Gaywood, JL; 88,
Thetford, HGG.
Still persists in the neighbouring county of Lincolnshire.

Pimpinella major (L.) Huds.   Greater Burnet Saxifrage
No mention in 1914 *Flora* but Miss A. Bell sent a West Norfolk specimen
to H. C. Watson who recorded it in his *Top. Bot.*, ed. 2, 192, 1883.

Andromeda polifolia L.   Marsh Andromeda
'Very rare. No recent records.'
West: 98, Larlingford = Larling, GRL in *Br. Fl.*, 1835; 99, Shropham,
SPW.

Centaurium littorale (D. Turner) Gilmour
Although reported from 'hollows of the sandhills between Wells and
Holkham' in 1900, its known distribution makes the record suspect and
we regard it as an error.

Mertensia maritima (L.) S. F. Gray   Northern Shore-wort
'Denizen or possibly native. Only one record. Sea-shores.'
East: 04, Blakeney, R. J. Pinchen, 1905.
One or more plants have been known to occur since 1905; one plant only
from 1921 until 1931.

Cuscuta europaea L.   Greater Dodder
'Native (?). Rare.'
Some twelve stations, chiefly in East Norfolk, are listed in the 1914 *Flora*
with gorse as the main host-plant. It would appear that there has been
some confusion, as we consider this species to be more or less confined to
Urtica dioica and but rarely to parasitise other species.

C. epilinum Weihe   Flax Dodder
'Alien. Rare.'
West: 59, Welney, E. T. Bennett, 1853; AB.
East: 18, Burston and Shimpling, KT.

Limosella aquatica L.   Mudwort
'Very rare. Stagnant muddy places.'
West: 63, Ingoldisthorpe, JL; 90, Scoulton Mere, FR, 1916.
East: 11, Swannington, HDG; 23, Aldborough, KT; 31, Hemblington, KT; 32, E. Ruston, KT.

Melampyrum arvense L.   Field Cow-wheat
'Weed of cultivation. Rare.'
Sixteen stations are listed in the 1914 *Flora*.
First British record: 'In the Corn on the right Hand just before you come to Lycham [*sic*] [Litcham] in Norfolk', J. Sherard in Ray's *Synopsis*, 3rd edit., 1724. Plate 1792 in *EB*, was drawn from a specimen sent by J. Pitchford from Costessey near Norwich.

Orobanche ramosa L.   Branched Broom-rape
'Very rare. Parasitic on hemp.'
West: 50, Outwell, CS in *BG*.
East: 38, Earsham, DS in KT; 39, Broome, TJW in *BG*; 49, Geldeston, DS in KT.
'Used to be found in Norfolk hemp-fields near Norwich, in the Waveney valley round about Bungay, and in the Fenland border at Outwell', EAE.

Orobanche rapum-genistae Thuill.   Greater Broom-rape
Last recorded from Heacham, W63, in 1920.

Teucrium scordium L.   Water Germander
'Very rare.'
West: 59, Welney, Hb. Salm.; 60, Stow Bridge, Bell.
East: 31, Horning, RW; KT.
Still persists in neighbouring Cambridgeshire.

Campanula patula L.
'Rare. Only once recorded.'
East: 03, near Holt, RBF; *Fl. Br.*, 1800.

Viburnum lantana L.   Wayfaring Tree
'Perhaps native. Rare. Woods and hedges on chalky soil.'
West: 61, E. Winch, GM; 72, Hillington, KT; 91, Mileham, KT; 92, Godwick, KT; 93, Walsingham, KT.
East: 23, Thorpe Market, KT; 28, Needham, *Fl. Pl. H.*

Valerianella rimosa Bast.   Sharp-fruited Lamb's-lettuce
'Rare. Cultivated land.'
West: 07, Blo' Norton, FR.
East: 13, Baconsthorpe, Miss Brenchley, 1912; 24, between Felbrigg Green and Cromer, BEC, 1891.

Senecio paludosus L.   Great Fen Ragwort
'Perhaps native in the Fens. Very rare, if not extinct.'

In *Trans. Nfk. and Norw. Nat. Soc.*, 1911–12, 438, Arthur Bennett reported that he had vainly tried to trace a Norfolk specimen but cited Redmore Fen, adjoining the Little Ouse at Brandon, as the station where it was found about 1833–40. In *Plant Extinctions since 1597*, Druce (BEC, 1919, suppt.) mentions 'Ranworth; Redmore Fen, 1835. Fen now turned into farm lands.' There is a Norfolk specimen in Sir J. E. Smith's herbarium.

S. palustris (L.) Hook.   Marsh Fleabane, loc. 'Trumpets'
'Native. Rare. Wet ditches.'
First record: 'In the way from Norwich to Yarmouth a little before you come to Oakley [*sic*] Bridge [? Acle Bridge], J. Sherard, 1724, and Herb. Mus. Brit., 1725.
'In the eighteenth and nineteenth centuries, marsh fleabane grew chiefly within a few miles of the Norfolk coast, in the valleys of the Ant and Thurne and the lower reaches of the Bure and Waveney, with the greatest concentration in the Fleggs and about Acle', EAE.
Last authenticated records:
West: 63, Dersingham, 1899, *Fl. Norf.*
East: 41, Fleggburgh, 1898, G. E. Harris.

Hypochoeris maculata L.   Spotted Cat's Ear
Always very rare. No recent records, the last being in 1904 from Carleton Forehoe, Eoo, MCHB.

Crepis foetida L.   Stinking Hawk's-beard
'Alien. Rare. Chalky places.'
West: 70, Beechamwell, DT; BG; Barton Bendish, JP; *WA*; 71, Narborough, KT; 94, between Binham and Longham, JC.

Fritillaria meleagris L.   Fritillary
'Rare. Pastures.'
East: 28, near Harleston, Rev. H. Tilney in *BG*; Needham, *Fl. Pl. H.*

Muscari atlanticum Boiss. & Reut.   Grape-Hyacinth
'Rare. Woods, etc.' Still occurs in the Suffolk Breckland but appears to have disappeared from the Harleston area, one of its last localities.

Colchicum autumnale L.   Meadow Saffron
'Native (?). Rare. Meadows.'
East: 28, Pulham, DS in KT; Starston, *Fl. Pl. H.*; 18, Burston, Rev. J. W. Millard, 1908.

Cephalanthera longifolia (L) Firtsch   Long-leaved Helleborine
'Native (?). Very rare. One record only.'
East: 02, Whitwell, KT.

Epipactis purpurata Sm.   Violet Helleborine
West: 92, Whissonsett, KT; 93, Fulmodeston, KT.
East: 11, Swannington, F. Norgate, 1910; 20, Howe, KT; 39, Hedenham, KT.

E. atrorubens (Hoffm.) Schult.   Dark-red Helleborine
'Very rare.'
KT's records for Docking must be regarded as errors.

Herminium monorchis (L.) R. Br.   Musk Orchid
'Native (?). Very rare. Chalky soils.'
West: 63, Snettisham, JC in *BG*; Heacham, JC; AB; 70, Marham,
TJW, 1779, in *Eng. Fl.*

Orchis ustulata L.   Burnt Orchid
'Very rare, only once recorded.'
West: 60, Shouldham, RF in *BG*.

Sparganium angustifolium Michx.   Floating Bur-reed
From its known distribution in the north and west, we consider the
records in the 1914 *Flora* are errors.

Eleocharis acicularis (L.) Roem. & Schult.   Slender Spike Rush
'Native. Rare. Damp heaths.'
Although recorded formerly from both vice-counties, there are no
recent records.

Scirpus sylvaticus L.   Wood Club-rush
'Native. Rare. Damp woods.' In Norfolk, HR, *c*. 1775, and JES.
East: 39, Ditchingham, TJW in *BG*.

Carex trinervis Degl.
'Native. (?). Very rare.'
H. G. Glasspoole first gathered the plant at Ormesby St Michael, E41, in
1869–70. Further gatherings were made in 1886. No ripe fruits were
found and there seems to be good reason to doubt its validity, an opinion
shared by both E. Nelmes and E. A. Ellis, the latter suggesting it is a
hybrid of which one parent is a member of the *C. nigra* complex.

Gastridium ventricosum (Gouan) Schinz & Thell.   Nit-grass
'Very rare, no recent records. Weed of cultivation.'
East: 04, Cley, JES in *Eng. Fl.*; 49, Gillingham, TJW in *BG*.

CASUALS

'Introduced species which are uncertain in place or persistence, i.e., not
naturalised or established' (Lousley in *The Recent Influx of Aliens into the
British Flora*, BSBI Conference Report, 1953).

The hunting of aliens, like fox-hunting, arouses very mixed feelings.
There are some who dismiss it with contempt and condemn it as 'dung-
hill botanising', whilst others consider casuals are potential newcomers
to the flora. In the future it might well be that some will exhibit an explosive increase such as has occurred with *Senecio squalidus*, *Epilobium
adenocaulon*, and more recently, *Veronica filiformis*.

We consider they should be noted and, where possible, both year of discovery and source of introduction recorded. Although some may in time become established, we would emphasise that no instances are known of persistence and spread. The low winter temperature appears to be the governing factor in Norfolk; central Norfolk is one of the coldest areas in the British Isles, equal to the east coast of Scotland, so that the persistence of many casuals, originating from America and the Mediterranean, is scarcely likely.

There have been two main sources of introduction. One, the extensive acreage of light, sandy soil of West Norfolk devoted to carrots towards the end of the Second World War when large supplies of seed were imported from the U.S.A., under Lend-Lease arrangements; some thirty-two adventives appeared, of which *Amaranthus* spp., and *Echinochloa crus-galli* were the most abundant. Whilst providing much interest for autumn botanising, we again know of no persistence, all later records being fresh introductions.

The other source is the very rich rubbish tip at Harford Bridges near Norwich in East Norfolk. The city of Norwich has always been famed for its large numbers of bird-fanciers ('Norwich canaries'), and this tip has provided many species from cage-bird seeds besides the usual garden throw-outs. The number of times Harford is mentioned reflects the assiduity of Miss Ruth M. Barnes (now Mrs Philip Race) during her tenure of office as Keeper of Natural History at the Castle Museum in that city.

Other agencies include imported grain used for milling, oil extraction, brewing and poultry food, e.g. *Bromus unioloides* in maize, *Centaurea melitensis* in linseed, *Anagallis foemina* in barley, and *Iva xanthifolia* in chicken food; with seeds imported for cultivation, e.g., *Amaranthus* spp., *Echinochloa crus-galli*; *Centaurea calcitrapa* in French lucerne seed; *Digitaria ischaemum* in annual blue lupin seed from Germany. With imported feedstuffs, e.g., *Rapistrum* spp., *Camelina sativa* and *Guizotia abyssinica* in bird-food. In imported timber, e.g., *Impatiens parviflora* in Russian timber from Archangel and *Euphorbia uralensis*; in ballast, *Chaenorhinum minus* (now established).

Where possible we give the year of discovery, and the coding represents source of introduction:

BF = cage-bird seeds **G** = grain and seed aliens
CF = carrot-fields **H** = horticulural origin.

RANUNCULACEAE

Delphinium ambiguum L. **G** Larkspur

West: 62, N. Wootton; 71, Marham, ELS, 1967; 79, Foulden, 1951; 99, Stow Bedon, FR, 1916.
East: 20, Norwich, WGC, 1915; 39, Ditchingham, MIB, 1950.

D. orientale Gay **H** Eastern Larkspur

West: 93, E. Barsham, ETD.
East: 20, Harford, JHS; 21, Spixworth, ETD.

D. consolida L. **G** Forking Larkspur

West: 72, Congham, 1947.
East: 04, Cley, FMD, 1956.

Anemone ranunculoides L.　　**H**　　Yellow Wood Anemone
West: 80, Ashill; Necton, ESE, 1954; 81, Little Dunham, ESE; 90, Ovington, 1916, FR.

A. apennina L.　　**H**　　Blue Anemone
East: 20, Framingham Pigot, ETD, 1966.

Adonis annua L.　　**H, G**　　Pheasant's-eye
West: 90, Saham Toney, ERN, 1960.
East: 20, Norwich, WGC, 1915.

PAPAVERACEAE
Papaver lateritium C. Koch　　**H**
East: 33, Mundesley, KHB, 1959.

Glaucium corniculatum (L.) Rudolph　　**H, G**　　Red Horned-poppy
Sent by Benj. Stillingfleet to Hudson in 1755 from Norwich. Found by J. J. Kidd in 1885 at King's Lynn.
West: 62, King's Lynn tip, JH, 1965.
East: 20, Norwich, WGC, 1915.

Eschscholzia californica Cham.　　**H**　　Californian Poppy
West: 88, Croxton tip, 1959.
East: 14, Sheringham, 1928, F. E. Sowter; 20, Harford tip, RMB.

FUMARIACEAE
Corydalis solida (L.) Swm.　　**H**
West: 88, Croxton, HDH, 1946.
East: 12, Reepham, E. Stimpson, 1957; 49, Norton Subcourse, 1961, Miss M. Brown in Herb. Norw. Mus.

C. cava (L) Schweigg. & Koerte　　**H**
East: 49, Norton Subcourse, RMB, 1961.

CRUCIFERAE
Brassica oleracea L.
Common as escaped cultivated races of the wild cabbage.

B. napus L.　　Rape, Cole, Swede
Frequent as an escape.

B. rapa L.　　Turnip
Occasional escape from cultivation.

B. juncea (L.) Czern.　　Chinese Mustard
West: 70, Cockley Cley, 1950; 88, Thetford, Miss M. Cobbe, 1919; 90, Watton, FR, 1919.
East: 20, Eaton Park, Norwich; Harford tip, RMB, 1963.
There are likely to be many more records in the future now that this species is being increasingly used in the manufacture of mustard.

Erucastrum gallicum (Willd.) O. E. Schultz
West: 79, banks of newly cut Relief Channel, Methwold Hythe, GT, 1966, conf. ELS.

Eruca sativa Mill.
West: 90, Watton, FR, 1915.

Raphanus sativus L.　　**H**　　Radish
East: 20, Harford tip, RMB, 1958.

Rapistrum perenne (L.) All.　　**G, BF**
West: 62, King's Lynn Docks, FR, 1917.
East: 20, Harford tip, RMB, 1958.

R. rugosum (L.) All., ssp. *rugosum*　　**BF**
East: 20, Harford tip, RMB, 1961; 28, Harleston, ELS, 1961; 41, Acle, ETD, 1964.

Ssp. *orientalis* (L.) R. & F.　　**BF**
East: 20, Harford tip, RMB, 1961.

Conringia orientalis (L.) Dum.　　**BF**　　Hare's-ear Cabbage
West: 83, Stanhoe, GT, 1966; 88, Thetford, AMS, 1958; 90, Watton, FR, 1915.
East: 11, Drayton, WGC, 1922.

Lepidium sativum L.　　**H**　　Garden Cress
East: 20, Harford tip, RMB, 1960.

L. virginicum L.
East: 11, Drayton, WGC, 1922; 20, Norwich, WGC, 1915; 32, E. Ruston.

L. densiflorum Schrad.
West: 90, Watton, FR, 1916; 88, Thetford, Mrs Wedgwood, 1921.

L. neglectum Thell.
West: 88, Thetford, FR, 1916.

L. perfoliatum L.
East: 11, Drayton, WGC, 1922.

Isatis tinctoria L.　　Woad
Cultivated in the Wisbech area as a dye plant up to 1913, the last crop from twenty acres. Now occurs rarely as a weed in gardens.
West: 64, Old Hunstanton, 1955.
East: 09, Attleborough, HDH, 1939; 21, Wroxham, PER, 1932.

Neslia paniculata (L.) Desv.
East: 20, Norwich, WGC, 1915.

Bunias orientalis L.
West: 71, Gayton, 1945, RPL.

Lobularia maritima (L.) Desv.　　**H**　　Sweet Alison
East: 14, Sheringham cliffs; 20, Harford tip, ETD; 24, Cromer, KHB, 1959; 49.

Barbarea verna (Mill) Aschers.
West: 61, Middleton, E. A. Atmore, 1922; 90, Carbrooke, FR, 1914.

Matthiola bicornis (Sibth.) DC.　　**CF**　　Night-scented Stock
West: 93, Sculthorpe, 1949.

Malcolmia maritima (L.) R. Br.   **H**   Virginia Stock
East: 20, Harford tip, RMB, 1958.

Sisymbrium loeselii L.
East: 20, Harford abattoir, JHS, 1963.

S. polyceratium L.
West: 62, King's Lynn, J. G. Gilbert, BSBI Exc., 1949.

Camelina sativa (L.) Crantz   **BF**   Gold of Pleasure
West: 84, Burnham Overy, ELS, 1967; 91, E. Dereham, PH, 1960; 94, Wells-next-the-Sea, JSP, 1954.
East: 20, Harford tip, RMB, 1958; 28, Pulham St Mary, CF.

RESEDACEAE

Reseda alba L.   **H**   Upright Mignonette
West: 70, Marham, FR, 1921; 99, Rockland St Peter, FR, 1916.
East: 20, Norwich, ETD, 1966; 39, Geldeston, PER, 1929.

CARYOPHYLLACEAE

Silene cretica L.
East: 28, Pulham St Mary, CF, 1960, det. JEL.

Vaccaria pyramidata Medic.   **BF**
West: 90, Watton, FR, 1915; 94, Wells-next-the-Sea, FR, 1916.
East: 20, Norwich, ETD, 1966; Keswick, WGC, 1915.

Kohlrauschia prolifera (L.) Kunth   Proliferous Pink
West: 79, Cranwich, up to 1950; gravel pit, Northwold, J. E. Little, 1927, in Herb. Mus. Brit.

K. nanteuilii (Burnat) P. W. Ball & Heywood
West: 79, Stoke Ferry, 1890; Northwold, 1889 – both gatherings in Herb. Mus. Brit., det. Ball & Heywood.

AMARANTHACEAE

Amaranthus caudatus L.   **H**   Love-lies-bleeding
East: 20, Harford tip, RMB, 1960, det. JMPB.

A. hybridus L., ssp. *hybridus*   **CF**
West: 61, Bawsey, 1945, det. JMPB; 69, W. Dereham Fen, beet-field, 1967, Mrs RCLH; 72, Appleton, 1949.
East: 20, Harford, 1953, RMB.

Var. *cruentus* Mansf.   **ÇF**
West: 72, Houghton, det. JMPB.

A. bouchonii Thell.   **CF, G**

West: 72, Appleton, 1949, det. P. Jovet, the first British record; 78, Weeting, 1961; 80, Hilborough, 1954.

A. quitensis K. B. K.

East: 11, Taverham, 1959; 20, Harford tip, JEL, in Herb. Norw. Mus.

A. retroflexus L., var. *retroflexus*   **CF, G**   Pig Weed

West: 72, Appleton, 1949, det. JMPB; 88, Thetford, 1916; FR; 89, W. Tofts, 1967.

East: 12, Marsham; Heydon; 19, Bunwell; 22, Felmingham; 28, Wortwell, CF, 1961; 32, E. Ruston; 33, Paston.

Sub-var. *rubricaulis* Thell.   **CF**

East: 14, Weybourne, ELS, 1967.

A. albus L.   **CF**

West: 61, Bawsey, 1945; 72, Appleton, 1949, det. JMPB; 93, Sculthorpe.

Sub-var. *rubicundus* Thell.

West: 72, Appleton, CEH, 1949, det. JMPB – new to the British Isles.

A. blitoides S. Wats.

West: 72, Appleton, CEH, 1949.

CHENOPODIACEAE

Axyris amarantoides L.

West: 61, Leziate, CPP; 72, Hillington, RPL, 1949.

Chenopodium opulifolium Schrad., ex Koch & Ziz

West: 88, Thetford, FR, 1916; 90, Watton; Ovington, both FR.

C. hircinum Schrad.

East: 20, Whitlingham, F. Long, 1917, in Herb. Norw. Mus.; 39, Geldeston, JEL.

C. pratericola Rydb.

West: 88, Thetford, FR, 1916.

C. murale L.   **CF, G**   Nettle-leaved Goosefoot

West: 61, Bawsey, 1945; 62, King's Lynn; 71, E. Walton, 1951; 82, beet-field, E. Rudham, ELS; 90, Cranworth, ALB 1962; 84, Burnham Market, KHB; 93, Sculthorpe, 1955.

East: 03, Saxlingham, GT; 08, Banham, DMM 1965; 49; 50, Gt. Yarmouth.

C. urbicum L.   Upright Goosefoot

Weed of cultivation. Formerly regarded as common locally in waste places; no recent records.

West: 90, Saham Toney, WCFN.

C. hybridum L.   **H**   Sowbane

West: 70, Langwade Green, ESE; 71, Westacre, 1951; 88, Thetford, Miss E. S. Todd, 1931; 99, Cranberry Farm, Hockham, ALB, 1966; Snetterton, FR.

Atriplex hortensis L.   **H**   Garden Orache

West: 62, tip at Roydon Common, JH, 1964.

East: 20, Norwich, Miss S. C. Puddy, 1963.

Salsola pestifer A. Nels.   **CF**   'Russian Thistle'
West: 62, King's Lynn Harbour; 73, Gt. Bircham, 1949; 79, Didlington, VML, 1967; 89, W. Tofts, JHS, 1967.
East: 14, Weybourne, JHS, 1967; 20, Whitlingham, F. Long, 1917. Both the Didlington and Weybourne populations contained plants with broad roseate perianth-segments = *S. kali* L., var. *caroliniana* (Walt.) Nutt., of Gray's *Manual*.

Kochia scoparia (L.) Schrad.   **H**   'Burning Bush'
East: 03, Briston, B. L. Palmer, 1952; 20, Harford tip, ELS.

K. densiflora Turcz.   **CF**
East: 14, Weybourne, JHS, 1967, det. and conf. JEL and Dr Aellen.

MALVACEAE

Malva parviflora L.   **CF**
West: 72, Houghton, ELS, 1950, sent to BSBI Exc.

Lavatera arborea L.   **H**   Tree Mallow
East: 13, 14, 24; 33, cliff tops from Sheringham to Happisburgh, EAE.

L. trimestris L.   **H**
East: 41, Acle New Road, ETD, 1964; JHS, det. CCT.

Althaea hirsuta L.   **G**   Hispid Mallow
West: 62, W. Newton, 1949, BSBI Excursion.

A. rosea (L.) Cav.   **H**   Hollyhock
West: 70, Shingham.
East: 00, Hackford, ALB; 04, Kelling Warren, 1961; 20, Harford tip, RMB.

Abutilon theophrasti Med.   **BF**   'Chinese Hemp', 'Indian Mallow'
West: 83, in field of mangolds, Stanhoe, GT, 1967.
East: 30, Thurton, EAE, in Herb. Norw. Mus.

LINACEAE

Linum usitatissimum L.   Cultivated Flax
A frequent escape from fields up to 1956 when cultivation ceased.
West: 62, King's Lynn Docks, 1939.
East: 00, Deopham, HDH, 1943; 14; 20, Harford tip, RMB.

GERANIACEAE

Geranium platypetalum F. v. M.   **H**
West: 79, Whittington, 1956; 91, Gressenhall, RMB.

G. sanguineum L.   **H**   Bloody Cranesbill
East: 03, Holt, 1949; 11, Attlebridge, W. A. Nicholson, 1920; 20, Dunston Common, ETD, 1966.

OXALIDACEAE

Oxalis europaea L.   **H**   Upright Yellow Sorrel

West: 62, King's Lynn Docks, JH, 1964; 88, Thetford churchyard, RCLH, 1959.

East: 12, Blickling, RCLH, 1961; 32, bulb farm, near Wayford Bridge, Stalham, CG & Mrs B. Russell, 1959, conf. D. P. Young.

O. articulata Savigny   **H**

West: 79, Whittington tip, RCLH, 1961.

East: 18; 20, Harford tip, RMB; 24, Between Cromer and E. Runton, JBE, 1957; 31.

O. corymbosa DC.   **H**

East: 12, Blickling, RCLH, 1961; 14, Weybourne, FMD, 1954; 20, weed in Messrs Daniels' Nursery, Norwich, RCLH, 1957; grounds of Carrow Works, RCLH, 1959.

VITACEAE

Parthenocissus quinquefolia (L.) Planch.   **H**   Virginia Creeper

East: 20, Harford tip, ELS, 1961.

PAPILIONACEAE

Lupinus polyphyllus Lindl.   **H**   Lupin

West: 90, edge of lake, Scoulton Mere, F. Robinson, 'I have known it for thirty years'.

L. luteus L.   Annual Yellow Lupin

West: 80, arable land, Gt. Cressingham, ELS, 1954.

Both this species and the sweet blue lupin are rich in protein and are grown occasionally on light, sandy soils for livestock feeding.

L. angustifolius L.   Sweet Blue Lupin

West: 79, near Feltwell, 1954; 80, Rowley Corner, Hilborough, 1954.

L. hirsutus L.   Bitter Blue Lupin

Sown occasionally as a green manure crop on soils deficient in nutrients.

West: 79, near Feltwell, 1954.

Melilotus indica (L.) All.   Small-flowered Melilot

West: 62, King's Lynn quayside and tip, 1943; 72, Little Massingham; 84, Burnham Overy, BT, 1962.

East: 14, Weybourne carrot-field, ELS, 1967; 20, Mousehold Heath, ETD, 1966.

Trifolium resupinatum L.   **G**

East: 11, Hellesdon, RPL, 1945.

T. incarnatum L.　　**G**　　Crimson Clover

West: 60, Shouldham Thorpe, 1945; 61, Blackboro' End, 1957; Totten-hill Row, 1956; 62, S. Wootton, 1950.

East: 27, Brockdish, R. E. Emms, 1954.

T. angustifolium L.　　**BF**

West: 71, Westacre, 1952.

Lotus angustissimus L.　　Slender Birdsfoot Trefoil

East: 04, High Kelling, garden weed, PHS, 1954.

Psoralea dentata DC

West: 94, Binham, R. Scott, 1959, det. B. M.

East: 20, Harford tip, RMB, 1961.

Galega officinalis L.　　**H**　　Goat's Rue

East: 20, Norwich, TCS, 1959; Eaton, ETD, 1967.

Colutea arborescens L.　　**H**　　Bladder Senna

East: 10, Ketteringham, ETD, 1964; 20, Mousehold Heath, RMB, 1958; 20, Harford tip, RMB, 1959.

Coronilla varia L.　　Crown Vetch

No mention in 1914 *Flora*.

East: 10, gravel pit, Wymondham, Miss Pomeroy, 1920; 20, Norwich, ETD, 1964, a dozen plants.

Vicia villosa Roth

East: 21, Frettenham gravel pit, JEL, 1949; 50, Gt. Yarmouth, EAE, 1930.

V. bithynica (L.) L.　　Bithynian Vetch

East: 20, Norwich, EAE, 1934.

V. faba L.　　Horse Bean

A frequent escape from cultivation.

Lathyrus aphaca L.　　Yellow Vetchling

East: 19, Forncett St Peter, MBA, 1966, garden weed; 20, Norwich, EAE, 1934; 28, Pulham St Mary, CF, 1965.

L. sativus L.

West: 62, King's Lynn quayside, 1942, det. A. E. Wade.

Pisum sativum L.　　Field Pea

A frequent escape from cultivation.

ROSACEAE

Spiraea salicifolia L.　　**H**　　Willow Spiraea

Both this species and *S. douglasii* Hook., are grown in gardens, the former escaping and becoming naturalised.

East: 03, Holt, FR, 1916; 20, Mousehold Heath, ETD, 1965; 21, Hainford, ETD.

106 PYRAMIDAL ORCHID   *Anacamptis pyramidalis*   *A. H. Hems*

107 BROAD-LEAVED COTTON GRASS   *Eriophorum latifolium*   *Miss G. Tuck*

108 SWEET FLAG   *Acorus calamus*   *E. G. Burt*

109 HARE'S-TAIL   *Eriophorum vaginatum*   *Mrs D. M. Dean*

110 LESSER TUSSOCK SEDGE   *Carex appropinquata*   *J. E. Lousley*

112 GREAT BROME   *Bromus diandrus*   *Miss G. Tuck*

111 REED SWEET-GRASS *Glyceria maxima* *A. C. Jermy*

113 GREY HAIR-GRASS *Corynephorus canescens* *W. H. Palmer*

114 DROOPING BROME *Bromus tectorum* *J. E. Lousley*

115 WOOD MILLET *Milium effusum* *Miss D. M. Maxey*

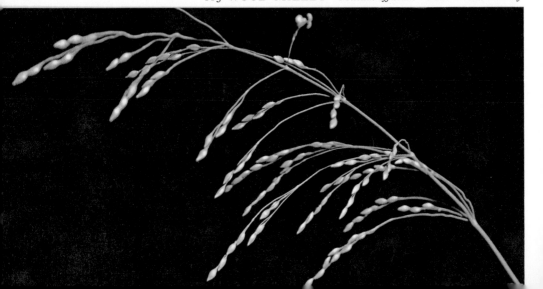

Potentilla norvegica L.
West: 88, Thetford, Miss M. Cobbe, 1920.
East: 20, Whitlingham, F. Long, 1918.

Pyracantha coccinea M. J. Roem.   **H**   Firethorn
East: 20, Harford tip, RMB, 1958.

CRASSULACEAE

Sedum spurium M. Bieb.   **H**
East: 33, Ridlington near Bacton, KHB, 1959.

SAXIFRAGACEAE

Tellima grandiflora (Pursh.) Dougl. ex Lindl.   **H**
East: 03, shady woodland, Stody, RPB-O, 1966, det. EAE.

UMBELLIFERAE

Astrantia major L.   **H**
East: 20, Norwich, Miss G. Taylor, 1961; 23, N. Walsham, Miss J. L. Brockwell, 1952; 32, Stalham, E. C. Crutwell, 1929.

Caucalis platycarpos L.   Small Bur-parsley
East: 20, Mousehold tips, TCS, 1959; Norwich, WGC, 1915.

Coriandrum sativum L.   Coriander
West: 80, Swaffham, E. M. Reynolds, 1919.
East: 04, Cley, E. Adye, 1953; 20, Norwich, L. A. W. Burder, 1930.

Bupleurum rotundifolium L.   Hare's-ear
West: 80, Saham Toney, 1954.
East: 20, Norwich, S. A. Manning, 1936; Thorpe, JHS, 1966.

B. lancifolium Hornem.   **BF**
West: 79, Didlington, VML, 1964; 92, N. Elmham, Miss R. Walton, 1960.
East: 11, Attlebridge, M. E. Smith, 1962; 20, Norwich, RMB, 1956.

Petroselinum crispum (Mill.) Nym.   **H**   Parsley
East: 30; 51, Caister, RCP, 1957.

Ammi majus L.   **CF**
West: 84, Burnham Westgate, RMB, 1963.

Apium leptophyllum (Pers.) F. v. Muell.
West: 90, Watton, FR, 1920.

Falcaria vulgaris Bernh.   **G**
West: 83, Quarles, BT, 1960.
East: 41–51, Hemsby to Caister, NYS, 1926.

Laser trilobum (L.) Borck.
West: 62, King's Lynn quayside, 1945.

CUCURBITACEAE

Cucurbita pepo L.   **H**   Vegetable Marrow
West: 62, King's Lynn tips, 1964; 88, Croxton tip, 1963.
East: 20, Harford tip, RMB, 1958.

EUPHORBIACEAE

Euphorbia virgata W. & K. × *E. esula* L.
(*E.* × *pseudovirgata* (Schur) Soó)
Hitherto regarded as *E. uralensis* Fisch.
West: 62, King's Lynn Docks, 1952; 80, Hilborough road, Swaffham, 1957, both det. A. R. Smith of Kew.

E. esula L.   Leafy-branched Spurge
West: 79, Feltwell, J. A. Ward det. EAE in Herb. Norw. Mus., 1954.
East: 20, Harford, RMB, 1964, det. T. Pritchard.

E. cyparissias L.   **H**   Cypress Spurge
West: 79, Feltwell, J. A. Ward, 1958; 81, Litcham, 1959; 83, Stanhoe, GT; 88, near Kilverstone, HDH, 1945.
East: 11, Taverham, ETD, 1966.

POLYGONACEAE

Polygonum patulum M. Bieb.   **CF**
West: 72, Houghton; Little Massingham, 1950.

P. sachalinense F. Schmidt.   **H**   Giant Knotweed
East: 14, W. Runton, PJB, 1962; 20, Norwich, RMB, 1964; 24, Overstrand, KHB, 1959.

P. campanulatum Hook. f.   **H**   Lesser Knotweed
East: 30, Ashby St Mary, R. Penton, 1963, det. EAE; 49, Geldeston, EAE.

P. senegalense Meisn.   **H**   Senegal Knotweed
East: 20, Keswick tip, EAE, 1963.

Fagopyrum esculentum Moench   Buckwheat
Occasionally sown for pheasant food and not always harvested.
West: 61, Wormegay, 1942; 70, Cockley Cley; 80, Hilborough; 84, Burnham Overy tip.
East: 10, Ketteringham, ETD, 1964; 20, Harford tip.

URTICACEAE

Helxine soleirolii Req.   **H**   Mind-your-own-business
East: 09, Spooner Row railway station, J. S. Rees, 1966; 39, colonising the brickwork of Geldeston Lock, R. Waveney, EAE, 1958; old wall overlooking stream, Ellingham Mill, ETD, 1964.

CANNABIACEAE

Cannabis sativa L.   **BF**   Hemp
West: 61, Setch, WGH, 1959; 84, Burnham Overy tip, 1967.
East: 20, Harford tip, RMB, 1959; Norwich, 1964; 24, Cromer, 1957; 30, Cantley, RMB, 1956; 31, Ludham, RMB, 1964.

EMPETRACEAE

Empetrum nigrum L.   Crowberry
Possibly introduced by birds passing along the coast on migration.
East: 24, cliffs at Cromer, C. D. Raby, 1967, det. EAE; 51, dunes at Winterton, JHS, 1967.

PRIMULACEAE

Cyclamen hederifolium Ait.   **H**   Cyclamen
West: 79, on island of Didlington Lake, EQB, 1950.
East: 19, orchard at Gt. Moulton, J. Goldsmith, 1967.

Lysimachia punctata L.   **H**   Loosestrife
West: 62, Dersingham Fen, 1949; 90, Scoulton Mere, FR, 1919.
East: 20, Thorpe, ETD.

Anagallis arvensis L., ssp. *foemina* (Mill.) Schinz. & Thell.   **G**
   Blue Pimpernel
West: 62, King's Lynn, 1949; 80, Little Cressingham, ETD, 1963; 88, Two Mile Bottom near Thetford, HDH, 1939.
East: oo, Crownthorpe, HDH, 1939.

BORAGINACEAE

Omphalodes verna Moench   **H**   Blue-eyed Mary
West: 60, Wallington Park wood, J. E. Little, 1922.

Amsinckia intermedia F. v. M.   **G**   Tarweed
Appears to be increasing in frequency. Little work has been done on the genus, which is badly in need of revision.
West: 79, Feltwell, J. A. Ward det. Kew; 90, Watton, FR, 1916; 94, Wells-next-the-Sea, FR, 1916; 99, E. Wretham, ALB, 1964.
East: 04, Cley, RSRF; 11, Costessey, RMB, 1958, det. Kew; 14, Sheringham, 1958; Weybourne, ETD, 1966; 20, Norwich, JHS; 33, Mundesley, JHS.

A. lycopsioides Lehm.   **G**
West: 61, Bawsey, CPP, 1930; 94, Wells-next-the-Sea, FR, 1916.
East: 04, Salthouse, E. S. Adye in Herb. Norw. Mus., 1954; 11, Drayton, WGC in Herb. Norw. Mus.; 14, Sheringham, H. E. S. Upcher; Weybourne, PHS.

Trachystemon orientalis (L.) G. Don **H**
East: 24, Overstrand, KHB, 1959.

CONVOLVULACEAE

Cuscuta campestris Yuncker **CF** Dodder
East: 21, Sprowston, RPL, 1930, det. Yuncker.

SOLANACEAE

Scopolia carniolica Jacq. **H**
East: 03, Brinton, weed in shrubbery, RPB-O, 1952.

Nicandra physalodes (L.) Gaertn. **G, H** Apple of Peru, Shoo Fly
West: 84, Burnham Overy tip, ELS, 1967.
East: 20, Harford tip, 1960; 23, N. Walsham, EAE, 1964; 30, Surlingham, EAE; 39, Ditchingham, in beet-field, MIB, 1946–52.

Physalis alkekengi L. **H** Bladder Cherry
West: 99, Rockland All Saints, 1963.

Solanum sarrachoides Sendtn. **CF**
West: 70, Cockley Cley, 1950; 72, Appleton, 1949; 73, Gt. Bircham, 1949; near Heacham, beet-field, ELS, 1966; 78, Weeting, in stubble field, ELS, 1961 and 1966.
East: 14, Weybourne, ELS, 1967; 30, Rockland St Mary, R. H. Sewell, in Herb. Norw. Mus., 1955.

S. rostratum Dunal.
East: 31, Upton, RMB, 1959, det. Kew.

S. tuberosum L. Potato
Appears annually on many rubbish tips and waste ground.

Lycopersicon esculentum Mill. **H** Tomato
Similar to the last species.

SCROPHULARIACEAE

Verbascum phlomoides L. **H** Woolly Mullein
West: 72, Harpley Dams, 1952; 78, near Chalk Hill Farm, Hockwold, 1964.
East: 20, Harford tip, RMB, 1960, conf. Kew; Mousehold, ETD, 1965; 30, Cantley, EQB, det. Kew.

V. lychnitis L. **H** White Mullein
East: 19, Forncett St Peter, RMB, 1962; 20, Mousehold, ETD, 1965.

V. sinuatum L.
West: 88, Thetford, HDH, 1943, det. JEL.

Antirrhinum siculum Ucria
West: 62, King's Lynn, garden weed, 1949.

Linaria repens (L.) Mill.　**H**　Pale Toadflax

East: 21, Wroxham churchyard wall, KHB, 1959, until 1965; 40.

L. maroccana Hook. f.

East: 33, cliff-top, Cliftonville near Mundesley, KHB, 1959.

Mimulus moschatus Dougl., ex Lindl.　**H**　Musk

West: 62, Dersingham Common, 1945.
East: 39, Ellingham near Bungay, RMB, 1962.

Erinus alpinus L.　**H**

East: 12, Cawston, walls of old barn, rectory and wall near church, 1943–64.

Veronica longifolia L.　**H**

East: 20, Harford tip, RMB, 1961.

Parentucellia viscosa (L.) Caruel　Yellow Bartsia

Appears to be a weed in grass-seed mixtures used for cricket pitches and golf greens but does not survive the mowing.

West: 61, grass heath, C. J. Cadbury, 1963; E. Winch Common cricket pitch, RPL, 1949; 94, Wells-next-the-Sea, Miss R. Carey, 1954; 64, Old Hunstanton golf links, TGT, 1936.

LABIATAE

Salvia reflexa Hornem.　**H**

East: 20, Harford tip, RMB, 1961, det. Kew; 29, Fritton, Miss Salmon, 1928.

S. verticillata L.　**H**

East: 13; 20, Harford tip, ELS, 1960, conf. Kew.

S. sylvestris L.　**H**

West: 79, Mid Warren Farm, Methwold, J. A. Ward, 1954; 84, dunes at Burnham Overy Staithe, 1953.

Stachys annua (L.) L.　**H**

East: 20, Harford tip, JHS, 1963; 28, Pulham St Mary, CF.

Lamium maculatum L.　**H**　Spotted Dead-nettle

East: 04, Kelling, CPP, 1928; 20, Harford tip; 08, ditch at Kenninghall Heath, HDH, 1936.

Leonurus cardiaca L.　**H**　Motherwort

Several stations listed in 1914 *Flora* but no recent records.

East: 09, Attleborough, FR, 1913.

PLANTAGINACEAE

Plantago indica L.　**BF**

East: 20, Harford tip, RMB, 1962; Mousehold Heath tip, TCS, 1959.

RUBIACEAE

Asperula taurina L.   **H**   Pink Woodruff
East: oo, swampy wood, Hardingham, EQB, 1958.

Galium tricornutum Dandy   **G**   Rough Bedstraw
East: 20, Norwich, WGC, 1915.

COMPOSITAE

Helianthus annuus L.   **CF, BF, H**   Annual Sunflower
West: E. Winch, 1940; 88, Thetford, RPL, 1954; 90, Cranworth, barley field, ALB, 1964.
East: 14, Weybourne, JHS, 1967; 20, Harford tip, 1961.

H. tuberosus L.   **H**   Jerusalem Artichoke
West: 60, W. Dereham Fen, 1964; 78, Hockwold, 1964; 79, Whittington, 1961.
East: 20, Harford tip, 1959.

H. rigidus (Cass.) Desf.   **H**   Perennial Sunflower
West: 62, King's Lynn tip, 1964.
East: 20, Harford tip, 1959.

H. decapetalus L.
West: 68, Hockwold Fen, 1958, Maps Scheme.

Guizotia abyssinica (L. f.) Cass.   **BF**   'Niger'
West: 62, King's Lynn, garden weed, 1964; 84, Burnham Overy tip, 1967.
East: 11, Attlebridge, 1961; 20, Harford tip, 1960.

Cosmos bipinnatus Cav.   **H**   Cosmea or Mexican Aster
West: 62, King's Lynn tip, 1964.
East: 20, Harford tip, RMB, 1959.

Hemizonia pungens Torr. & Gray   **CF**   Californian Spikeweed
West: 79, Didlington, ELS, 1967.
East: 04, Salthouse, EAE, 1967.

Iva xanthifolia Nutt.   **G**
West: 62, King's Lynn, chicken run, 1960, det. Kew.

Ambrosia artemisiifolia L.   **G, H**   Ragweed, Roman Wormwood
West: 80, Saham Toney, ERN, 1964, det. Kew; 92, N. Elmham, EAE, 1958.
East: 02, Foxley, ALB, 1967; 23, Antingham, EAE, 1958.

A. trifida L.   **H**   Great Ragweed
East: 11, Drayton, WGC; 23, N. Walsham, WGC, 1915.

Xanthium spinosum L.   Spiny Cocklebur
West: 83, Whitehall Farm, Syderstone, 1959, field dressed with shoddy in 1955; 70, Cockley Cley, kale field, EAE, 1958.

Tagetes pumila L.   **H**   French Marigold
East: 20, Harford tip, RMB, 1959.

Senecio doria L. **H**
East: 32, Ingham, garden weed, RMB, 1951.

S. fluviatilis Wallr. Broad-leaved Ragwort
East: 30, Wheatfen Broad, where it was introduced and has become well established, EAE.

Calendula officinalis L. **H** Garden Marigold
West: 63, Heacham tip, 1959; 79, Whittington tip, RCLH, 1961; 84, Burnham Overy Staithe tip, 1959; 88, Thetford tip, 1959; oo, Reymerston tip, ALB, 1962.

Dimorphotheca pluvialis Moench **H** Star of the Veldt
West: 72, Houghton, 1950.

Anaphalis margaritacea (L.) Benth. **H** Pearly Everlasting
West: 94, Warham churchyard, 1950; probably introduced by E. K. Robinson, founder of the Brit. Empire Nat. Assoc.

Solidago canadensis L. **H** Golden-rod
A frequent garden throw-out on tips and waste places.

Aster novi-belgii L. **H** Michaelmas Daisy
Probably the most frequent of the alien asters which are to be found on most of the rubbish tips and dumping grounds in the county.

Callistephus chinensis (L.) Nees **H** Chinese Aster
East: 20, Harford tip, RMB, 1959.

Erigeron mucronatus DC. **H**
East: 14, walls of Beeston Regis church, KHB, 1959; 21, wall of Belaugh church, RMB, 1961.

E. bonariensis L. **H**
East: 20, Harford tip, RMB, 1959.

Anthemis tinctoria L. **H** Yellow Chamomile
West: 71, W. Bilney, 1947–9.
East: 20, Harford tip, RMB, 1962; roadside, Harford, ETD, 1965.

Artemisia biennis Willd.
West: 88, Thetford, Miss M. Cobbe, 1920.
East: 51, Caister-on-Sea, Miss M. Cobbe, 1920.

Echinops commutatus Jur. **H** Globe Thistle
West: 79, Didlington; 80, Cockley Cley, VML.

Carduus tenuiflorus Curt. Slender Thistle
West: 62, King's Lynn Docks, 1950; 80, between Swaffham and Cockley Cley, ESE, 1950.
East: 24, 29, 31, 39, 40 (Maps Scheme records).

Centaurea aspera L. Rough Star Thistle
West: 89, Tottington, FR, 1914.

C. calcitrapa L. Star Thistle
West: 83, Stanhoe, GT, 1960; 94, Stiffkey, FR, 1915.
East: 11, Drayton, WGC, 1922.

C. solstitialis L.   **G, CF**   Yellow Star Thistle

West: 61, Bawsey sand-mill, CPP, 1930; 62, N. Wootton, in sown grass field, CPP, 1957; 72, Houghton, 1950; 80, Hilborough, ALB, 1963; 83, Stanhoe, GT, 1960; 88, Thetford, FR, 1916; 89, W. Tofts, 1967; 94, Stiffkey, FR, 1915.

East: 10, Hethersett, WGC, 1921; 11, Drayton, WGC, 1922; 14, Weybourne, JHS, 1967; 21, Gt. Plumstead, JHS, 1950; 49, Aldeby, L. Lloyd, 1932.

C. melitensis L.   Maltese Star Thistle

West: 62, King's Lynn Harbour, ELS, 1940.

Lactuca sativa L.   **CF**   Cultivated Lettuce

West: 72, Harpley, 1949.
East: 14, Weybourne, JHS, 1967.

L. serriola L.   Prickly Lettuce

West: 62, King's Lynn tip and Docks, 1939, 1962; 82, Gt. Massingham carrot-field, 1951.

East: 14, Weybourne, carrot-field, 1967; 18; 20, Harford tip, RMB, 1959; 21, Spixworth, ETD, 1965; 28, 39, 49.

Cicerbita macrophylla (Willd.) Wallr.   **H**   Blue Sow-thistle

West: 93, Gt. Walsingham, near the ford, BT, 1961.

LILIACEAE

Muscari comosum (L.) Mill.   **H**   Tassel Hyacinth

West: 98, W. Harling Heath, Lady Fitzroy, 1920.
East: 14, Sheringham cliffs, 1954.

Allium carinatum L.

East: 20, Keswick, by the ford, EAE, in Herb. Norw. Mus., 1955, formerly introduced by James Crowe from Westmorland in the eighteenth century, EAE.

A. roseum L.

East: 22, Hautbois, ETD, 1965.

IRIDACEAE

Crocosmia × *crocosmiflora* (Lemoine) N.E. Br.   **H**   Montbretia

West: 79, Whittington dump, RCLH, 1961.
East: 20, Harford tip, RMB, 1959.

PALMACEAE

Phoenix dactylifera L.   Date Palm

East: 20, Harford tip, many plants, ELS, 1959.

FOR CASUAL GRASSES SEE MAIN LIST.

INDEX